Biochemistry and Neurological Disease, edited by A. N. Davison, PhD, DSc, FPS, 325 pp, 80 illus, $45, Philadelphia, JB Lippincott Co, 1976.

This is a well-written and extensively documented volume that consists of six monographs reviewing recent developments in neurochemistry. While there is little evidence of coordination between the different authors, much of the overlap in the subject material is useful, since the focus of each chapter varies greatly.

The major bias a neuropathologist has to overcome in evaluating the efforts of neurochemists is that his chemical colleagues know too little about the pathologic features of the tissues they are analyzing. While stones can be thrown from either side, this point is embarrassingly evident in the first chapter, where obvious plaques in slices of brain from a patient with multiple sclerosis are ignored, and sections of cortical gray matter at the depths of sulci are labeled "plaques" instead. The types of studies reported in this chapter depend on many variables, including accuracy of diagnosis, the microscopic heterogeneity of the nervous system, and postmortem changes, so that the specter of inaccuracy in sampling renders this whole approach suspect.

The lucid and critically written chapter on retinal degeneration is based on animal models of retinitis pigmentosa and correlations are made with the findings in humans. Less extensive coverage is given to toxic damage to the retina and diabetic retinopathy.

Current information about the biochemical alterations associated with epilepsy is also critically reviewed. But here, too, ambiguity surrounds discussions of morphology, particularly the distinction between collagen fiber and glial scars. The former, associated with abscesses and penetrating wounds, is much more frequently associated with seizures than the latter.

The chapter on transmitter amines in brain disease is well illustrated and skillfully constructed. My only reservation is that insufficient weight is given to the possibility that, in many degenerative CNS disorders, decreases in enzyme or transmitter amine levels may not be causally related, but rather secondary to neuronal degeneration. This chapter is particularly careful to point out the limitations of inferring what is happening in the brain from analyses of urine, blood, or cerebral spinal fluid.

The chapter on coma is deceptive, for it is really about hypoxia, ischemia, and ammonia toxicity. Here the correlation between structural and biochemical alterations is particularly well drawn.

The final monograph on brain-specific antigens is complementary to the rest of the book, tying together many structural and functional threads that were previously loose. Excellent use is made of tables. This chapter also contains a very useful technical section outlining many of the pitfalls awaiting a newcomer to this field.

DOYLE G. GRAHAM, MD, PHD
Durham, NC

Biochemistry and Neurological Disease

Biochemistr

EDITOR: A. N. DAVISON
B.Pharm., Ph.D., D.Sc., F.P.S.
Professor of Neurochemistry,
Institute of Neurology,
University of London

nd Neurological Disease

BLACKWELL SCIENTIFIC PUBLICATIONS
OXFORD LONDON EDINBURGH MELBOURNE

© 1976 Blackwell Scientific Publications
Osney Mead, Oxford,
85 Marylebone High Street, London, W1M 3DE
9 Forrest Road, Edinburgh,
P.O. Box 9, North Balwyn, Victoria, Australia.

All rights reserved. No part of this publication
may be reproduced, stored in a retrieval system,
or transmitted, in any form or by any means,
electronic, mechanical, photocopying, recording
or otherwise without the prior permission of
the copyright owner.

ISBN 0 632 00377 4

First published 1976

Distributed in the United States of America by
J. B. Lippincott Company, Philadelphia,
and in Canada by
J. B. Lippincott Company of Canada Ltd., Toronto.

Printed and bound in Great Britain by
William Clowes & Sons, Limited
London, Beccles and Colchester

Contents

	Contributors	vii
	Preface	ix
1	Biochemistry of Brain Degeneration *D. M. Bowen and A. N. Davison*	2
2	Biochemistry of Retinal Degeneration and Blindness *H. W. Reading and A. J. Dewar*	52
3	Biochemistry and Basic Mechanisms in Epilepsy *H. F. Bradford and P. R. Dodd*	114
4	Transmitter Amines in Brain Disease *G. Curzon*	168
5	Biochemistry of Coma *H. S. Bachelard*	228
6	Brain Specific Antigens: Biochemical Role in Selective Pathogenesis *E. J. Thompson*	278
	Index	317

Contributors

D. M. BOWEN and A. N. DAVISON

Miriam Marks Department of Neurochemistry, Institute of Neurology, Queen Square, London WC1N 3BG, England

H. W. READING and A. J. DEWAR

M.R.C. Brain Metabolism Unit, Department of Pharmacology, University of Edinburgh, 1 George Square, Edinburgh EH8 9JZ, Scotland

H. F. BRADFORD and P. R. DODD

Department of Biochemistry, Imperial College of Science and Technology, South Kensington, London SW7 2AZ, England

G. CURZON

Miriam Marks Department of Neurochemistry, Institute of Neurology, Queen Square, London WC1N 3BG, England

H. S. BACHELARD

Department of Biochemistry, Institute of Psychiatry, De Crespigny Park, Denmark Hill, London SE5 8AF, England

E. J. THOMPSON

Department of Chemical Pathology, Institute of Neurology, Queen Square, London WC1N 3 BG, England

Preface

The very considerable advance in basic biochemical knowledge has been accompanied by recent progress in our understanding of the metabolism of specialized tissues of the nervous system.

In this book we aim to show how neurochemistry is relevant to the scientific basis of neurology. Since progress in research on brain tumours and on the aetiology of cerebral vascular disease is related to basic work on the larger fields of cancer and cardiovascular disease, these major topics are only incidentally referred to in this book. We have also restricted ourselves to dealing with problems of the mature nervous system, and the interesting questions of the developing brain and paediatric neurology are not dealt with.

Apart from certain notable exceptions, routine biochemical analysis of body fluids, including even cerebrospinal fluid, makes a disappointingly minor contribution to diagnosis and management of neurological disease. Research on experimental animals should supplement this work but, unfortunately, there are few exact parallels of diseases of the human nervous system. In the past, chemical analysis of pathological tissues, largely centred on rare inborn errors of metabolism and other equally obscure conditions, has resulted in the accumulation of a great deal of basic data. However, an understanding of the biochemistry of neurological disorders demands much more than study of the molecular disorganization and metabolism of diseased nervous tissue. Not only must account be taken of the physiology and morphology of the nervous system, but the pathology (of the disease) must also be related to biochemical data. This demands a new approach to the subject. It necessitates a deliberate attempt by those trained in the traditional basic scientific disciplines to learn and, as far as possible, to utilize other newer subjects of relevance to their research. Thus the use of stereotatic surgery has become almost a routine experimental tool to the biochemical neuropharmacologist. Radioimmune and fluorimetric assay has dramatically altered the sensitivity of biochemical analyses, and a knowledge of immunological principles is an essential prerequisite for interpretation of many of the molecular changes found in degenerative disorders. Another change in attitude has come from the realization of the much wider utility of human post-mortem tissue than had, hitherto, been considered. The identification and quantitation of brain proteins by electrophoresis supplements the now routine studies of lipid analysis. Quite unexpectedly, assay of free amino acid in the brain post-mortem has been shown of value. Some of the transmitter synthesizing enzymes have been found to retain activity for long enough to be of practical importance. The finding of striking and specific tissue decarboxylase deficiencies in the dyskinesias (p. 188) suggests a selective loss of specific tracts linked to either GABA (in Huntington's chorea) or to dopamine (in Parkinson's disease). Indeed, this principle of *selective vulnerability*, whether it be at the molecular, cellular or regional level, emerges as a general theme throughout the book. For instance, Bachelard refers to the varying susceptibility of different brain regions to ischaemia and Reading and Dewar to specific changes in the retina as a result of inherited factors or following poisoning. Epilepsy is another condition in which there is frequently an association with the temporal lobe. There are many other examples of such vulnerability such as the selective effect of dihydrostreptomycin on minute regions of the cochlea or the action of 6-hydroxydopamine on the peripheral nerve noradrenaline-containing neurones. It will be made evident, in this book, that we understand relatively little about the biochemical basis of selective vulnerability. Such changes are customarily ascribed to variations in blood supply to a brain region or to differences in the metabolism of various groups of nerve cells. Hardly any data are available which might explain their selective vulnerability. Certainly it appears that nerve cells have become so specialized that, with the possible exception of microneurones, they have lost the ability to regenerate. Moreover, there is a growing realization of the possibility of structural changes at a distance from the initial lesion (diaschisis). It is vital, therefore, to understand not only the pathogenesis of neurological disease but to turn, as well, to possible alternatives to anatomical regeneration of cells and tracts as mechanisms for functional recovery. This poses the question, 'Is there restitution after irreversible lesions in the central nervous system, or merely substitution of original functions by alternative remaining pathways?' Appreciation of these issues is intertwined with the concepts of cerebral organization. A start on the problems of functional recovery after lesions of the nervous system has been made by experimental psychologists (Eidelberg & Stern, 1974). Perhaps they will soon be joined by the biochemist to tackle this most practical problem at the molecular level.

1 Biochemistry of brain degeneration

D. M. Bowen

A. N. Davison

I Introduction, 4
   A. Abiotrophy and diaschisis, 4
   B. Pathoclisis, 5

II Histopathology, 6
   A. Chromatolysis, 6
   B. Wallerian degeneration, 7
   C. Gliosis and CNS phagocytes, 8
   D. Inflammatory response, 8
   E. 'Dying back', 9
   F. Demyelination, 10
   G. Senile changes, 10
      1. Nerve cell loss, 10
      2. Intra-neuronal changes, 14
      3. Changes in neuropil, 17

III Experimental models, 19
   A. Oxygen deprivation and selective vulnerability, 19
      1. Ischaemia, 19
      2. Cellular pathoclisis, 20
   B. Hyperammonaemia, 22
   C. Experimental demyelination, 23
      1. Agents directly acting on myelin, 23
      2. Agents acting on the oligodendrocytes, 23
      3. Cell-mediated reactions, 24
   D. Metals, 25
   E. Viruses, 26

IV Human neurological diseases, 27

    A. Problems associated with investigating human tissue, 28
    B. Multiple sclerosis, 28
        1. Chemical pathology, 28
            Plaques, 28
            Normal appearing white matter, 29
            Cerebrospinal fluid, 31
        2. Conclusion, 31
    C. Senile dementia, 31
        1. Biochemical correlates of morphological change, 32
        2. Biochemical changes, 33
            Proteins and decarboxylases, 33
            Other indices, 37
        3. Conclusions, 38
            Pathogenesis of senile dementia, 38
            Comparison with other abiotrophies, 39

V Summary, 41

1 Introduction

In contrast to other types of cells in the nervous system (e.g. oligodendrocytes, astrocytes, microglia and pericytes) the mature nerve cell is especially vulnerable to damage, for it relies on the continuous supply of substrates supported by virtually no energy reserve to maintain function and sustain the axon and nerve terminals. This is especially true of long nerve fibres where the energy requirement must be relatively large. Interruption in oxygen or glucose supply for more than a few seconds causes marked functional changes. Permanent alterations to the central nervous system follow several minutes of ischaemia, for neurones have very restricted power of recovery. Thus, compression of the abdominal aorta can result in permanent paralysis with injury to the ventral horn cells, although cells of other organs such as the kidney are spared. The neuronal cell body is primarily affected by insufficient supply of nutrients. Oligodendroglia are less affected than neurones in ischaemia, while astrocytes and especially endothelial cells appear to be better adapted to resist diminished blood supply. Susceptibility to ischaemia differs in various parts of the central nervous system; cells of the frontal and occipital regions may die before injury to the basal nuclei or the more resistant vital areas of the brain. In man, the brain regions that are especially affected in anoxia are the cerebral cortex (particularly in the depths of sulci following a single anoxic episode), basal ganglia (particularly following multiple anoxic episodes), Ammon's horn (especially Sommer's sector) and the Purkinje cells of the cerebellum, whereas the hypothalamus is usually not damaged and the brain stem and parts of the thalamus are of intermediate sensitivity.

A. ABIOTROPHY AND DIASCHISIS

Damage to the nervous system may result from many other well-recognized causes, such as bacterial and viral infections, neoplasia, the action of various toxic agents or from vitamin deficiency. However, there are many examples of neurological disease where the causative factor is so far unrecognized. This was realized even at the beginning of the century when in 1907 Gowers (see Anon 1974) classified certain brain degenerative disorders as *abiotrophies*, in which, it was postulated, there was an unidentified essential defect of vital endurance leading to cellular death. While it is clear that damage in a brain infarct is related to the territory previously sustained by the now defective blood supply, disruption of function may affect neuronal complexes well removed from the site of structural damage. Thus for example in this process of *diaschisis* (introduced

Abbreviations: DOPAD, aromatic acid decarboxylase; GAD, glutamic acid decarboxylase.

by Monakow in 1914, see Teuber, 1974) posterior cortical lesions could affect visually guided behaviour not because of the direct damage to the cortex but rather from the side effects emanating from the focal lesion. Recovery of function may sometimes occur following brain damage but little is understood about the underlying cellular mechanism of neuronal and synaptic plasticity (Cragg, 1974) necessary for return of function. The functional interrelationship between different areas of the brain is also evident in higher mental activity (e.g. memory) where localized areas are subserved by other parts of the brain (Iversen, 1973).

B. PATHOCLISIS

In 1922, Vogt and Vogt introduced the concept of *pathoclisis* to explain the bilateral symmetry of selective vulnerable regions (topistic units). It was proposed that specific physicochemical properties within the topistic units are responsible for the varying predispositions to noxious agents and to genetic defects (Pentschew, 1969). The topistic unit may include more than one nucleus (e.g. substantia nigra and locus caeruleus in Parkinson's disease) or a specific group of neurones (e.g. motor neurones in amyotrophic lateral sclerosis). In some cases damage may be primarily associated with reduction in specific transmitter tracts. Thus, in the inherited disease, Huntington's chorea, the choreiform movements and possibly mental symptoms may be associated with progressive loss of GABA-mediated neurones, particularly from the caudate nucleus. Similarly, marked neurophagia especially in the pigmented DOPA-containing cells of the substantia nigra is associated to the typical dyskinesia found in Parkinson's disease. Relatively little is known about the pathogenesis of this group of abiotrophic diseases, but one exception is that of hepatolenticular degeneration or Wilson's disease. Therefore, this disease can no longer be considered a true abiotrophy for the abnormal and toxic accumulation of copper has been traced to deficiency in a specific protein (Evans, Dubois & Hambridge, 1973). Another example of selective vulnerability is seen in demyelinating diseases where there is primary destruction of the myelin sheath. In the case of multiple sclerosis, the disease is associated with an inflammatory response, and demyelination which often has a distribution similar to that of the venous drainage. The periventricular predilection seen in this disease may have some connection with the cerebro spinal fluid.

Although we lack knowledge of the aetiology of many of these 'degenerative' disorders, it is, nevertheless, possible to study the biochemical changes occurring in the damaged nervous tissue in relation to the pathology especially where model systems are available. Before passing on to discuss possible biochemical mechanisms of the pathogenesis of different neurological and experimental conditions, we shall briefly describe some histopathological changes accompanying damage to the nervous system.

II Histopathology

In this section morphological hallmarks are reviewed that occur in degenerative brain diseases. At the macroscopic level the anatomy of normal human brain is particularly well illustrated in the atlas by Roberts and Hanaway (1970).

Neuroanatomy at the light and electron microscopic levels has been usefully summarized by Cragg (1968) while for a systematic description of neuropathology the reader is referred to Greenfield's Textbook of Neuropathology.

A. CHROMATOLYSIS

Within a few hours following stress, observable changes occur in the perinuclear rough endoplasmic reticulum of the neurone. This so-called Nissl substance first disappears from the perikaryon nearest to the nucleus, subsequently extending to the rest of the cytoplasm except for the periphery where a thin rim remains identifiable (Fig. 1.1). In this process of Nissl degeneration or central chromatolysis, the nucleus moves close to the cell plasma membrane. Swelling of the neuronal perikaryon may occur. In addition, there is an increase in nucleolar size and an increase in total RNA so that the cell may be regarded as

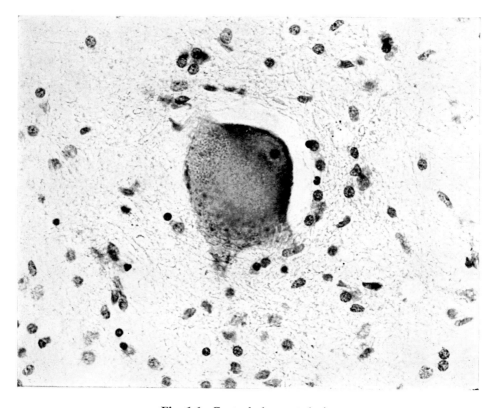

Fig. 1.1. Central chromatolysis.

responding to injury by heightened metabolic activity. If the damage is not too severe, recovery may take place within 7 days. New RNA moves from the nucleolus to appear in the cytoplasm at Nissl substance.

Death of the neurone may be followed by neuronophagia in which microglia (p. 8) serve a phagocytic function together with infiltrating blood-borne macrophages. In cases such as the lipidoses there is abnormal accumulation of lipid (e.g. ganglioside G_{M2} in Tay-Sachs disease) within neurones (often in residual bodies, p. 17) or the concentration of other products increases (e.g. lipofuscin in ageing) which may eventually kill the cell.

B. WALLERIAN DEGENERATION

Following nerve section axonal degeneration and loss of myelin occurs above the point of damage up to the next internode, that is to the uppermost part of the Schwann cell severed by the injury. Distal loosening of myelin lamellae may be seen within 24 h. Retraction of myelin from the nodes of Ranvier occurs and characteristic sudanophilic lipid globules appear in about a week. Scavenging macrophages engulfing myelin debris are apparent at about the same time. Primary attack on the myelin sheath such as is thought to occur in multiple sclerosis initially spares the axon so that this disease may solely involve the sheath and its satellite oligodendroglial cells. Chromatolysis may accompany severing (axotomy) or damage to the axon when the process is known as axonal reaction or retrograde degeneration (Fig. 1.1). In addition, a primary attack on the neuronal perikaryon (e.g. rabies, virus encephalitis and poliomyelitis) will lead to axonal degeneration. Injection of actinomyosin D prior to axotomy prevents the dispersion of Nissl bodies which suggests that new RNA might be necessary for the initiation of the retrograde changes (Torvik & Skjörten, 1974).

The sequence of biochemical and pathological changes following section of peripheral nerve is now well-established. Within a day there is an increase of neurofilaments and accumulation of axonal mitochondria. By 48 h the entire axis cylinder undergoes degeneration, after which physical fragmentation of the myelin sheath occurs. During the next stage (4–10 days) there is an increase of cell population (either Schwann cells or macrophages) paralleling a marked increase in total lysosomal hydrolase activity (Hallpike, 1972). Later, 8–32 days after transection there is a slow loss of myelin lipid and an accumulation of cholesterol esters. Breakdown of myelin occurs more slowly in the central as compared with the peripheral nervous system. Disrupted myelin with many individual myelin lamellae persists for long periods in degenerating central tracts (Lampert & Cressman, 1966). Lipid droplets are found late in the degenerative process within macrophages (Bignami & Ralston, 1969) and cholesterol esters are not demonstrable until 7–8 weeks after severance of the posterior roots. In rats, following removal of the eye, Bignami and Eng (1973) showed

60% loss of myelin by 54 days after enucleation, but the isolated myelin appeared morphologically and chemically normal. A floating fraction consisting largely of cholesterol ester was found 54, 76 and 90 days after enucleation (Bignami & Eng, 1973). Since there appeared to be no loss of basic protein from the remaining myelin such as found in multiple sclerosis plaques and in peripheral nerve Wallerian degeneration it is unlikely that endogenous proteolytic activity is responsible for the breakdown of the myelin. Thus, breakdown of myelin appears to occur mainly in macrophages.

C. GLIOSIS AND CNS PHAGOCYTES

Destruction of neurones and demyelination frequently leads to formation of glial scar tissue at the site of damage. The deposited fibrous material is provided by proliferation of the fibrillary astrocytic processes (gliosis). In this process protoplasmic astrocytes of the grey matter may be transformed to fibrous astrocytes (present in white matter) whose long thin fibrillary processes form the glial fibrils of the scar tissue. Eng and his colleagues (1971) have isolated an acidic protein from fibrous astrocytes. They suggest that the water soluble glial fibrillary protein is composed of non-aggregated subunits of the water insoluble glial filament material. Microglia also react and become reactive microglia or gitter cells. Since some microglia may originate from blood monocytes (Langevoort et al., 1970), and as monocytes are transformed into macrophages (Gordon & Cohn, 1973), microglia are probably best considered, along with other phagocytic cells such as the Kupffer cells of liver, as tissue macrophages. In addition to fibrous astrocytes and microglia, endothelial cells called pericytes (which are associated with cerebral blood vessels) have a phagocytic role in the CNS. Generally speaking, fibrous astrocytes are usually involved in minor neural degeneration, while in Wallerian degeneration of the optic nerve microglia and astrocytes react. In stab wounds, encephalitis and experimental allergic encephalomyelitis (EAE) (p. 24), pericytes and circulating leucocytes can also play a phagocytic role (Vaughn & Skoff 1972).

D. INFLAMMATORY RESPONSE

The defensive reactions of the body's tissues to injury may result in acute (up to a few days), subacute (of intermediary duration) or chronic inflammation (lasting months or years). At the site of acute damage, blood polymorphs and some mononuclear cells are found (perivascular cuffing, Fig. 1.2) while the chronic inflammatory response differs in containing primarily mononuclear cells accompanied by proliferating astrocytic tissue. The mononuclear cells include lymphocytes, derived plasma cells and macrophages. The plasma cells are the antibody forming cells derived from B (bone-marrow)-lymphocytes and the macrophages under chemotactic influence are phagocytic. A key factor in

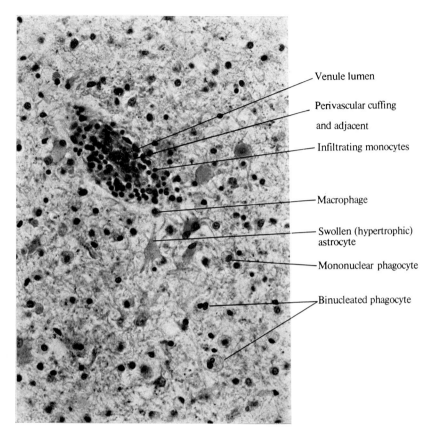

Fig. 1.2. Perivascular cuffing in multiple sclerosis. (High-power section of white matter showing loss of both organized oligodendroglia and myelinated nerve fibres.)

the immunological reaction are the T (thymus)-lymphocytes which as 'killer' lymphocytes are involved in the cellular immune response as well as influencing B-cell transformation to plasma cell and antibody synthesis (McGregor & Mackaness 1974).

E. 'DYING BACK'

The process of degeneration of axon and nerve cell present in subjects poisoned by tri-ortho cresyl phosphate has been recognized since the turn of the century. Although myelin degeneration may occur, this phenomenon is distinct from Wallerian degeneration, for in the latter the axon degenerates at the site of section while in organophosphorus ester poisoning the distal portions of the nerve fibre degenerate and neuronal cell bodies may only show minimal morphological changes. This so-called 'dying back' process occurs more severely in the largest and longest fibres and is said to be consistent with the view that this type of selectivity is related to the amount of work the cell body has to do (Cavanagh, 1973).

F. DEMYELINATION

Demyelination, or loss of myelin, occurs in both the 'demyelinating diseases' and as a phenomenon secondary to axonal damage (secondary demyelination). If the axon is spared, as in multiple sclerosis, the myelin degeneration is said to be primary.

On section of the cerebral hemisphere from a case of multiple sclerosis the plaques or lesions of demyelination are usually obvious in white matter as greyish areas in fixed material, or as pinkish gelatinous areas in the fresh specimen (Fig. 1.3a). When examined by commonly used cell stains, such as haematoxylin eosin, old or fibrous plaques exhibit an almost complete disappearance of oligodendroglial cells which gives the plaques a less cellular appearance than the surrounding tissue. This loss may be incomplete in shadow or early plaques. In recent lesions, perivascular cuffing and microglia (often in the early stages of lipid phagocytosis, and detected by the Marchi stain) and hypertrophic or swollen astrocytes are seen; binucleated or paired astrocytes (Cajal stain) may also be present, which indicates that astroglial proliferation has occurred (Fig. 1.2).

G. SENILE CHANGES

In the ageing, or senile brain, loss of brain weight, thinning of the cortical ribbon, widening of the sulci and enlargement of the ventricles can be quite obvious (Fig. 1.3b). From a cursory gross macroscopic examination of normal human brain it is not clearly apparent that the amount of grey matter exceeds the white (Ogata & Feign, 1973), with a ratio of grey matter to white matter volume which approaches 2 in children, decreases to a minimum of about 1·1 at age 60 years, then rises again to about 1·3 at age 85 years (Miller A.K.H. & Corsellis J.A.N. personal communication).

The light and electron microscopic findings in senile brain and related disorders are well-documented (Table 1.1).

1. NERVE CELL LOSS

It has been generally accepted that there is a steady loss of neurones from the human brain throughout life. Brody estimated that by the eighth decade only half the neuronal population remains in the cortex. In a more recent study (Cragg, 1975) araldite instead of paraffin (which can cause a large and variable volumetric shrinkage) has been used as the embedding medium. Thus in this study although most of the elderly subjects exhibited cerebral atherosclerosis,

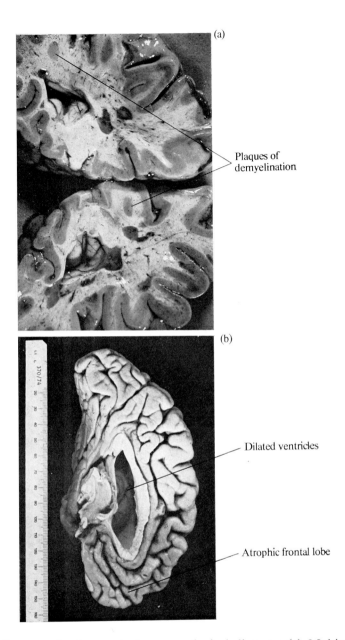

Fig. 1.3. Macroscopic changes in two neurological diseases: (a) Multiple sclerosis; (b) Senile dermentia.

neither the density of neurones nor synapses in cortex were found to be reduced by the seventh decade (Table 1.2). Similarly, Konigsmark & Murphy found no significant age regression in counts of nerve cells in the ventral cochlear nucleus (Prestige, 1974). Since the total weight of brain decreases with age, the question

Table 1.1. Clinical and morphological features of the progressive dementias[a] and related disorders

Disease	Probable aetiology	CLINICAL FEATURES		MORPHOLOGICAL CHANGES[b]			
		Dementia	Parkinsonian signs	Tangles[c]	Senile plaques	Granulo-vacuoles[d]	Excessive Lipofuscin
		CHILDHOOD					
Batten's disease	Defective autoxidation?	+	—	—	—	—	+
Subacute sclerosing panencephalitis	Measles virus	+	—	±	—	—	?
Down's syndrome[e]	Defective neural development	±	?	+	+	+	?
		MIDDLE AGE					
Creutzfeldt-Jacob disease	'Slow virus'	+	+	±	+	?	?
Alzheimer's disease		+	+	+	+	+	?
Pick's disease		+	—	?	±	+	?
Huntington's chorea	Abiotrophies[i]	+	+[f]	—	—	—	+
Guam–Parkinsonism dementia		+	+	+	—	?	?
Parkinson's disease		±	+	+[h]	—	—	?
Motor neurone[g] diseases							
Common		±	—	—	—	—	?
Guamanian		—	—	+	—	+	?
		OLD AGE					
Normal elderly human adults	—	—	—	±	±	±	+
Senile dementia	Abiotrophy	+	?	+	+	+	?

[a] Dementia is also present in metabolic diseases (e.g. Wernicke–Korsakoff syndrome due to B_1 deficiency) and as a result of vascular disease (e.g. 'arteriosclerotic senile dementia'), hydrocephalus and cerebral tumours (particularly of the frontal lobe).
[b] Symbols indicate: +present; ±can occur; −absent; ? unknown or unclear.
[c] Neurofibrillary degeneration.   [d] Granulo-vacuolar degeneration in hippocampus.   [f] In rigid form of the disease.
[e] Senile changes only occur in brains from patients aged over 35–40.
[g] Amyotrophic lateral sclerosis.   [h] In post-encephalitic type; Lewy bodies occur in idiopathic Parkinsonism.
[i] Aluminium ion may be an aetiological factor in at least two diseases of this group; see p. 25 for Alzheimer's disease and Yase et al. (1974) for the Kii Peninsula type of Amyotrophic lateral sclerosis.
Key references: Wisniewski & Terry (1973); Zeman (1974); Ellis et al. (1974); Corsellis (1969); Bernheimer et al. (1973).

Table 1.2. Effect of age on neuronal and synaptic population in human cortex[a] (from Cragg 1975)

Mean age[b]	Synapses			Neurones		Ratio synapses/ neurones
	Count[c]	Length (μ)	Density[d]	Diameter (μ)	Density[e]	
38	1·41	0·30	5·9	19	15·6	39,360
73	1·43	0·29	6·1	16	15·6	40,460

[a] From frontal and temporal lobe.   [b] 7 specimens/group.
[c] In screen area of 7·56 μ^2.   [d] $\times 10^{11}/cm^3$.   [e] $\times 10^6/cm^3$.

of whether neuronal loss occurs in aged humans still remains unanswered, for conventional histological methods do not measure the overall change.

By counting neurones in suspensions of entire mouse brain it has been estimated that the number of neurones are decreased by half in extreme old age. This is in contrast to glial cells for although the number of astrocytes and oligodendrocytes remain constant, microglia increase slowly with increasing age in rodent brain (Table 1.3).

Table 1.3. Effect of age on neural cell numbers in rodent brain

Age (months)	Neurones[a] (millions per brain)	Glia[b]		
		Microglia	Astrocytes (Cells/field 55 μm × 270 μm)	Oligodendrocytes
3	5·6	0·38	1·06	0·61
12	5·1	0·39	0·99	0·67
24	4·2	0·53	0·97	0·64
27	2·5	0·63	0·98	0·68

[a] Mouse (Johnson & Ferner, 1972).
[b] Rat auditory cortex (Vaughan & Peters, 1974).

2. INTRA-NEURONAL CHANGES

Changes within neuronal cells are quite numerous and include neurofibrillary degeneration (Figs. 1.4, 1.5) and the presence of Hirano bodies; the latter occur in various and unrelated pathological conditions. Two other intra-neuronal changes, detected at the light microscopic level, occur in ageing neurones.

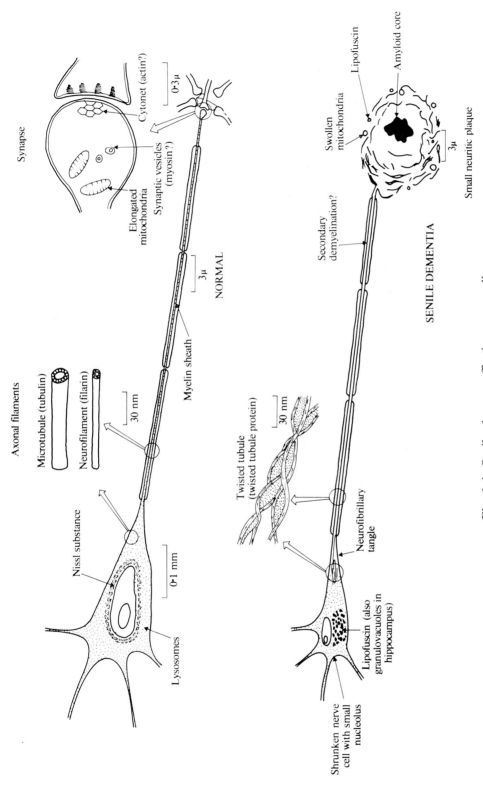

Fig. 1.4. Senile changes affecting nerve cells.

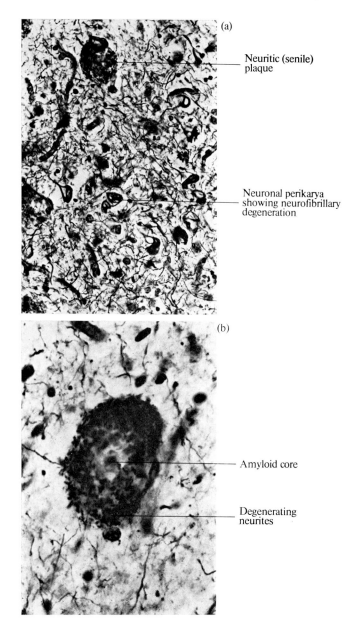

Fig. 1.5. Some histological changes that occur in some progressive dementias and in normal elderly human adults: (a) low power, cerebral cortex; (b) high power, neuritic plaque.

Firstly, a granulovacuolar degenerative change (Fig. 1.4) occurs, with accumulation of lipofuscin-like (Wisniewski & Terry, 1973) haematoxylinophilic material in the hippocampus that is said to be almost diagnostic of senile dementia (Tomlinson, 1972). Secondly, there is the deposition of the fluorescent pigment granules of lipofuscin (Fig. 1.4; Bowen *et al.*, 1974a). The rate of lipofuscin increase that occurs with advancing years in neurones is concomitant with a decrease in cytoplasmic RNA. The pigment granules are rarely evenly distributed in the cell body but accumulate in an area devoid of RNA staining or oxidative enzyme activity (Mann & Yates, 1974). Although lipofuscin accumulates in some cases of senile dementia (Nishioka *et al.*, 1968) and in Huntington's chorea (Tellez-Nagel *et al.*, 1974) this is a widespread phenomenon occurring even in childhood in the inferior olives (Braak, 1971). Pigment deposition is also seen, with increasing age, in the cortex in normal people (Mann & Yates, 1974) and in many cell-types from organs with a low capacity for cell division. Lipofuscin may arise from lysosomes (Fig. 1.4, 1.6) due to autoxidation. It is, therefore, of interest that a deficiency in paraphenylene diamine-linked peroxidase activity in granulocytes has been reported to occur in children with neuronal ceroid-lipofuscinosis (Armstrong *et al.*, 1974). Since this enzyme has not been described in brain and as neither catalase nor glutathione peroxidase occur in brain (De Marchena *et al.*, 1974) the relevance of this finding, in Batten's disease, to lipofuscin in the brain is unclear. The reaction catalysed by superoxide dismutase does, however, furnish the brain (Fried & Mandel, 1974) with a convenient method of disposing of the highly reactive and potentially damaging superoxide ion ($O_2^{\bullet-}$). Results in this laboratory show that extracts of dialysed human brain homogenates inhibit the superoxide ion-dependent autoxidation of pyrogallol (Marklund & Marklund, 1974). This should provide a simple and rapid method for the assay of superoxide dismutase in post-mortem brain specimens.

Apart from lipofuscin, other types of electron dense subcellular organelle, enclosed by a single limiting membrane, occur in neural cells (Fig. 1.6). These include lysosomes and possibly also peroxisomes (Bowen *et al.*, 1974a). Lysosomes contain many of the hydrolytic enzymes found in brain cells, while peroxisomes contain enzymes that generate certain toxic substances, such as hydrogen peroxide. A defect in the functioning of either type of organelle can, therefore, have a potentially devastating effect on the brain. This can lead to the deposition of characteristic types of dense bodies that may be collectively called 'residual' bodies (Fig. 1.6) and these phenomena may be associated with or cause cell death.

3. CHANGES IN NEUROPIL

Perhaps the most striking of the several well-defined changes that occur in neuropil are the neuritic or senile plaques. (Figs. 1.4, 1.5). Recent findings (Wisniewski *et al.*, 1974) indicate that plaques retain connections with the perikaryonal cytoplasm. Thus under these circumstances, plaque formation does not appear to reflect a typical 'dying back' process. Plaques are observed in the cortex and more rarely in the basal ganglia of normal old people. An increase

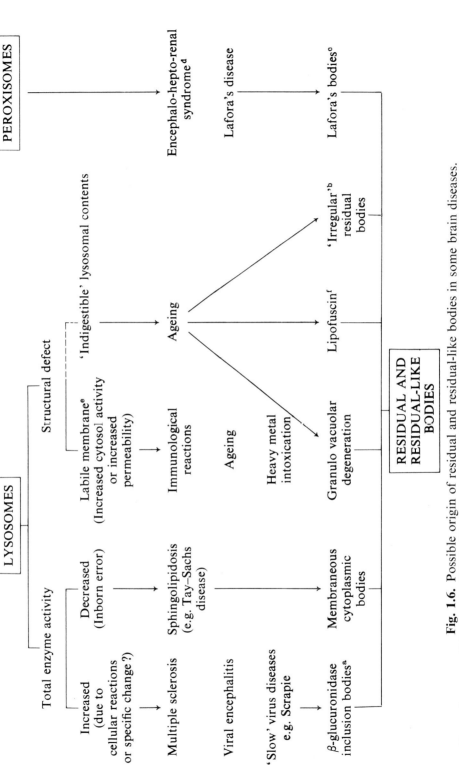

Fig. 1.6. Possible origin of residual and residual-like bodies in some brain diseases. Key references: [a]Mackenzie (1971); [b]Brunk et al. (1973); Brunk (1973); [c]Carpenter et al. (1974) may be similar to Bielschowsky bodies (de Leon 1974); [d]Goldfischer et al. (1973); [e]Bowen & Radin (1969); Brunk & Ericsson (1972); [f]Bowen et al. (1974a); Sekhon & Maxwell (1974).

in the number of plaques does, however, correlate with a deterioration of behavioural and intellectual functioning.

III Experimental models

The study of experimental models of the degenerating nervous system provides a necessary basis for our understanding of the pathology of human disease. Thus, the use of animal material allows examination of the time course of the disease process. The availability of fresh tissue eliminates postmortem artifact and the numbers of animals studied can reduce errors due to biological variability in response. Unfortunately, few exact parallels of human disease exist so that great care has to be exercised in the interpretation of experimental findings.

The literature on experimental models is voluminous. We have, therefore, been selective and briefly describe oxygen deprivation, hyperammonaemia, cuprizone toxicity and experimental demyelination as examples of experimental methods for inducing changes in neurones, astrocytes, microglia and oligodendroglia, respectively. The effects of metal poisoning and viral encephalitides are also described for studies in animals and may be very relevant to certain human degenerative diseases.

A. OXYGEN DEPRIVATION AND SELECTIVE VULNERABILITY

1. ISCHAEMIA

Various methods are available for reducing oxygen supply to the brain, for example, exposure to nitrogen, lowering atmospheric pressure, cyanide poisoning (Hirano et al., 1967) and interference with either pulmonary exchange or cerebral circulation. The effects of ischaemia have been studied in the brain of decapitated rodents by Lowry and his colleagues. Changes may be quite rapid, for example, following a vascular occlusion the oxygen content of the cortex drops to zero within 30–150 s. In the very early stages of ischaemia they found changes in the concentration of high energy phosphate compounds and of glycolytic pathway intermediates (p. 245). There is a rapid loss in the small reserves of glycogen in the brain and a parallel increase in lactate concentration; creatine phosphate and ATP concentrations drop within a few seconds to low levels a minute after complete severance of blood supply. Increased concentration of ADP, AMP and P_i may modulate regulatory enzyme activity—phosphorylase, hexokinase and phosphofructokinase (Lowry, 1966). However, with hypoxia, even accompanied by serious depression of brain function, cerebral levels of ATP may almost be normal (Duffy et al., 1972). Perhaps permanent changes are not due to a failure in the 'power supply' but rather to the marked fall in the pH which accompanies prolonged anoxia.

A convenient method for induction of ischaemia *in vivo* is by ligature of the carotid artery. In most ligated animals either a redistribution of the blood to the brain can occur via the circle of Willis or a high mortality rate results. However, the Mongolian gerbil has an anomalous circle of Willis, lacking a major posterior communicating artery, and provides a most suitable model of ischaemia (Harrison et al., 1973). Fatal cerebral ipsilateral ischaemic infarction follows unilateral carotid artery ligation in over 50% of cases (Levine & Payan, 1966; Harrison et al., 1973). The success of surgery can be assessed before the animals regain consciousness by examination of retinal blood flow. Neurological signs, following successful surgery, include rapid circling, twisting of the head and trunk and rolling (all in the direction to the side ligated). In affected animals infarctions can be first detected ultramicroscopically within only 20 min after ligation as mitochondrial swelling (in neurones) and as distention of the endoplasmic reticulum in the pericapillary foot processes of astrocytes (Hartman, et al., 1973). At the light microscopic level nuclear pyknosis and cytoplasmic vacuolation is present at 2 h but not prominent until 24 h after surgery; invasion by leucocytes does not occur before 24 h (McMartin et al., 1972). Since lysosomal acid hydrolases have been implicated at the onset of cellular autolysis, the activities of representative enzymes of this type have been measured in cerebral infarcts produced in gerbils. Ten days after ligation the total activities of cathepsin D and β-glucuronidase are increased about 5 and 10 fold respectively. This phenomenon appears to correlate with the invasion of macrophages into the lesion (McMartin et al., 1972). Up to 24 and 3 h following ischaemia no clear change is seen in total and free activities of lysosomal hydrolases respectively (McMartin et al., 1972; Clendenon et al., 1971). Since most lysosomal hydrolases are enriched at least 2–5 fold in neuronal perikarya (Sinha & Rose, 1972) in comparison with the neuropil fraction, one might conclude that neuronal lysosomes are relatively insensitive to ischaemia.

2 CELLULAR PATHOCLISIS

A day after ligation of the gerbil carotid artery a 46% reduction in dopamine concentration occurs in the infarcted side while the level of noradrenaline remains unchanged (Zervas et al., 1974). This suggests that there are differences in the susceptibility of noradrenergic and dopaminergic neurones to anoxia. Another example of biochemical evidence of selective vulnerability are the studies of Ferrendelli & McDougal (1971) on regional differences in energy consumption following the abnormally high metabolic demands associated with audiogenic seizures. Phosphocreatine levels in the thalamus were reduced before those in the cerebellum but the latter was finally the most markedly affected. No change was apparent in the cortex. This selective vulnerability may be due to vascular factors or alternatively attributed to minor metabolic differences between cells in various brain regions (pathoclisis p. 5). Passonneau & Lowry (1971) have compared early changes in fractions enriched in two types of neural element from a control and anaesthetized mouse. The immediately frozen-dried cerebral tissue was sectioned and pieces of individual anterior horn cell bodies and adjacent neuropil were removed by dissection for analysis. Phosphocreatine concentrations in the cell bodies and neuropil from the control animal were 80 and 34% of the value obtained in the animal anaesthetized before decapitation. As judged by the greater difference in phosphocreatine levels, the expenditure of energy resulting from the intense motor discharge following decapitation is more in the neuropil. This phenomenon is attributed to the much greater ratio of surface area of the neuropil processes compared with the cell bodies. Ionic fluxes in the relatively larger surface area of the neuropil fraction would require a greater expenditure of energy. This phenomenon of selective vulnerability at the regional and cellular level can clearly complicate the interpretation of biochemical data, particularly in experiments where whole brains are analysed. Another difficulty is that damage following an anoxic episode may be accompanied by secondary axonal degeneration. Following hypoxia the sequence of pathological change (Fig. 1.7) does not begin to overlap with those of Wallerian degeneration until after the first day. In this laboratory, changes in neurotransmitter related enzymes have been examined within 24 h of acute ischaemia in gerbil and rat. In addition to ischaemia, carbon monoxide poisoning and prolonged coma appear to reduce the activity of several neurotransmitter biosynthetic enzymes (Table 1.4). The results obtained with the rodents and human post-mortem tissue suggest that the decarboxylases are more affected by insult than choline acetyl transferase activity. One implication of

Table 1.4. Effect of carbon monoxide poisoning and prolonged coma on neurotransmitter biosynthetic and related enzymes

Brain region	Enzyme activity (% of control; control = 100%)[a]				
	Carbon monoxide poisoning[b]		Prolonged coma before death[c]		
	DOPAD	GAD	GAD	Choline acetylase	Acetylcholine esterase
Putamen	12	25	13	64	109
Caudate nucleus	19	36	12	42	116
Cortex	—	52	9	33	186

[a] Calculated from data expressed as enzyme activity/g fresh wt.
[b] 34-year-old male with severe anoxic changes in nerve cells of cortex and deep grey matter, total neurosis of the putamen (Corsellis J.A.N., personal communication); control 50-year-old male dying of coronary thrombosis (Bowen D.M. & White P. unpublished).
[c] 69-year-old male; control 45-year-old male accident victim (McGeer et al., 1973). Although the control was younger, GAD activity in normal subjects does not fall precipitously until after 70 years of age.

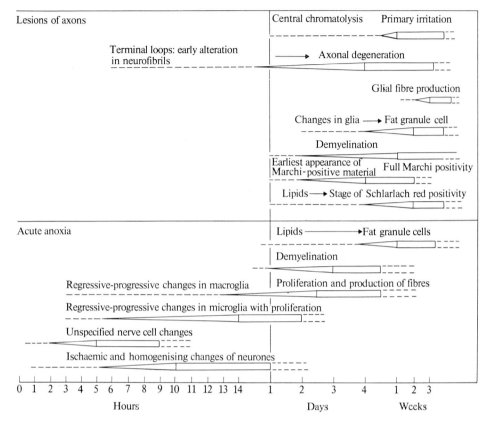

Fig. 1.7. Timing of morphological changes in human anoxia compared with changes due to axonal lesioning (after Jacob, 1963). From Schadé and McMenemy (Eds.) *Selective Vulnerability of the Brain in Hypoxaemia* (C.I.O.M.S.) Oxford, Blackwell Scientific Publications.

this is that the vulnerability of neurones may differ. This concept of '*neuronal pathoclisis*' is not unexpected from a teleological view, for cholinergic pathways may be most primitive in evolutionary terms and their loss would presumably be incompatible with life.

B. HYPERAMMONAEMIA

Cellular abnormalities (Cavanagh, 1972, 1974; Norenberg & Lapham, 1974), particularly in the astrocytes in the brain, are often associated with chronic liver disease. It is possible to produce similar changes experimentally in rats by means of a portocaval shunt (Eck fistula) which causes chronic hyperammonaemia, suggesting that astrocytes play a key role in the control of ammonium ions entering the brain. A glial hypertrophy may be detected in this disorder which taken with the appearance of glycogen granules, is believed to be an expression of extensive denervation in the adjacent neuropil and to reflect a 'dying back' type of neuropathy (p. 9). A spongiform degeneration also occurs in cerebellum in both experimental hyperammonaemia and in cerebral cortex in chronic hepatic disease in humans. The spongiform change is characteristic of an important group of subacute encephalopathies of man which include Kuru and Jacob–Creutzfeldt disease (Gibbs & Gajdusek, 1972).

Biochemical studies on rats with portocaval anastomoses show that the glutamine content of brain is increased up to 4 times normal while glutamate and GABA concentration remain unchanged. Alterations in labelling of cerebral lipids, proteins and amino acids following injection of radioactive acetate or glucose indicates an overall depression of synthetic activity in brain in these animals (Cavanagh, 1974). The effect of portocaval anastomosis on the activities of cerebral lysosomal hydrolases and enzymes involved in the metabolism of neurotransmitters is being studied (Bowen D.M., Flack R.H.A. & White P. unpublished). The most striking change is a 2 to 2·5 fold increase in activity of β-glucuronidase. Since the overall synthetic capacity of the brain is reduced in portocaval anastomosis the increases detected in activities of β-glucuronidase, β-galactosidase and cathepsin A are presumably due to the astrocytic reaction. The activities of acetylcholinesterase, GAD and DOPAD are essentially normal. These biochemical observations appear to be consistent with morphological observations for the overriding histological change is the glial reaction.

C. EXPERIMENTAL DEMYELINATION

1. AGENTS DIRECTLY ACTING ON MYELIN

The chronic administration of triethyl tin causes a demyelinating encephalopathy that begins with severe oedema and in which demyelination occurs without the intervention of macrophage activity. The degeneration that occurs results in an accumulation of fat-rich material which may be isolated on centrifugation as a 'floating fraction' (Smith, 1973) similar to that found in certain human brain degenerations (Ramsey et al., 1974). Hexachlorophene induces a demyelinating process similar to the one elicited by triethyl tin (Matthieu et al., 1974) and a particularly fascinating model has been studied in vivo using tadpoles (Webster et al., 1974).

In peripheral nerve diphtheria toxin induces fragmentation of myelin lamellae as the primary event. Demyelination follows local injection of the toxin in the CNS and thereby provides a useful model for studying disturbance to nerve conduction produced by demyelination (Harrison et al., 1972).

2. AGENTS ACTING ON THE OLIGODENDROCYTES

Demyelination occurs in some strains of laboratory mice following subacute administration of the copper chelating agent, cuprizone (biscyclohexanone oxaldihydrazone) and is probably due to degeneration of oligodendrocytes (Blakemore, 1972). The microglial reaction seen in this disorder follows a predictable course and it is possible to isolate what appear to be reactive microglia from diseased animals (Pattison & Jebbett, 1973). When mice are withdrawn from the diet they soon recover from functional and histological abnormalities and remyelination may occur. Cuprizone-induced encephalopathy resembles the non-inflammatory subacute spongiform encephalopathies (e.g., scrapie, Kuru and Jacob–Creutzfeldt disease).

3. CELL-MEDIATED REACTIONS

When whole brain, myelin suspensions or small amounts of myelin basic protein, together with complete Freuds adjuvant are injected into experimental animals an acute inflammatory reaction in the brain results. Paralysis and other neurological signs usually follow some weeks later, depending on the species affected. In contrast to other demyelinating disorders (p. 28) in experimental allergic encephalomyelitis (EAE) the myelin is ultrastructurally normal before the appearance of cells.

In monkeys histological examination shows widespread inflammation of the central nervous system with predominantly mononuclear cells and some polymorphonuclear leucocytes, particularly in areas of severe perivascular infiltration. Some lesions are haemorrhagic and necrotic. Lesions in monkeys may be macroscopically visible, so that combined histochemical and biochemical estimations may be made. Loss of myelin in the damaged areas is reflected in decreased cyclic nucleotide phosphohydrolase activity and reduced concentration of encephalitogenic basic protein (Govindargan et al., 1974). In the lesions there is increased acid proteinase (cathepsin A, D and B) as well as neutral proteinase activity.

Unlike monkeys and guinea pigs, the Wistar rat is relatively resistant to damage, only about 5% of the brain undergoes demyelination (Smith et al., 1974) preceded by an increase in water content due to enhanced vascular permeability. Necrosis is absent and commonly there is only a mononuclear infiltration (Rauch et al., 1973).

Similar less marked changes are seen in the central nervous system of the rat and, with the onset of acute paralysis, there is enhanced cathepsin A activity in the brain stem (Smith et al., 1974). The increased activity of the acid proteinases can be correlated with the digestion of the myelin basic protein—for the concentration of this protein is decreased in lesions. The role of the physiologically more interesting neutral proteinase (also capable of digesting basic protein) is not clear.

In the lymph nodes of EAE rat cathepsin A activity is increased 11–14 days after injection during the period of the acute response. The lymph nodes increase in weight (3–4 times) during the immune response and there is a comparable increase in total acid proteinase activity—so that infiltrating lymphocytes could account for the raised acid proteinase in the lesion. These changes seem to be related to the inflammatory process not to the demyelination—for no change in proteolytic enzymes occurs in triethyl tin induced demyelination (Smith *et al.*, 1974).

It should be noted that the pathology of EAE is related to human rabies post-vaccinal encephalomyelitis and the resemblance to multiple sclerosis is slight, although EAE may serve as an experimental system for studying some of the secondary changes occurring in an acute inflammatory reaction. A relapsing form of EAE offers promise of a closer model of multiple sclerosis.

D. METALS

Environmental pollution by heavy metals, such as lead and mercury, has assumed notoriety for their devastating neurotoxic effect. It is therefore not surprising that inorganic lead and organic derivatives of mercury are useful agents for retarding neuronal maturation (Patel *et al.*, 1974; Krigmann *et al.*, 1974) for changing metabolic compartmentation and inhibiting protein synthesis (Cavanagh, 1973) in the brain of experimental animals. In contrast to the heavy metals, ingestion of even large doses (0·5 g/day) of aluminium is not harmful (Bailey, 1972) while cerebral injection of trace amounts induces a chronic encephalopathy. A neurofibrillary change is associated with this disorder and clinical symptoms and altered electrical activity occur that are somewhat similar to those seen in pre-senile dementia (Alzheimer's disease). Furthermore, in tissue culture, aluminium salts affect neurofilaments and in pre-senile dementia the levels of aluminium are markedly elevated in parts of the cortex (Crapper *et al.*, 1973). Although the mechanism by which aluminium acts is unknown the metal is a potent inhibitor (Harrison *et al.*, 1972) of brain hexokinase, a key enzyme in aerobic metabolism.

Neuronal loss and an associated chronic epileptic focus can be produced by application of either cobalt or nickel (Hartman *et al.*, 1974) to the cortex of experimental animals. Histological examination of the action of other metals in comparison with cobalt indicates that damage occurs to the 'blood-brain barrier'. It seems possible that calcification, vascularization and probably a specific attack by cobalt on the nerve cell proteins are responsible for the epileptogenic effect (Payan, 1967). The epileptic focus, produced by application of a cobalt-gelatine pellet to rat frontal cortex, is particularly amenable to biochemical study for a discrete secondary and selective cortical degeneration develops in the cortex contralateral to the implant (Emsen & Joseph, 1975). In this model synaptic terminal degeneration is abundant in the ipsilateral caudate nucleus and is also noticeable in the anterior end of the caudate nucleus. Although GAD activities are reduced in both the contralateral cortex and ipsilateral caudate, choline acetyl transferase activity is reduced only in the cortex. This is one example of the value of neurotoxins in establishing neuronal pathways.

E. VIRUSES

At the morphological and clinical level a number of virus infections in animals resemble human degenerative brain conditions (Table 1.5). However, apart from extensive studies on the scrapie agent (Hunter, 1972; Kimberlin, 1973) and the synthesis of virus protein, neurotropic virus infections have been largely ignored

Table 1.5. Animal virus infections that resemble some human neurological disorders

Disease/Malformation	Virus	Species
Subacute sclerosing panencephalitis	Measles-type	Rats, hamster
Multiple sclerosis	Neurotropic hepatitis	Mouse
	Canine distemper	Dog
Schilder's disease	Visna	Sheep
	Parainfluenza type 1	Chick embryo
Creutzfeldt–Jacob } Kuru	Scrapie Transmissible encephalopathy	Sheep, mice, Mink, squirrel monkey
Micrencephaly	Influenza and Newcastle-disease	Chick embryo
Hydranencephaly	Blue tongue	Sheep, mouse
Microcephaly	Hog cholera	Pig
Cerebellar hypoplasia	Lymphocytic Choriomeningitis	Rat

Key references:
Johnson (1972); Johnson & Weiner (1972); Byington & Johnson (1972); Gibbs & Gajdusek (1972); ZuRhein & Echroade (1974).

by neurochemists. However, Bowen et al. (1974b) have studied the biochemistry of examples of acute experimental encephalitis in which the histopathological changes were typical of those seen in degenerative neurological disorders. In these disorders acid hydrolase activity was measured, for changes in the activities of these enzymes were expected as a result of the proliferation of brain phagocytic cells (p. 8), immune reactions (Dingle et al., 1967) and the virus host-cell interaction (Allison, 1967; Buening & Gustafson, 1971). The impetus to much of this work was the finding of increased cathepsin D activities in multiple sclerosis plaque tissue (p. 29). Measurement of the several typical lysosomal enzymes clearly demonstrates that the changes in enzyme activities are characteristic of each disorder and that of the enzymes studied β-glucuronidase is, in general, the most affected. Increased β-glucuronidase activity is often associated with astroglial reactions but this is probably not a completely reliable marker for an increase in activity may reflect glial (presumably oligodendrocytic) proliferation during myelination (Robins *et al.*, 1961) and possibly also a reaction of pericytes or the accumulation of intraneuronal bodies rich in β-glucuronidase (Mackenzie, 1971). Although β-galactosidase activity is enriched in neuronal perikarya (Sinha & Roxe, 1972) the activity does increase in several encephalopathies (p. 23; Emson & Joseph, 1975) which seriously limits its usefulness as an index of neuronal damage. Similarly, the finding of increased cathepsin A activity in portocaval anastomosis (p. 23) may invalidate this enzyme as a marker of mononuclear phagocytes in white matter (Bowen & Davison, 1974; Smith *et al.*, 1974a). An increase in cathepsin D activity does not appear to correlate with any one type of morphological change and therefore appears to have a more ubiquitous distribution in neural cells. In summary, although certain acid hydrolases may be selectively increased or appreciably low in activity in certain neural cells from normal brain, the activity of these enzymes in unfractionated brain homogenates from animals infected with viruses no doubt represents the balance between enzyme loss (due to neuronal and oligodendroglial damage, or both) and an increase in activity due to the hyperactivity of central nervous system phagocytes. As judged by the effect of the scrapie agent on the activity of lysosomal and glycosyl transferase enzymes (Hunter, 1972; Kimberlin, 1973; Suckling & Hunter, 1974) virus-host cell interactions might be expected to further complicate interpretation by stimulating (and also possibly suppressing) the activity of *specific* enzymes.

IV. Human neurological diseases

Although astrocytes have been reported to show early changes in an acute virus infection, it is generally assumed that the neuroglial reactions or 'gliosis' (p. 8) seen in brain diseases, such as senile dementia and multiple sclerosis, are phenomena secondary to more primary changes in neurones or oligodendroglia. Apart from rare microgliomas there is no evidence that microglia initiate any type of brain degeneration. Thus, in this section senile dementia and multiple sclerosis are described as examples of brain diseases that primarily involve damage to neurones and oligodendroglia, respectively. The myelin sheath, which degenerates in multiple sclerosis, is formed in brain as a projection from the oligodendroglia, and is somewhat analogous to the synaptic terminals, for the latter are an appendage of the neurone. In multiple sclerosis it is not clear whether the pathological changes occur indirectly as a result of damage to the oligodendrocyte perikaryon or are due to a direct attack on myelin. Similarly, the changes that occur in senile dementia may be due to either primary damage to the neuronal cell body or to the portion of the nerve cell distal to the perikaryon.

A. PROBLEMS ASSOCIATED WITH INVESTIGATING HUMAN TISSUE

Many cerebral enzymes (Lazarus et al., 1962; Swanson et al., 1973; Cuzner & Davison, 1973; Stahl & Swanson, 1974; Bird & Iversen, 1974) and subcellular organelles (Wannamaker et al., 1973; Garey & Heath, 1974; Kornguth et al., 1974) are remarkably resistant to post-mortem change. Nevertheless, it is clearly desirable that the time intervals between death, body refrigeration (Bird & Iversen, 1974) and post-mortem are not significantly different in comparing cases. The possibility that post-mortem changes are exacerbated in diseased brain cannot be ruled out (but see Robins et al., 1958; Bird & Iversen, 1974). For this reason, we believe that it is difficult to interpret small changes in GABA levels, for the concentration of this neurotransmitter increases very rapidly after death (Alderman & Shellenberger, 1974). Since oxygen deprivation (Bowen et al., 1975a) prolonged terminal coma (Table 1.4) and drug treatment (Bird & Iversen, 1974) may affect the level of neurotransmitter related enzymes, it is essential to have complete clinical records. This should include neuropsychiatric evaluation shortly before death. It is particularly helpful if patient and control material originate from the same hospital, for this reduces variability due to differences in post-death procedures. Morphological examination of both diseased and control specimens is essential (Bowen et al., 1975a).

B. MULTIPLE SCLEROSIS

With the decline of neurosyphilis, multiple sclerosis has become, at least in north-west Europe and the U.S.A., the commonest neurological disease of young adults. It has been established that an inflammatory reaction (p. 9) occurs in multiple sclerosis and is associated with the distribution of plaques (p. 30). Fog found over 80% of plaques to be associated with venules. Thus it is clear that extra-neural factors are involved in the aetiology of multiple sclerosis. Recent studies on the epidemiology of multiple sclerosis throughout the world have shown major differences in the prevalence of multiple sclerosis among people of the same genetic stock, which shows that it is primarily a disease of the environment. There is evidence that individuals with the HLA 3 and 7 and LD-7a tissue antigen type are more susceptible to multiple sclerosis and there are epidemiological data to suggest involvement of an infectious agent in the aetiology (Arnason et al., 1974).

1. CHEMICAL PATHOLOGY

Plaques. Extensive studies on the chemical composition of multiple sclerosis plaques have shown a reduction in overall lipid content. Some investigators claim, however, that cerebrosides are selectively reduced, but the temporal relationship of such changes in the pathogenesis is not clear. As the loss of lipid proceeds cholesterol esters accumulate, possibly as a mechanism for the detoxication of potentially poisonous long chain fatty acids. Of particular interest is the finding that total acid proteinase (cathepsin D) activity is increased (Cuzner & Davison, 1974) in plaques, for this enzyme actively degrades myelin protein releasing on encephalitogenic fragment. The significance of this change in enzyme activity in the pathogenesis of the disease is difficult to both assess and investigate. For example, in early plaques the source of the increased activity may be the infiltrating white blood cells, while, in more mature areas of demyelination, acid proteinase activity from both reactive microglia and astrocytes may predominate. As judged by studies on demyelination in canine distemper the physical presence of large numbers of infiltrating lymphocytes does not account for the bulk of the increased cathepsin D activity in this disease (McMartin et al., 1972). The changes in hydrolase activity in multiple sclerosis and canine distemper are similar, for cathepsin D activity is appreciably increased while β-glucuronidase activity is essentially unchanged. Therefore, it seems unlikely that lymphocytes account for the bulk of the increased cathepsin D activity seen in most multiple sclerosis plaques. This is substantiated by the fact that more lymphocytes are seen in areas of demyelination in distemper than in multiple sclerosis plaques. Furthermore, in multiple sclerosis appreciable numbers of lymphocytes are present only in relatively early plaques while the enzyme change can be detected in most types of plaques. It has not been established, in either canine distemper or multiple sclerosis, whether perivascular cuffing precedes primary demyelination. We have presented evidence (Bowen & Davison, 1974) consistent with the view that macrophages or monocytic phagocytes (originating from either microglia or blood monocytes or both, p. 8) are one source of the increased cathepsin D activity.

Adams et al., (1975) believe that the macrophages seen in plaque tissue are derived locally, for they react negatively for catalase while typical reticuloendothelial cells are catalase positive. Although microglia and possibly also astrocytes may be quantitatively most important in changing the lysosomal enzyme content of multiple sclerosis tissue, other phenomena that cause changes in the activity of these enzymes may be more closely involved in the primary processes of the disease. Some possible sites of involvement of lysosomal enzymes in multiple sclerosis are shown in Fig. 1.8. It would not be unexpected if the causative or initiating factor(s) in this disease selectively affect the oligodendrocyte for primary demyelination, with sparing of axons, is a characteristic feature of multiple sclerosis.

Normal appearing white matter. Some reports indicate that the chemical composition of grossly normal appearing white matter from multiple sclerosis brain may be abnormal with respect to content of cerebroside, acid proteinase, basic protein and proportion of polyunsaturated fatty acid. These findings are difficult to interpret, for it is impossible to establish whether the specimens of tissue analysed were devoid of histological change, and it has been concluded

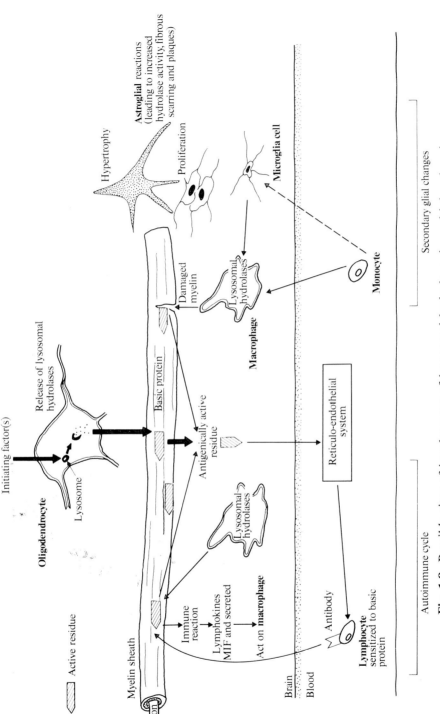

Fig. 1.8. Possible sites of involvement of lysosomal hydrolases in multiple sclerosis.

Lysosomal enzymes from at least 6 types of cell may be involved. In the figure the various cell types are in bold type. The heavy arrows, in the upper portion of the figure, indicate a possible primary sequence of events in which the initiating factor acts upon the oligodendrocytic lysosomes. (See review by Allison, 1971, on labilization of lysosomes in virus-infected cells.) An antigenically active residue, released from myelin basic protein by action of cathepsin D from oligodendrocytes, could initiate the auto immune cycle outlined in the left-hand portion of the figure. As a result of the perivascular reaction and initial damage to myelin the secondary glial reactions, outlined on the right, occur.

that there is no evidence for a primary lipid defect of myelin in multiple sclerosis (Alling et al., 1971). When changes in lipid composition of the apparently normal white matter are found, cholesterol esters accumulate and there is an increase of lysosomal hydrolase activity (Cuzner & Davison, 1974). It thus appears that the brain is normal in composition in regions remote from plaques before an attack of multiple sclerosis and where changes are found these are due to microscopic areas of demyelination.

Cerebrospinal fluid. Several changes have been reported to occur in the cerebrospinal fluid of cases of multiple sclerosis as compared with normal controls. The predominant characteristic of the cerebrospinal fluid profile indicative of multiple sclerosis is that approximately two-thirds of patients have a significant elevation of IgG expressed as a percentage of the total protein. Tourtellotte (1975) believes that most of the increased IgG is probably synthesized at the edges of plaques of demyelination by extra and perivascular plasma cells. Another change is that a number of studies have indicated that higher antibody titres to measles antigen are present in the cerebrospinal fluid (and serum) of patients compared with normal control subjects.

2. CONCLUSION

The association of plaques with venules and the resultant perivascular cuffing with infiltrating monocytes suggest that blood borne factors are of importance in the disease process. However the nature of the initiating factor remains elusive. Our attention should therefore be directed at finding such factors in plasma and cells. In addition, immunochemical studies on desensitization by the encephalitogenic protein and on the nature of the immunological reaction to this protein are relevant to an understanding of multiple sclerosis. Although it is likely that the causative agent will be a virus (see however Nemo et al., 1974) it may be that control of the secondary autoimmune response will prevent the irreversible scarring of nerve fibres and allow some remyelination to occur.

C. SENILE DEMENTIA

The signs and symptoms of senile dementia differ from those seen in multiple sclerosis, for in the latter they result from the destruction of the myelin sheaths that insulate nerve fibres and, in the former, the analogous hallmark may prove to include loss or dysfunction of synaptic terminal. (As the course of multiple sclerosis progresses, however, destruction of nerve fibres passing through areas of demyelination occurs. This no doubt accounts for the incidence of the dementia in multiple sclerosis patients.) Neurological diseases presenting with dementia include the cerebral atrophies (of which senile dementia is a prominent example) chronic inflammatory diseases, metabolic disorders, hydrocephalus and tumours (footnote to Table 1.1). By investigating the chemistry of abiotrophies (p. 4), it is hoped to elucidate the pathogenesis and to ultimately effect a therapeutic treatment, such as has so successfully been achieved in Parkinson's syndrome (p. 189).

In the western world, elderly patients account for nearly half of the total mental hospital population and, perhaps, at least a third of these patients have senile dementia. Therefore, this disease is the commonest organic nervous disease of mature adults. Organic brain syndromes of old age are associated with unmistakeable pathological changes in the brain, and common clinical features are memory disturbances, particularly of new impressions, impairment of other intellectual functions, disorientation and a coarsening of personality. Senile dementia *per se* is associated with diffuse loss of brain tissue and global impairment of intellect, ending in advanced dementia. There is little or no atheroma (Corsellis, 1962). The findings of some studies (Larsson *et al.*, 1963) are consistent with a genetic aetiology for the disease, although twin studies are inconclusive on this point. Other possible aetiological factors include transmittable agents, similar to those implicated in the subacute spongiform encephalopathies of man (p. 23, Table 1.5), aluminium toxicity (p. 25) and abnormal autoxidation processes. Autoimmune reactions (Nandy, 1973) may occur in normal ageing brains and this could be another aetiological factor for according to the 'ageing theory' of senile dementia the disease is an exaggerated form of the usual decline that occurs as people grow old. However, in contrast to multiple sclerosis (in which an autoimmune cycle probably occurs, Fig. 1.8) the neuropathological lesions in senile dementia (p. 15) are not associated with an inflammatory reaction. Furthermore, although the distribution of neuritic plaques may appear to be related to the distribution of vessels (Migakawa *et al.*, 1974) this may be coincidental due to the prolific capillary supply.

1. BIOCHEMICAL CORRELATES OF MORPHOLOGICAL CHANGE

Several attempts have been made to analyse the chemical composition of the abnormal structural elements (p. 33) seen in senile dementia. Thus, methods have been developed for the bulk isolation from brain in dementia of nerve cell bodies and a fraction enriched in twisted tubules (Iqbal *et al.*, 1974). Methods also exist for isolation of lipofuscin (Siakotos *et al.*, 1970) and neuritic plaques (Nikaido *et al.*, 1972). Using some of these techniques, neuronal perikarya have been isolated from hippocampal cortex from cases of senile dementia and then further sub-fractionated to yield a fraction enriched in twisted tubules. The fraction contains a unique protein band (M.W. 50,000) that migrates on SLS gels just ahead of tubulin; this new protein is similar to but not identical to neurofilament protein (Iqbal *et al.*, 1974).

Apart from the bizarre finding of silica in isolated neuritic plaques and neurofibrillary tangle bearing neurones (Nikaido *et al.*, 1972) there appears to be no other precise information on the chemical composition of the other abnormal structures that are seen in senile dementia.

It is possible that the onset of the clinical symptoms of at least pre-senile dementia is not primarily due to any of the pathological changes mentioned above (for example, see Ferrendelli et al., 1971 and Mann & Yates, 1974 for views on lipofuscin) but rather to an exaggerated loss of nerve cells (Colon, 1973) over that which may occur in normal ageing (p. 14). Based on morphological data, Wisniewski and Terry, (1973) conclude that neuronal fibrous proteins and changes in synaptic mitochondria (such as swelling, Fig. 1.4) are particularly important features in senility.

Although the presence of large numbers of neuritic plaques, granulovacuoles and neurofibrillary degeneration is characteristic of senile dementia, one or all of these changes occur in other progressive dementias (Table 1.1). Furthermore, since neuritic plaques but not neurofibrillary degeneration occur in aged dogs, while only the latter is found in Guam–Parkinsonism it seems unlikely that the neurofibrillary change is a prerequisite for the formation of neuritic plaques.

2. BIOCHEMICAL CHANGES

Apart from the work in this laboratory, and the investigations mentioned above, only limited studies on cerebral lipids (Bowen et al., 1973) and metabolites of catecholamines (Gottfries et al., 1969) have been carried out on the chemistry of senile dementia. Although most authorities agree (briefly reviewed by Wisniewski & Terry, 1973) that there is no major difference between the morphological changes seen in Alzheimer's disease and senile dementia, at least one epidemiological study suggests that these diseases are distinct entities (Larsson et al., 1963). Thus, it is difficult to assess how far the results of the metabolic studies in Alzheimer's disease (reviewed by Embree et al., 1972 and critically assessed by Appel & Festoff, 1971) can be extrapolated to senile dementia. The salient finding of these studies emphasized the similarity, rather than the difference, between biopsy tissue from Alzheimer's disease and normal patients with respect to protein synthetic activity, oxygen uptake and glucose utilization.

Proteins and decarboxylases. Perhaps the most striking of our findings in senile dementia are a marked deficiency in the concentration of a water-soluble acidic protein (neuronin S-6, Fig. 1.9) and the reduction in the activity of GAD and DOPAD (Table 1.6).

The change in the concentration of neuronin S-6 (M.W. about 47,000 on SDS polyacrylamide gels) appears to be characteristic of a chronic degenerative process for the change was not detected in acute experimental encephalopathies (Bowen et al., 1974b). As judged by criteria, such as peptide mapping, molecular weight and lack of reactivity towards specific antisera, the protein does not appear to be identical in properties to S-100, antigen α (14.3.2), actin, tubulin or neurofilament protein (Bowen et al., 1975a). Since the loss of neuronin

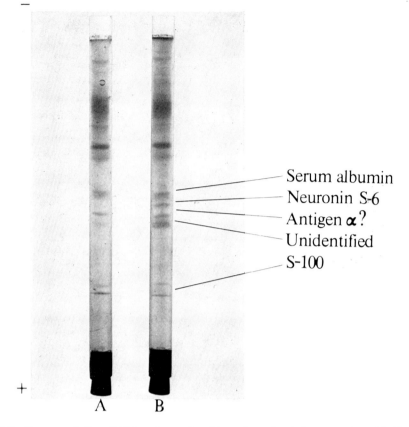

Fig. 1.9. Representative 7·5% polyacrylamide gels stained for protein with Procion Brilliant Blue.

Water-soluble proteins were extracted from frontal grey matter and subjected to electrophoresis. A. Senile dementia (22 cases; neuronin was within the normal range in only one brain), Alzheimers' disease (5 cases; neuronin S-6 was markedly depleted in 4 cases). B. Normal brain (20 cases). Modified from Smith et al. (1974b).

S-6 (a water-soluble protein) occurs in brain regions that rarely exhibit neurofibrillary degeneration (Smith et al., 1974b) the phenomenon does not appear to be directly related to the appearance of an SLS-soluble protein (M.W. 50,000) in the twisted tubule enriched fraction. It is of interest that two other water-soluble acidic proteins have been reported to be in a water-insoluble form in frontal cortex from cases of senile dementia and Alzheimer's disease (Hariguchi et al., 1969). One of these proteins is called B-4 (M.W. 52,000–55,000) and may be a constituent of senile plaques and/or neurofibrillary tangles (Nishimura et al., 1975).

Thirty brains from patients where the primary clinical diagnosis did not stress neurological disorder were studied. Where low levels of neuronin S-6 and GAD were found these invariably correlated with evidence of either some types

Table 1.6. Brain decarboxylase activities in senile dementia and related diseases (from Bowen et al., 1974c, 1975a and b)

Region	Activity in senile dementia[a] (% of age matched controls; control = 100%)	
	GAD	DOPAD
1. Putamen[b]	37·0 ($<0·001$)	2·6 ($<0·05$)
2. Caudate[b]	31·4 ($<0·001$)	15·0 ($<0·05$)
3. S. nigra[c]	27·9 ($<0·02$)	3·7 (N.S.)
4. G. pallidus[d]	20·6 ($<0·01$)	1·8 (N.S.)
5. Frontal g.m.[e]	34·0 ($<0·001$)	—
6. Thalamus[e]	22·4 ($<0·001$)	—
7. Temporal lobe[e,f]	36·7 ($<0·02$)	—

One case each of *Parkinson's disease* (P.D., 78 years old, with a degree of dementia but morphological evidence for senile degeneration was slight) and *Huntington's chorea* (H.C., 72 years old) were also examined. GAD activities (% of controls) for regions 1–7, respectively, were for P.D., 20·1, 4·0, 5·6, 8·1, 25·2, 21·2 and 86·9; and for H.C., 88·3, 56·3, 65·8, 62·0, 88·2, 108·5 and 123·5. DOPAD activities (% of controls) for regions 1–4, respectively, were for H.C., 122·3, 177·8, 333·0 and 374·0. Numbers in parenthesis refer to P (significance, Students t); N.S., non-significant decrease.

[a] calculated from the original data which were expressed as enzyme activity per gram fresh weight of tissue, [b-e] DOPAD activity is high ([b]), variable ([c]), moderately low ([d]), absent or too low to quantitate ([e]), [f] defined in Table 1.7.

of hypoxia or abnormal brain morphology (Bowen et al., 1975a). None of the cases of senile dementia (Table 1.6) exhibited post-mortem evidence of profound cerebral anoxia or, as far as could be judged, had blood defects or evidence of prolonged terminal coma. (This conclusion is substantiated by biochemical data for the levels of total protein were normal in senile dementia but were reduced in at least two regions from patients with signs of profound terminal hypoxia, Bowen D. M. & Smith C.B. unpublished.) Furthermore, if cyanosis was present this was only of the peripheral type immediately before death. However, one important difference between our cases was that the typical causes of death were cardiac failure for the controls and bilateral bronchopneumonia for the cases of senile dementia. Preliminary findings (Bowen D.M., Smith C.B. & White P., unpublished) indicate that the levels of neuronin S-6, GAD and DOPAD are preserved in the frontal cortex but are depleted in parts of the basal ganglia in non-demented patients dying with bronchopneumonia. However,

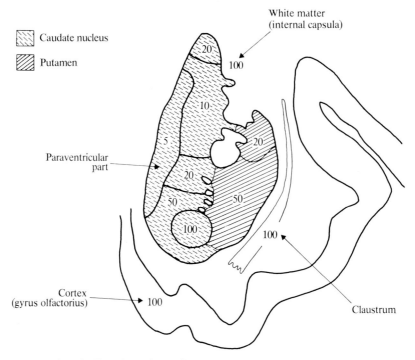

Fig. 1.10. Brain Choline Acetyltransferase activity in Huntington's Chorea (modified from Aquilonius, S.-M., Eckernas, S.-A. & Sundwall, A. (1975) Regional distribution of choline acetyltransferase in the human brain: changes in Huntington's chorea. *J. Neurol. Neurosurg. Psychiat.* **38**, 669 677). (Numbers are % of control brain values, where control = 100%, in a section 5 mm rostral to anterior commissure.)

although cardiac arrest was usually given as the actual cause of death in the controls (Table 1.6) the autopsy examination frequently showed that they exhibited pulmonary congestion, thickening of air passages, peribronchial fibroses and emphysema; in some instances there was also a clinical history of chronic bronchitis. Moreover, the case of Huntington's chorea (Table 1.6) died of bronchopneumonia but the GAD activities (and concentrations of neuronin S-6) were in most regions almost identical to the levels in the controls. The case of Parkinson's disease with dementia (Table 1.6) died suddenly with evidence of bronchopneumonia; GAD activities were markedly depleted in cortex, basal ganglia and substantia nigra. One of the cases of senile dementia (Table 1.6) and a patient with depressive illness both died at age 85 years within two weeks of each other in the same hospital. The cause of death in both cases was pulmonary thromboembolism. The clinical history and post-mortem examination indicated that the respiratory system of the depressive patient was the more chronically diseased. The levels of neuronin S-6, GAD and DOPAD in this patient were typical of the controls while the levels in the dement were conspicuously within the range for cases of senile dementia dying of bronchopneumonia. In summary, we conclude that the effect of bronchopneumonia and

Table 1.7. Biochemical indices of brain degeneration in senile dementia[a]

Parameter	Measured as potential index of:	% of control (control = 100%)	N[b]	P. (Students 't' test)
SELECTIVE LOSSES				
Neuronin	unknown	15·6	5	<0·01
GAD	synaptic terminals; GABA-ergic neurones	34·8	4	<0·05
Succinic acid dehydrogenase	mitochondria	67·4	6	<0·01
2′, 3′ cyclic nucleotide 3′ phosphohydrolase	myelinated axons	66·8	6	<0·02
Fresh wt. temporal lobe	gross atrophy	81·3	6	<0·05
UNCHANGED				
Antigen α (14.3.2)?	neuronal perikarya	74·3	5	N.S.
Total protein	—	81·5	6	N.S.
Carbonic anhydrase	'glial cells'	88·3	6	N.S.
β-galactosidase/ β-glucuronidase	cellularity[c]	84·8	6	N.S.
DNA	cellularity	96·6	6	N.S.
RNA/DNA	neurones (rel. number)	100·0	6	N.S.

[a] Bowen D.M., Smith C.B., White P., & Davison A.N. (unpublished). Results are based on analysis of the entire 'temporal lobe' (tissue ventral to the lateral cerebral fissure extended to meet the occipital lobe). Data (corrected for wt. differences in lobes between sexes) from which these results were calculated were expressed/lobe.
[b] Number of cases of senile dementia, mean age 85 years (range 79–97), compared with 4–5 controls, mean age 83 years (range 70–100).
[c] Ratio of neuronal perikarya/reactive glia.

other anoxemic conditions may contribute to the terminal biochemical changes particularly in the region of the basal ganglia.

Other indices. In an on-going study (Table 1.7) an attempt is being made to establish which neural element(s) is quantitatively most affected in senile dementia. This approach is clearly fraught with difficulty for, apart from post-mortem changes, the whole question of definitive biochemical indices of cell type and subcellular structures is tentative. Nevertheless, with these reservations the parameters listed in Table 1.7 have been used. To circumvent expressing the results in terms of unit mass, the entire temporal lobe (which includes the hippocampus) was homogenized for the assay of the potential markers. Although the mean weight of the temporal lobes in dementia was decreased by about 20%, no changes were found in the values of the potential indices of neuronal perikarya and the total number of neurones and glial cells. In contrast, parameters that may reflect the number or state of myelinated axons and synaptic terminals were reduced to about a third or half normal. Neuronin S-6 is a soluble protein that is present in the primary cytosol fraction of brain homogenates but is conspicuously absent from bulk-isolated neuronal perikarya (Smith, 1975). The temporal lobes from the cases of senile dementia were markedly depleted in this protein (Table 1.7). In contrast, another soluble protein, (which may be 14.3.2), is found in bulk-isolated neuronal perikarya but was not changed in concentration in senile dementia (Table 1.7). From these results it is tempting to conclude that the pathological process in senile dementia selectively affects the nerve cell distal to the perikaryon.

In our attempt to establish more precisely the inter-relationships between each of the biochemical parameters the data have been analysed using a new technique for analysing complex biological data (Gedye J.L. & Lee M., personal communication). This zygological method is based on the Q analysis technique (Atkins, 1974) and can be thought of as a generalization of the principles of multidimensional chromatography (the 'mixture' being the set of brains under study while the indices form the different developing solvents which run orthogonally to each other). In brief, the analysis of the data expressed per temporal lobe shows that most of the parameters measured (Table 1.7) run together as two separate groups. Group I (includes GAD, DOPAD, and neuronin) links very closely with diagnosis via GAD activity. Group II (includes most of the other parameters) is weakly, but directly, linked to diagnosis through succinate dehydrogenase and ganglioside NANA. Zygological analysis of the inter-relationship between GAD activity in the various brain regions examined (Table 1.6) shows that the activity in frontal grey matter is linked most closely to the diagnosis of senile dementia. This relationship is not apparent from the 't' test data but is not unexpected for of the regions studied the frontal grey matter exhibits the most marked morphological damage and is relatively unaffected by bronchopneumonia.

3. CONCLUSIONS

Pathogenesis of senile dementia. The zygological analyses suggest that the changes in the activities of neurotransmitter related enzymes (particularly GAD activity in frontal cortex) and concentration of neuronin may play a key role in explaining the pathogenesis of senile dementia. The biochemical observations thus appear to suggest that there is a major disability in at least non-cholinergic synapses in senile dementia. Although glutamic acid decarboxylase has been shown to be enriched in synaptosomes from rat brain it is not clear whether the changes in brain decarboxylases in senile dementia are due to a degeneration of nerve-terminals or whether they result from a more specific terminal disturbance of neurotransmitter metabolism.

The reduction in activity of the brain decarboxylases occurs in the thalamus and basal ganglia (Table 1.6) which are regions that usually exhibit little evidence of light microscopic changes in senile dementia (Corsellis J.A.N., personal communication). However, there is other biochemical evidence that dopamine metabolism, at least, is disturbed in the basal ganglia in both senile dementia (Gottfries *et al.*, 1969) and Alzheimer's disease (Gottfries *et al.*, 1974). It has also been reported that the neurofibrillary changes seen in brain stem and hypothalamus in Alzheimer's disease have a distribution that is strikingly similar to that of monoamine containing nerve cells (Ishii, 1966). Furthermore, Parkinsonian signs are apparently present in over 60% of patients with Alzheimer's disease (Pearce, 1974). One explanation for the findings of enzyme and protein changes in regions that have so far been found to show relatively little morphological abnormality is that these changes are non-disease specific effects, secondary to chronic oxygen deprivation, and brought about by terminal broncho-pneumonia and reduced blood flow. Another possibility is that the structural changes or atrophy in all parts of the cerebral cortex may lead to secondary enzymatic changes in the fibres that are known to connect the basal ganglia and thalamus. However, chronic simultaneous unilateral pre-frontal (Cowey & Bozek, 1974) and dorsal hippocampal (Stevens & Cowey, 1973) lesions do not affect the activities of the neurotransmitter related enzymes in discrete regions of the basal ganglia and thalamus (Bowen D.M. & White P., unpublished).

Comparison with other abiotrophies. Pioneer studies by Hornykiewicz and Perry and associates have demonstrated that neurotransmitter metabolism is defective in Parkinson's disease (Table 1.8) and Huntington's chorea (Table 1.8). Other studies extend this work for they indicate that in Huntington's chorea the concentration of muscarinic receptor is decreased (Hiley & Bird, 1974), and choline acetyltransferase activity is selectively affected in parts of the striatum (Fig. 1.10).

The magnitude of the fall in GAD in the putamen and the caudate nucleus in senile dementia is intermediate between the effect in Parkinson's disease and in Huntington's chorea (Table 1.8). Although Parkinsonian signs were not conspicuous in most of these cases of senile dementia, the extent which DOPAD activity is reduced in the caudate nucleus and the putamen is similar to that reported in Parkinson's disease (Table 1.8). However, the reduced levels of DOPAD were not unexpected, for in dements homovanillate levels in the caudate nucleus are reduced to almost half normal (Table 1.8) which is not dissimilar to the fall that occurs in this region in Parkinson's disease (Table 1.8). The results also agree with other findings in Parkinson's disease (Lloyd *et al.*, 1973) for the mean reduction in DOPAD activity in senile dementia is greater in the putamen than the caudate nucleus (Table 1.8) and the marked decrease in DOPAD activities in brain areas such as the substantia nigra and the globus pallidus is not statistically significant presumably due to the larger scatter in the elderly control group. Contrasting with these similarities, preliminary findings (footnotes

Table 1.8. Summary of metabolism of dopamine and γ-aminobutyric acid in some neurological diseases

Parameter	UNITS IN DISEASED TISSUE (% control; control = 100%)					
	Parkinson's disease		Huntington's chorea		Senile dementia	
	Putamen	Caudate	Putamen	Caudate	Putamen	Caudate
Homovanillate	13 (S)[a]	41 (S)[a]	90 (NS)[a]	63 (S)[a,g]	64 (S)[f]	58 (S)[f]
Tyrosine hydroxylase	48 (3)[b]	47 (3)[b]	107 (3)[b]	165 (3)[b]	nd	nd
Dopa decarboxylase	8 (S)[a]	14 (S)[a]	122 (1)[c,h]	178 (1)[c,h]	3 (S)[c]	15 (S)[c]
γ-Aminobutyric acid	110 (1)[d]	82 (1)[d]	38 (S)[d]	60 (S)[d]	nd	nd
Glutamic acid decarboxylase	48 (S)[a]	49 (S)[a]	27 (S)[e]	20 (S)[e]	37 (S)[c]	31 (S)[c]

Symbols in parenthesis indicate either number of brains or significance (S, significant where P = at least <0·05, usually <0·01; NS, non-significant); nd, not determined.
a–h = references:
[a] Bernheimer et al., 1973; Lloyd et al., 1973. [b] McGeer et al., 1973. [c] Bowen et al., 1975b. [d] Perry et al., 1973. [e] Bird & Iversen, 1974. [f] Gottfries et al., 1969. [g] Dopamine level is normal in non-rigid form but is unusually high in the rigid form. [h] Non-rigid form.

to Table 1.6) indicate that in comparison to the widespread changes in senile dementia, decarboxylase activities are more selectively affected in Parkinson's disease and Huntington's chorea. For example, in the case of Huntington's chorea, DOPAD activities were not decreased in any part of the brain and in comparison with the changes in senile dementia the GAD activities in the case of Parkinson's disease (with dementia) were reduced more in the substantia nigra but less in the temporal lobe. It is difficult to draw conclusions from comparisons such as these for, although they support the concept of pathoclisis at the regional and neuronal level (p. 5), they are based on differences in enzyme activities expressed per unit mass and take no account of, for example, the selective atrophy of caudate nucleus in Huntington's chorea. Thus the data in Table 1.8 may over-emphasize the degree of abnormality in the basal ganglia for in senile dementia thinning of the cortical ribbon is greater than atrophy of the basal ganglia; furthermore no correction was made for agonal effects.

V Summary

This chapter highlights recent advances made in the neurochemistry of the abiotrophic neurological diseases; particular emphasis is placed on the concept of pathoclisis. Thus, the selective changes that occur in the activities of neurotransmitter related enzymes (DOPAD, GAD and choline acetyl-transferase) in many disorders (e.g., senile dementia, Huntington's chorea, Parkinson's disease, prolonged coma, CO poisoning and ischaemia), may reflect differences in the vulnerability of different types of neurones to different damaging agents. Alternatively, this selectivity may be due to specific effects on the individual enzymes (e.g. dissociation of the polymeric GAD molecule into enzymatically inactive subunits) or to differences in compartmentization (e.g., most of the DOPAD and choline acetyl-transferase activities in fractionated brain homogenates are recovered in the primary cytosol and synaptosomal fractions, respectively. In dementia DOPAD activity in putamen is reduced from normal by over 90% while unpublished data show that choline acetyl-transferase activity is reduced by less than 50%). As judged by morphological and functional criteria relatively well-preserved synaptosomes can be isolated from human postmortem tissue (p. 28). Therefore, it should be possible to measure the uptake of specific neurotransmitters into the particulate fraction of human brain homogenates. Experiments of this type (for a particularly elegant example using rat brain, see Cuello *et al.*, 1973) may indicate whether the changes in neurotransmitter-related enzymes are due to loss of specific types of synapses.

In this chapter we have also shown how biochemical measurements of specific macromolecules, including the activity of certain enzymes, can be correlated with the pathological changes in degenerative diseases. There are many difficulties in interpretation of the molecular changes occurring in the damaged tissue. Much of the information relates to the end result of a degenerative process which originated many years earlier. Thus the failure to detect a virus in the CNS of multiple sclerosis patients may be simply because it is only present at the onset of the disease, in late childhood. As a result biochemists would necessarily examine the secondary effects of the infective agent rather than studying the mechanism of its initial attack. We have referred to other problems in dealing with post-mortem tissues. The question of adequate controls is one such difficulty, even those dying without clear evidence of neurological disease may, in our experience, be found to have clear histological evidence of neural involvement. This could well account for the low GAD activity reported in brain regions from some control patients by Urquhart et al. (1975). Pre-mortem conditions (e.g., broncho-pneumonia) are other possible complicating factors. Despite these restrictions, neurochemical observations can be of value. For example, the discovery of reduced GAD activity in the abiotrophic diseases focuses attention on the possible vulnerability of the GABA-ergic neurone (Roberts, 1974) and its synaptic connections. Changes seen in the concentration of the encephalitogenic basic protein in plaque-tissue from multiple sclerotic brains indicates the possible importance of lysosomal cathepsins and neutral proteinase, at least in the later stages of the demyelinating process. Further research on acute material from diseased brains should help us to find the source of the destructive proteinases.

In the future, there is a clear need to continue and extend the work described in this chapter, using carefully evaluated post-mortem and biopsy material. Release of brain proteins (e.g., tubulin or neuronin) or their products into the CSF should provide the basis of a valuable diagnostic method for evaluation of the degenerative disorders (see chapter 6). There is also the likelihood that on-going pathological changes in the CNS may be accompanied by immunological and other biochemical changes in the plasma and leucocytes. For example, increased polymorphonuclear neutral proteinase activity is related to exacerbation of the disease in patients with multiple sclerosis . Apart from such clinical studies, progress must come from the use of better animal models for the neurological diseases and from the more extensive use of experimental brain lesions as a means of following degenerative change.

Acknowledgments

We wish to thank all our colleagues whom we have consulted for advice during the preparation of this chapter. In particular, we are grateful to Professor W. Blackwood and Dr. J.A.N. Corsellis for critical advice, and the supply of both post-mortem specimens and many of the photographs that appear in this chapter.

References

ADAMS C.W.M., BAYLISS G.B. & TURNER D.R. (1975) Phagocytes, lipid removal and regression of atheroma. *J. Path.*, in press.

ALDERMAN J.L. & SHELLENBERGER M.K. (1974) γ-Aminobutyric acid (GABA) in the rat brain: re-evaluation of sampling procedures and the post-mortem increase. *J. Neurochem.* **22**, 937–940.

ALLING C., VANIER M.T. & SVENNERHOLM L. (1971) Lipid alterations in apparently normal white matter in multiple sclerosis. *Brain Res.* **35**, 325–336.

ALLISON A.C. (1967) Lysosomes in virus infected cells. *Perspectives in Virology* **5**, 29–61.

ALLISON A.C. (1971) The role of membranes in the replication of animal viruses. *Int. Rev. Expt. Pathol.* **10**, 182–235.

ANON (1974) Abiotrophies. *Br. med. J.* **1**, 337–338.

APPEL S.H. & FESTOFF B.W. (1971) Biochemical dysfunction and dementia, in Wells C.E. (Ed.) *Dementia*, pp. 133–149. Oxford, Blackwell Scientific Publications.

ARMSTRONG D., DIMMITT S. & VAN WORMER D.E. (1974) Studies in Batten disease 1. Peroxidase deficiency in granulocytes. *Arch. Neurol.* **30**, 144–152.

ARNASON B.G.W., FULLER T.C., LEHRICH J.R. & WRAY S.H. (1974) Histocompatibility types and measles antibodies in multiple sclerosis and optic neuritis. *J. neurol. Sci.* **22**, 419–428.

ATKINS R.H. (1974) *Mathematical structure in human affairs*. London, Heinemann.

BAILEY R.R. (1972) Aluminium toxicity in rats and man. *Lancet* **ii**, 276–277.

BERNHEIMER H., BIRKMAYER W., HORNYKIEWICZ O., JELLINGER K. & SEITELBERGER F. (1973) Brain dopamine and the syndromes of Parkinson and Huntingdon. *J. neurol Sci.* **20**, 415–455.

BIGNAMI A. & ENG L.F. (1973) Biochemical studies of myelin in Wallerian degeneration of rat optic nerve. *J. Neurochem.* **20**, 165–173.

BIGNAMI A. & RALSTON H.J. (1969) Myelination of fibrillary astroglial processes in long term Wallerian degeneration. The possible relationship to 'status marmoratus'. *Brain Res.* **11**, 710–713.

BIRD E.D. & IVERSEN L.L. (1974) Huntington's Chorea, post-mortem measurement of glutamic acid decarboxylase, choline acetyl-transferase and dopamine in basal ganglia. *Brain* **97**, 457–472,

BLAKEMORE W.F. (1972) Observations on oligodendrocyte degeneration, the resolution of status spongiosus and remyelination in cuprizone intoxication in mice. *J. Neurocytol* **1**, 413–426.

BOWEN D.M. & DAVISON A.N. (1974) Macrophages and cathepsin A activity in multiple sclerosis brain. *J. neurol. Sci.* **21**, 227–231,

BOWEN D.M., DAVISON A.N. & RAMSEY R.B. (1974a) Dynamic role of lipids in the nervous system, in Goodwin T.W. (Ed.) *Biochemistry of Lipids (MPP International Review of Sciences, Biochemistry series ONE)*, pp. 141–179. London, Butterworths.

BOWEN D.M., FLACK R.H.A., MARTIN R.O., SMITH C.B., WHITE P. & DAVISON A.N. (1974b) Biochemical studies on degenerative neurological disorders 1. Acute experimental encephalitis. *J. Neurochem.* **22**, 1099–1107.

BOWEN D.M., FLACK R.H.A., WHITE P., SMITH C.B. & DAVISON A.N. (1974c) Brain-decarboxylase activities as indices of pathological change in senile dementia. *Lancet* **ii**, 1247–1249.

BOWEN D.M. & RADIN N.S. (1969) Cerebroside galactosidase: A method for determination and a comparison with other lysosomal enzymes in developing rat brain. *J. Neurochem.* **16**, 501–511.

BOWEN D.M., SMITH C.B. & DAVISON A.N. (1973) Molecular changes in senile dementia. *Brain* **96**, 849–856.

BOWEN D.M., SMITH C.B., WHITE P. & DAVISON A.N. (1975a) Senile dementia and related abiotrophies: in *The Neurobiology of Ageing*, Raven Press, New York, in press.

BOWEN D.M., WHITE P. & DAVISON A.N. (1975b) Glutamic acid and L-DOPA decarboxylase activities in senile dementia. *Proc. VIIth Int. Cong. Neuropath*, in press.

BRAAK H. (1971) Uber das Neurolipofuscin in der unteren Olive und dem dentatus Cerebelli in Gehirn des Menschen. *Z. Zellforsch.* **121**, 573–592.

BRAY D. & BROWNLEE S. (1973) Peptide mapping of proteins from acrylamide gels. *Anal. Biochem*, **55**, 213–221.

BRUNK U. (1973) Distribution and shifts of ingested marker particles in residual bodies and other lysosomes. *Expt. Cell Res.* **79**, 15–27.

BRUNK U. & ERICSSON J.L.E. (1972) Electron microscopical studies on rat brain neurons. Localization of acid phosphatase and mode of formation of lipofuscin bodies. *J. Ultrastructure Res.* **38**, 1–15.

BRUNK U., ERICSSON J.L.E., PONTEN J. & WESTERMARK B. (1973) Residual bodies and 'aging' in cultured human glia cells. *Expt. Cell Res.* **79**, 1–14.

BUENING G.M. & GUSTAFSON D.P. (1971) Growth characteristics of scrapie-infected mouse brain cell cultures. *Am. J. vet. Res.* **32**, 953–958.

BYINGTON D.P. & JOHNSON H.P. (1972) Experimental subacute sclerosing panencephalitis in the hamster: correlation of age with chronic inclusion cell-encephalitis. *J. Infectious Dis.* **126**, 18–25.

CARPENTER S., KARPATI G., ANDERMANN F., JACOB J.C. & ANDERMANN E. (1974) Lafora's disease: Peroxisomal storage in skeletal muscle. *Neurology* **24**, 531–538.

CAVANAGH J.B. (1972) Cellular abnormalities in the brain in chronic liver disease. *Scientific Basis of Med. Ann. Revs.* 238–247.

CAVANAGH J.B. (1973) Peripheral neuropathy caused by chemical agents. *Critical Rev. in Toxicology* **2**, 365–417.

CAVANAGH J.B. (1974) Liver bypass and the glia, *Res. Publ. Ass. nerv. ment. Dis.* **53**, 13–38.

CLENDENON N.R., ALLEN N., KOMATSU T., LISS L., GORDON W.A. & HEIMBERGER K. (1971) Biochemical alterations in the anoxic-ischemic lesion of rat brain. *Arch. Neurology* **25**, 432–448.

COLON E.J. (1973) The cerebral cortex in presenile dementia: A quantitative study. *Acta neuropath. (Berl.)* **23**, 281–290.

CORSELLIS J.A.N. (1962) Mental illness and the ageing brain. *Maudsley Monograph, No. 9.* Oxford University Press.

CORSELLIS J.A.N. (1969) The pathology of dementia. *Brit. J. Hospital Med.* **10**, 695–703.

COWEY A. & BOZEK T. (1974) Contralateral neglect after unilateral dorsomedial preferential lesions in rats. *Brain Res.* **72**, 53–63.

CRAGG B.G. (1968) Gross, microscopical and ultramicroscopical anatomy of the adult nervous system, in Davison A.N. & Dobbing J. (Eds.) *Applied Neurochemistry* pp. 3–47. Oxford, Blackwell Scientific Publications.

CRAGG B.G. (1974) Plasticity of synapses. *Brit. Med. Bull.* **30**, 141–144.

CRAGG B.G. (1975) The density of synapses and neurones in normal, mentally defective and aging human brains. *Brain*, **98**, 81–90.

CRAPPER D.R., KRISHNAM S.S. & DALTON N.J. (1973) Brain aluminium distribution in Alzheimer's disease and experimental neurofibrillary degeneration. *Science* **180**, 511–513.

CUELLO A.C., HORN A.S., MACKAY A.V.P. & IVERSEN L.L. (1973) Catecholamines in the median eminence: new evidence for a major noradrenergic input. *Nature (London)* **243**, 465–467.

CUZNER M.C. & DAVISON A.N. (1973) Changes in cerebral lysosomal enzyme activities and lipids in multiple sclerosis. *J. neurol. Sci.* **19**, 29–36.

DE LEON G.A. (1974) Bielschowsky bodies: Lafora-like inclusions associated with atrophy of the lateral pallidum. *Acta neuropath. (Berl.)* **30**, 183–188.

DE MARCHEMA O., GUARNIERI M. & MCKHANN G. (1974) Gluthatiane peroxidase levels in brain. *J. Neurochem.* **22**, 773–776.

DINGLE J.T., FELL H.B. & COOMBS R.N.A. (1967) The breakdown of embryonic cartilage and bone cultivated in the presence of complement-sufficient antiserum, Part 2 (Biochemical changes and the role of the lysosomal system). *Int. Arch. Allergy* **31**, 293–303.

DUFFEY T.E., NELSON S.R. & LOWRY O.H. (1972) Cerebral carbohydrate metabolism during acute hypoxia and recovery. *J. Neurochem.* **19**, 959–978.

EMBREE L.J., BASS N.H. & POPE A. (1972) Biochemistry of middle and late life dementias, in Lajtha A. (Ed.) *Handbook of Neurochemistry* **1**, pp. 329–369.

EMSON P.C. & JOSEPH M.H. (1975) Neurochemical and morphological changes during development of cobalt induced epilepsy in the rat. *Brain Res.*, **93**, 91–110.

EVANS G.W., DUBOIS R.S. & HAMBIDGE K.M. (1973) Wilson's disease; identification of an abnormal copper-binding protein. *Science* **181**, 1175–1176.

ENG L.F., VANDERHAEGHEN J.J. & GERSTL B. (1971) An acidic protein isolated from fibrous astrocytes. *Brain Res.* **28**, 351–354.

FERRENDELLI J.A. & MCDOUGAL D.B. (1971) The effect of audiogenic seizures on regional CNS energy reserves, glycolysis and citric acid cycle flux. *J. Neurochem.* **18**, 1207–1220.

FERRENDELLI J.A., SEDGWICK W.G. & SUNTZEFF V. (1971) Regional energy metabolism and lipofuscin accumulation in mouse brain during aging. *J. Neuropath. Exp. Neurol.* **30**, 638–649.

FRIED R. & MANDEL P (1974) Superoxide dismutase of mammalian nervous system. *J. Neurochem.* **23**, in press.

GAREY R.E. & HEATH R.G. (1974) Uptake of catecholamines by human synaptosomes. *Brain Res.* **79**, 520–523.

GIBBS C.J. & GAJDUSEK D.C. (1972) Isolation and characterization of the subacute spongiform virus encephalopathies of man: kuru and Creutzfeldt-Jakob disease. *J. clin. path* **25**, Suppl. (*Roy Coll. Path.*) **6**, 84–96.

GORDON S. & COHN Z.A. (1973) The macrophage. *Int. Rev. Cytol.* **36**, 171–212.

GOLDFISHER S., MOORE C.L., JOHNSON A.B., SPIRO A.J., VALSAMIS M.P., WISNIEWSKI H.K., RITCH R.H., NORTON W.T., RYPIN I. & GARTNER L.M. (1973) Peroxisomal defects in cerebro-hepato-renal syndrome. *Science* **182**, 62–64.

GOTTFRIES C.G., GOTTFRIES R. & ROOS B.E. (1969) The investigation of homovanillic acid in the human brain and its correlation to senile dementia. *Brit. J. Psychiat.* **115**, 563–574.

GOTTFRIES C.G., KJALLQUIST A., PONTEN U., ROOS B.A. & SUNDBAIG G. (1974) CSF pH and monoamine and glucolytic metabolites in Alzheimer's disease. *Br. J. Psychiat.* **124**, 280–287.

GOVINDARAJAN K.R., RAUCH H.C., CLAUSEN J. & EINSTEIN E.R. (1974) Changes in cathepsins B, and D, Neutral Proteinase and 2', 3'-cyclic nucleotide-3-phosphohydrolase activities in monkey brain with experimental allergic encephalomyelitis. *J. neurol. Sci.* **23**, 295–306.

HALLPIKE J.F. (1972) Enzyme and protein changes in myelin breakdown and multiple sclerosis, in Graumann W., Lojda Z., Pearse A.G.E. & Schiebler T.H. (Eds.)

Progress in Histochemistry and Cytochemistry, Vol. 3, pp. 179–216. Stuttgart, G.F. Verlag.

HARIGUCHI S., NISHIMURU T., ICHIMARU S. & KANEKO J. (1969) The cerebral water soluble proteins in patients with presenile and senile dementia. *Shinkei Kagaku* **8**, Suppl. 9, 9.

HARRISON W.H., CODD E. & GRAY R.M. (1972) Aluminium inhibition of hexokinase. *Lancet* **ii**, 277.

HARRISON M.J.G., BROWNBILL D., LEWIS P.D. & ROSS RUSSELL R.W. (1973) Cerebral edema following carotid artery ligation in the gerbil. *Arch. Neurol.* **28**, 389–391.

HARRISON B.M., McDONALD W.I. & OCHOA J. (1972) Central demyelination produced by diphtheria toxin: an electron microscopic study. *J. neurol. Sci.* **17**, 281–291.

HARTMANN J.F., BECHER K.A. & COHEN M.M. (1973) Cerebral ultrastructure in experimental hypoxia and ischemia, in Cohen M.M. (Ed.) *Monog. in Neurol. Sciences 1*, 50–64 in '*Biochemistry, Ultrastructure and Physiology of Cerebral anoxia, Hypoxia and Ischemia*'. Basel, S. Karger.

HARTMAN E.R., COLASANTI B.K. & CRAIG C.R. (1974) Epileptogenic properties of cobalt and related metals applied directly to cerebral cortex of art. *Epilepsia* **15**, 121–129.

HILEY C.R. & BIRD E.D. (1974) Decreased muscarinic receptor concentration in post-mortem brains in Huntington's chorea. *Brain Res.* **80**, 355–358.

HUNTER G.D. (1972) Scrapie: a prototype slow infection. *J. infect. Dis.* **125**, 427–440.

IQBAL K. & TELLEZ-NAGEL I. (1972) Isolation of neurons and glial cells from normal and pathological human brain. *Brain Res.* **45**, 296–301.

IQBAL K., WISNIEWSKI H.M., SHELANSKI M.L., BROSTOFF S., LIWNICZ B.H. & TERRY R.D. (1974) Protein changes in senile dementia. *Brain Res.* **77**, 337–343.

ISHII T. (1966) Distribution of Alzheimer's neurofibrillary changes in the brain stem and hypothalamus of senile dementia. *Acta Neuropath.* **6**, 181–187.

IVERSEN S.D. (1973) Brain lesions and memory in animals, in Deutsch J.A. (Ed.) *The Physiological Basis of Memory*, pp. 305–356. New York, Academic Press.

JACOB H. (1963) Patterns of CNS vulnerability—CNS tissue and cellular Pathology in Hypoxaemic states, in Schadé J.P. & McMenemey W.H. (Eds.) *Selective Vulnerability of the Brain in Hypoxaemia*, pp. 153–163. Oxford, Blackwell Scientific Publications.

JOHNSON H.A. & ERNER S. (1972) Neuron survival in the aging mouse. *Exp. Geront.* **7**, 111–117.

JOHNSON R.T. & WEINER L.P. (1972) The role of viral infections in demyelinating diseases, in Wolfgram F., Ellison G.W., Stevens J.G. & Andrews J.M. (Eds.) *Multiple Sclerosis, Immunology, Virology and Ultrastructure*, pp. 245–264. New York, Academic Press.

JOHNSON R.T. (1972) Effects of viral infections on the developing nervous system. *N.E. J. Med.* **287**, 599–604.

KIMBERLIN R.A. (1973) Subacute spongiform encephalopathies in domestic and laboratory animals. *Biochem. Soc. Trans.* **1**, 1058–1061.

KORNGUTH S., WANNAMAKER B., KOLODNY E., GEISON R., SCOTT G. & O'BRIEN J.F. (1974) Subcellular fractions from Tay-Sach's brain: ganglioside lipid and protein composition and hexosaminidase activities. *J. neurol. Sci.* **22**, 383–406.

KRIGMAN M.R., DRUSE M.J., TRAYLOR T.D., WILSON M.H., NEWELL L.R. Lead encephalopathy in the developing rat: effect on cortical ontogenesis. Hogan E.L. (1974) *J. Neuropath. Exp. Neurol.* **33**, 671–686.

LAMPERT P.W. & CRESSMAN M. (1966) Fine structural changes of myelin sheaths after axonal degeneration in the spinal cord of rats. *Am. J. Path.* **49**, 1139–1155.

LANGEVOORT H.L., COHN Z.A., HIRSCH J.G., HUMPHREY J.H., SPECTOR W.G. & VAN FURTH R. (1970) Introduction, in Van Furth R. (Ed.) *Mononuclear Phagocytes*, p. 5. Oxford, Blackwell Scientific Publications.

LARSSAN T., SJOGREN T. & JACOBSON G. (1963) Senile dementia, a clinical sociomedical and genetic study. *Acta Psychiat. Scand. Suppl. 167*, **39**, 1–259.

LAZARUS S.S., WALLACE B.J., EDGAR G.W.F. & VOLK B.W. (1962) Enzyme localization in rabbit cerebellum and effect of post mortem autolysis. *J. Neurochem* **9**, 227–232.

LEVINE S. & PAYAN H. (1966) Effects of ischemia and other procedures on the brain and retina of the gerbil (*meriones unguiculatus*). *Exp. Neurol.* **16**, 255–262.

LLOYD K.G., DAVIDSON L. & HORNYKIEWICZ O. (1973) Metabolism of levodopa in the human brain, in Calne D.B. (Ed.) *Advances in Neurology, Vol. 3: Progress in the Treatment of Parkinsonism*, pp. 173–188. New York, Raven Press.

LOWRY O.H. (1966) Metabolic levels as indicators of control mechanisms. *Fed. Proc.* **25**, 846–849.

MCGEER P.L., MCGEER E.G. & FIBIGER H.C. (1973) Choline acetylase and glutamic acid decarboxylase in Huntington's chorea. *Neurology* **23**, 912–917.

MCGREGOR D.D. & MACKANESS G.B. (1974) Lymphocytes, in Zweifach B.W., Grant L., McLuskey R.T. (Eds.) *The Inflammatory Process*, Vol. III, pp. 1–32. New York, Academic Press.

MCMARTIN D.N., KOESTNER A. & LONG J.F. (1972) Enzyme activities associated with the demyelinating phase of canine distemper. *Acta Neuropath (Berl.)* **22**, 275–287.

MACKENZIE A. (1971) Applied enzyme histochemistry in scrapie. *Ph.D. Thesis*, University of Reading.

MANN D.M.A. & YATES P.O. (1974) Lipoprotein pigments—their relationship to ageing in the human nervous system I. The lipofuscin content of nerve cells. *Brain* **97**, 481–488.

MARKLAND S. & MARKLAND G. (1974) Involvement of superoxide anion radical in the autotoxidation of pyrogallol and a convenient assay for superoxide dismutase. *Eur. J. Biochem.* **47**, 469–474.

MATTHIEU J.M., ZIMMERMAN A.W., WEBSTER H. DE F., ULSAMER A.G., BRADY R.O. & QUARLES R.H. (1974) Hexachlorophene intoxication: characterization of myelin and myelin related fractions in the rat during early postnatal development. *Expl. Neurol.* **45**, 558–575.

MIYAKAWA T., SUMIYOSHI S., MURAYAMA E. & DESHIMARU M. (1974) Ultrastructure of capillary plaque-like degeneration in senile dementia. *Acta Neuropath (Berl.)* **29**, 229–236.

NANDY K. (1973) Brain-reactive antibodies in serum of aged mice. *Prog. Brain Res.* **40**, 437–454.

NEMO G.J., BRODY J.A. & WATERS D.J. (1974) Serological responses of multiple-sclerosis patients and controls to a virus isolated from a multiple-sclerosis case. *Lancet* **ii**, 1044–1046.

NIKAIDO T., AUSTIN J., TRUEF L. & RHINEHART R. (1972) Studies in aging of the brain II Microchemical analysis of the nervous system of Alzheimer disease patients. *Arch. Neurol.* **27**, 549–554.

NISHIMURA T., HARIGUCHI S., TADA K. & KANEKO Z. (1975) Changes in brain water soluble proteins in presenile and senile dementia. *Proc. VIIth Intern. Cong. Neuropath*. in press.

NISHIOKA N., TAKAHATA N. & IZUKA R. (1968). Histochemical studies on the lipopigments in the nerve cells. A comparison with lipofuscin and ceroid pigment *Acta Neuropath.* **11**, 174–181.

NORENBERG M.D. & LAPHAM L.W. (1974) The astrocyte response in experimental portal-systemic encephalopathy: an electron microscopic study. *J. Neuropath. Expt. Neurol.* **33**, 422–435.

OGATA J. & FEIGIN I. (1973) The relative weight of grey and white matter of the normal human brain. *J. Neuropath. Expt. Neurol.* **32**, 585–589.

PASSONNEAU J.V. & LOWRY O.H. (1971) Metabolite flux in single neurons during ischemia and anesthesia, in Dubach U.C. & Schmidt U. (Eds.) *Recent Advances in Quantitative Histo- and Cytochemistry*, pp. 198–212. Bern, Stuttgart, Vienna, Hans Huber.

PATEL A.J., MICHAELSON I.A., CREMER J.E. & BALÁZS R. (1974) Changes within metabolic compartments in the brains of young rats ingesting lead. *J. Neurochem.* **22**, 591–598.

PATTISON I.H. & JEBBETT J.N. (1973) Clinical and histological recovery from the scrapie-like spongiform encephalopathy produced in mice by feeding cuprizone. *J. Path. Bact.* **109**, 245–250.

PAYAN H.M. (1967) Cerebral lesions produced in rats by various implants; epileptogenic effect of cobalt. *J. Neurosurg.* **28**, 146–152.

PEARCE J. (1974) The extra pyramidal disorder of Alzheimer's disease. *Europ. Neurol.* **12**, 94–103.

PENTSCHEW A. (1969) Walsh F.B. & Hoyt W.I.H. (Eds.) *Clinical Neuro-Ophthalmology*, 3rd Edition, p. 2738. Baltimore, Williams and Wilkins.

PERRY T.L., HANSEN S. & KLOSTER M. (1973) Huntington's chorea, deficiency of γ-aminobutyric acid in brain. *New England J. Med.* **288**. 337–342.

PRESTIGE M.C. (1974) Axon and cell numbers in the developing nervous system. *Br. med. Bull.* **30**, 107–111.

RAMSEY R.B., BANIK N.L., BOWEN D.M., SCOTT T. & DAVISON A.N. (1974) Biochemical and ultrastructural studies on subacute sclerosing panencephalitis and demyelination. *J. neurol. Sci.* **21**, 213–225.

RAUCH H.C., EINSTEIN E.R. & CSEJTEY J. (1973) Enzymatic degradation of myelin basic protein in central nervous system lesions of monkeys with experimental allergic encephalomyelitis. *Neurobiology* **3**, 195–205.

ROBERTS E. (1974) γ-Aminobutyric acid and nervous system function—a perspective. *Biochemical Pharmacology* **23**, 2637–2649.

ROBERTS M. & HANAWAY J. (1970) *Atlas of the Human Brain in Section*. Philadelphia, Lea & Febiger.

ROBINS E., FISHER H.K. & LOWE I.P. (1961) Quantitative histochemical studies of the morphogenesis of the cerebellum II. Two β-glycosidases. *J. Neurochem.* **8**, 96–104.

ROBINS E., SMITH D.E., DAESCH G.E. & PAYNE K.E. (1958) The validation of the quantitative histochemical method for use on post-mortem material II. The effects of fever and uraemia. *J. Neurochem.* **3**, 19–27.

SEKHON S.S. & MAXWELL D.S. (1974) Ultrastructural changes in neurons of the spinal anterior horn of ageing mice with particular reference to the accumulation of lipofuscin pigment. *J. Neurocytol.* **3**, 59–72.

SIAKOTOS A.N., WATANABE I., SAITO A. & FLEISCHER S. (1970) Procedures for the isolation of two distinct lipopigments from human brain: lipofuscin and ceroid. *Biochem. Med.* **4**, 361–375.

SINHA A.K. & ROSE S.P.R. (1972) Compartmentization of lysosomes in neurones and neuropil and a new neuronal marker. *Brain Res.* **39**, 181–196.

SMITH C.B. (1975) Neurochemical correlates of brain degeneration in senile dementia. *Ph.D. Thesis*, University of London.
SMITH C.B., BOWEN D.M. & DAVISON A.N. (1974b) Loss of a specific protein from brain in dementia. *Trans. Biochem. Soc.* **2**, 661–663.
SMITH M.E. (1973) Studies on the mechanism of demyelination: triethyl tin-induced demyelination. *J. Neurochem.* **21**, 357–372.
SMITH M.E. (1965) Lipid biosynthesis: the central nervous system in experimental allergic encephalomyelitis. *Ann. N.Y. Acad. Sci.* **122**, 95–103.
SMITH M.E., SEDGEWICH L.M. & TAGG J.S. (1974a) Proteolytic enzymes and experimental demyelination in the rat and monkey. *J. Neurochem.* **23**, 965–972.
STAHL W.L. & SWANSON P.D. (1974) Biochemical abnormalities in Huntington's chorea brains. *Neurology* **24**, 813–819.
STEVENS R. & COWEY A. (1973) Effects of dorsal and ventral hippocampal lesions on spontaneous alternation, learned alternation and probability learning in rats. *Brain Res.* **52**, 203–224.
SUCKLING A.J. & HUNTER G.D. (1974) Glycosyl transferase activity in normal and scrapie-affected mouse brain. *J. Neurochem.* **22**, 1005–1012.
SWANSON P.D., HORNEY F.H. & STAHL W.C. (1973) Subcellular fractionation of post-mortem brain. *J. Neurochem.* **20**, 465–476.
TELLEZ-NAGEL I., JOHNSON A.B. & TERRY K.D. (1974) Studies on brain biopsies of patients with Huntington's chorea. *J. Neuropath. exp. Neurol.* **33**, 308–332.
TEUBER H.L. (1974) Recovery of function after lesions of the central nervous system: history and prospects. *Neurosciences Res. Prog. Bull.* **12**, 197–211.
TOMLINSON B.E. (1972) Morphological changes in non-demented old people, in Van Praag H.M. & Kalverboer A.F. (Eds.) *Ageing of the Central Nervous System, Biological and Psychological Aspects*, pp. 38–58. Haarlem, De Erven F., Bohn N.V.
TORVIK A. & SKJORTEN F. (1974) The effect of antinomycin D upon normal neurons and retrograde nerve cell reaction. *J. Neurocytol.* **3**, 87–97.
TOURTELLOTTE W. (1975) Multiple sclerosis. *HMSO*, in press.
URQUHART N., PERRY T.L., HANSEN S. & KENNEDY J. (1975) GABA content and glutamic acid decarboxylase activity in brain of Huntington's chorea patients and control subjects. *J. Neurochem.* **24**, 1071–1075.
VAUGHAN D.W. & PETERS A. (1974) Neuroglial cells in the cerebral cortex of rats from young adulthood to old age: an electron microscope study. *J. Neurocytology* **3**, 405–429.
VAUGHN J.E. & SKOFF R.P. (1972) Neurologia in experimentally altered central nervous system, in Bourne G.H. (Ed.) *The Structure and Function of Nervous Tissue V*, pp. 39–72. New York, Academic Press.
VOGT C. & VOGT O. (1922) Erkankungen der Grosshirnrinde im Lichte der Topistik, Pathoklise und Pathoarchitektonik. *J. Psychol. Neurol. (Leipzig)* **28**, 9–171.
WANNAMAKER B.B., KORNGUTH S.E., SCOTT G., DUDLEY A.W. & KELLY A. (1973) Isolation and ultrastructure of human synaptic complexes. *J. Neurobiol.* **3**, 543–555.
WEBSTER H.F., ULSAMER A.G. & O'CONNELL M.F. (1974) Hexachlorophene induced myelin lesions in the developing nervous system of *Xenopus* tadpoles: morphological and biochemical observations. *J. Neuropath. exp. Neurol.* **33**, 144–163.
WISNIEWSKI H.M., GHETTI B., SPENCER P.S. & TERRY R.D. (1974) Neuritic (senile) plaque—An expression of a cortical form of axonal dystrophy. *J. Neuropath. exp. Neurol.* **31**, 187.
WISNIEWSKI H.M. & TERRY R.D. (1973) Morphology of the aging brain, human and animal. *Prog. Brain Res.* **40**, 167–186.

YASE Y., YOSHIMASU F., UEBAYASHI Y., IWATA S. & KIMURA K. (1974) Amyotrophic lateral sclerosis. Interaction of divalent metals in CNS tissue and soft tissue calcification. *From Proceedings of the Japan Academy, Vol. 50, Nos. 5, 6 (1974).*

ZERVAS N.T., HORI H., NEGORA M., WURTMAN R.J., LARIN F. & LAVYNE M.H. (1974) Reduction in brain dopamine following experimental cerebral ischaemia. *Nature* **247**, 283–294.

ZURHEIN G.M. & ECKROADE R.J. (1974), in Zeman W. & Lennette E.H. (Eds.) *Slow Virus Diseases*, Chap. 3. Baltimore, Williams and Wilkins, in press.

2 The biochemistry of retinal degeneration and blindness

H. W. Reading

A. J. Dewar

I **Introduction**, 54

II **Inherited retinal dystrophy**, 54
    A. Introduction, 54
    B. Carbohydrate metabolism, 56
    C. Lipid metabolism, 57
    D. Nucleic acid metabolism, 58
    E. Total protein and amino acid metabolism, 59
    F. The visual cycle, 61
    G. Hexose monophosphate shunt (HMP) pathway 63
    H. Retinal alcohol dehydrogenase, 64
    I. Changes in ATPase, 65
    J. Abnormalities in visual pigment metabolism, 66
    K. Vitamin A metabolism, 68
    L. Visual cycle dysfunction, 70
    M. The nature of the primary biochemical lesion, 71
    N. The effects of light deprivation and anti-inflammatory drugs, 77
    O. Other enzyme defects in retinal dystrophy, 82
        1. Lactate dehydrogenase, 82
        2. Phosphodiesterase, 83
    P. Conclusions to be reached from the studies of retinal degeneration in animals, 88
    Q. The Cortical effects of retinal dystrophy, 88
    R. Behavioural studies in dystrophic animals, 91

III **Retinotoxic substances**, 92
    A. Introduction and classification of toxic substances, 92

B. Involvement of thiol groups, 92
C. Di-(p-aminophenoxy)-pentane (M and B 968A), 93
D. Retinal sensitivity to toxic agents, 96
E. Substituted phenothiazines and the quinoline derivatives (chloroquine), 96

IV Biochemical aspects of diabetic retinopathy, 97

A. Introduction, 97
B. Action of insulin in diabetes, 97
C. Retinal biochemistry and diabetes, 98
D. Effect of glucose and insulin in normal retina, 99
E. Diabetic retina—animals spontaneously diabetic or diabetic by induction, 99
F. Fatty acid metabolism in the diabetic retina, 99
G. Mucopolysaccharides of the diabetic retina, 100
H. Changes in basement membrane of kidney, lens and retina in diabetes, 101
I. Platelet abnormalities, 103
J. Drug therapy of diabetic retinopathy, 103

V The present position: an overview, 104

I Introduction

Blindness may result from abnormalities such as atrophy, or damage to the optic nerve, or other parts of the optic tract. (See diagrammatic representation of optic pathway, Fig. 2.1.) Degenerative changes associated with the optic nerve and secondary changes in the retina are well documented (Sorsby, 1972). In this chapter we will deal with the question of some specific degenerative changes in the retina which are considered to be due to metabolic abnormalities manifested through the specific metabolic characteristics of the retina itself. In our review we will exclude retinal atrophy following optic nerve damage and deal with inherited retinal dystrophy and with biochemical changes associated with retinal degeneration caused by toxic substances and diabetes. In the retina it is of interest to note that the pathological changes in inherited pigmentary degeneration and toxicity are changes largely confined to the neuroepithelium (i.e. the photoreceptor cells) and the pigment epithelium. This lies in close apposition to the retina and which is thought to serve the retina in functional exchange processes supplying nutrients and removing waste products.

Three forms of retinal degeneration have been the subject of extensive biochemical investigation; inherited retinal dystrophy, retinal degeneration produced by administration of retinotoxic drugs and the degenerative changes associated with diabetes known as diabetic retinopathy. Inherited retinal dystrophy (or primary pigmentary degeneration of the retina) is a true degeneration of the neuroepithelium involving actual loss of visual cells and is under the control of genetic determinants, whereas diabetic retinopathy of unknown aetiology, is primarily associated with capillary changes in the retina which eventually produce loss of vision, presumably by anoxia.

II Inherited retinal dystrophy

INTRODUCTION

Inherited retinal dystrophy in the rat, mouse and dog closely resembles the process of retinitis pigmentosa in man; both conditions are inherited, that in the rat as an autosomal recessive characteristic. In man the situation is more complicated in that at least seven different forms of inheritance have been recognized. Pigmentary retinal degeneration is often associated with a number of other conditions such as deafness, mental retardation, disorders of lipid metabolism, endocrine dysfunction (e.g. pituitary dysfunction in the Laurence–Moon–Biedl syndrome) and hereditary renal disease. These various manifestations of retinal degeneration cannot bear a common gene abnormality, although the pathological end result may be the same in all cases, whatever the genetic determinants.

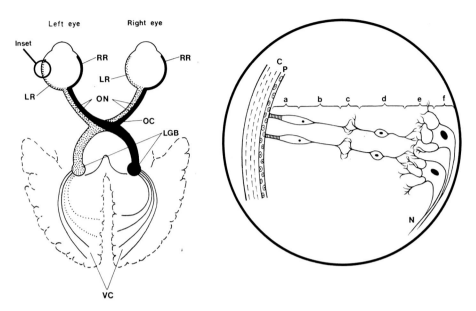

Fig. 2.1. Diagrammatic representation of the optic pathway. Key: LR, left side retina; RR, right side retina; ON, optic nerve; OC, optic chiasma; LGB, lateral geniculate bodies; VC, visual cortex (striate area); C, choroid; P, pigment epithelium; a, rod cell layer; b, outer nuclear layer (nuclei of rods); c, outer plexiform layer; d, inner nuclear layer (bipolar cells); e, inner plexiform layer; f, ganglion cell layer; N, nerve fibres going to optic nerve.

Biochemical research into pigmentary degeneration of the retina has been pursued for over 30 years in various laboratories in Europe and America. The obvious difficulty has always been, and will continue to be, obtaining human material for research purposes. One or two cases have occurred where excision of an eye has taken place for reasons other than retinitis pigmentosa but most of the material has been from older patients where the retinal degeneration has progressed too far for useful information to be obtained. However, the biochemist is fortunate in having a number of strains of dystrophic animals at his disposal. In these, as in man, blindness occurs some time after birth as a result of degeneration of the photoreceptor cells.

Keeler (1927) was the first to record the finding of a strain of 'rodless' mice in which the visual cells were almost completely absent. Bruckner (1951) presented the first ophthalmoscopic description of the fundus of albino and pigmented wild mice affected with retinal dystrophy. The incidence was as high as 50% in some of the strains examined.

Sorsby *et al.* (1954) carried out histological investigations on Bruckner's mice and established that the degeneration process in the retina developed subsequent to cellular differentiation but before the final stages of development of the rods. Similar changes were observed in the rat by Bourne *et al.* (1938) and by

Lucas *et al.* (1955), in the Irish setter by Lucas (1954) and in other breeds of dog by Barnett (1966). In affected mice, retinal development is normal until around the 10th postnatal day, whereupon early degenerative changes are seen. In the rat and dog, retinal degeneration starts about 12 days after birth, though full differentiation of the rods does not occur until 21–28 days. The pattern of changes in the rat broadly resembles those occurring in human retinitis pigmentosa (Cogan, 1950).

The condition was shown to have a simple autosomal recessive inheritance in the three species mentioned. Sorsby *et al.* (1954) stressed the fact that affected retinae were degenerate before the neuroepithelium had reached the stage of full differentiation, and suggested the term dystrophy, rather than abiotrophy, to describe such disorders.

Lasansky and De Robertis (1960) examined the dystrophic mouse retina by electron microscopy, confirming the failure of the full development of the rod cells, and showed that the outer segment of the rod in the normal mouse proceeds from a primitive cilium to the formation of membranous material which is reorientated into sacs, finally building up into a definite regularly layered structure. In the dystrophic animals, changes occur in the reorientation stage and the regularly layered structure never appears; in its stead, degeneration sets in. Dowling and Sidman (1962) in a similar examination of the eyes of rats affected with retinal dystrophy, reported an overproduction of visual pigment associated with the appearance of swirling sheets of 'extracellular' lamellae lying between the outer segments of the rods and the pigment epithelium. These authors also described the electroretinogram (ERG) pattern in the dystrophic rat and showed that it developed almost to the normal adult state but thereafter showed abnormalities.

Virtually all the biochemical studies on inherited retinal degeneration have been carried out either on the strain of dystrophic pink-eyed piebald agouti rats (known as 'Campbells') descended from the original strain of Bourne *et al.* (1938) or on the strain of dystrophic mice known as C3H/HeJ (Noell, 1958). The sighted strains generally used as controls have been the Piebald Virol Glaxo (PVG) strain of rat and the DBA/IJ strain of mice. The former are remarkable in that they, unlike the albino Wistar strain, very rarely show signs of spontaneous retinal degeneration.

B. CARBOHYDRATE METABOLISM

The dependence of the retina, like brain, on carbohydrate as a fuel for the supply of energy prompted early investigators to study retinal glucose metabolism in detail.

In the case of the dystrophic rat various groups of workers, amongst them Graymore *et al.* (1959), Walkers (1959) and Brotherton (1962), found decreases in the rate of anaerobic glycolysis but these were not pronounced until after

16–25 days of age. In a study of aerobic glucose metabolism using radioisotope methods, Reading and Sorsby (1962) found no striking differences in the overall pattern of glucose metabolism between normal and affected retinae during the age range 10–21 days after birth. Later, however, affected retinae showed obvious decreases in lactic acid, carbon dioxide and amino acid production accompanied by decreases in respiratory rate and glucose utilization. However, before the age of 21 days, affected retinae did not retain intracellularly amino acids formed from glucose. Aspartate, glutamate and GABA 'leaked' into the fluid medium bathing the tissue. This leakage did not occur from normal retinae.

Lolley (1972) demonstrated that in the dystrophic mouse the content of glucose, lactate, ATP and phosphocreatine within the photoreceptor layer were within normal limits for the first 10 postnatal days but after 15 days there was a decreased utilization of glucose. In normal mice glucose utilization is enhanced as visual function develops.

Thus in both the dystrophic rat and mouse the changes in energy metabolism are secondary to the retinal degeneration and it is therefore unlikely that a disorder of glucose metabolism is involved in the primary aetiology of the disease.

C. LIPID METABOLISM

In humans, abnormal lipid metabolism has been associated with pigmentary degeneration in such conditions as abetalipoproteinaemia (Bassen–Kornzweig disease) and phytanic-acid storage disease (Refsum's disease).

A survey by Dawson and Newell (1974) indicated that a defect in arachidonate metabolism may be involved in certain types of retinitis pigmentosa in man. They estimated that plasma arachidonic acid concentration was elevated in 15 out of 23 patients with retinitis pigmentosa. Both neutral lipid and phospholipid fractions showed a considerable enrichment in arachidonic acid content, but triglycerides and esterified cholesterol were primarily affected. Patients with other eye disorders did not exhibit any abnormality in plasma arachidonic acid levels.

In contrast, Fujiwara (1969) in a study involving 41 retinitis pigmentosa cases, actually found a decrease in plasma arachidonic acid but an increase in palmitic, palmitoleic and oleic acid concentrations. He also found a decrease in plasma phophatidylethanolamine and an increase in phosphatidylcholine. The total plasma lipid, triglyceride, phospholipid and cholesterol content was unchanged.

The significance of these findings is not really understood but there have been attempts to treat cases of retinitis pigmentosa by administration of certain phosphatides. Weiss and Kosmath (1971) claimed that intramuscular injections of a phosphatide preparation from animal retinae (known as Etaretin) had a beneficial effect on 11 out of 17 cases of retinitis pigmentosa but no effect on cases of diabetic retinopathy. This preparation had a phosphatide composition of 10·2% lysolecithin + monophosphoinositide, 7·6% sphingomyelin + lysophosphatidylethanolamine + phosphatidyl serine, 41% lecithin + phosphatidylserine, 20·5% phosphatidylethanolamine and 6·3% cardiolipin and a fatty acid composition of 27·8% palmitic acid, 22·7% stearic acid, 24·8% oleic acid, 1·3% linoleic acid and 6·7% arachidonic acid.

There have been few studies of lipid metabolism in dystrophic animals. The retina of adult dystrophic rats has been shown to have an abnormally high lipid phosphorus content but the dystrophic retina does not incorporate labelled CDP choline into lecithin at a higher rate than normal (Swartz & Mitchell, 1970). This study was undertaken on rats which had lost virtually all their photoreceptor cells. In addition the extracellular lamellae had appeared between the pigment epithelium and the area of the rod outer segments. However, although we have no evidence for a disorder of lipid metabolism preceding the onset of the degeneration, the above evidence is indicative of a defect in the pigment epithelium of adult dystrophic rats which interferes with the catabolism of retinal lipids.

D. NUCLEIC ACID METABOLISM

In the normal rat, retinal RNA content rises from birth until 1–2 weeks and then decreases markedly until the age of 3–4 weeks and remains approximately constant thereafter (Yates et al., 1974). However, in the Campbell rat, although the initial increase is observed, the retinal content falls steadily until about 14 weeks where it stabilizes approximately 70% of the value observed in the sighted PVG strain. The content of retinal DNA in normal and dystrophic retinae is maximal at 1–2 weeks. In the sighted rat it falls until 4–5 weeks and remains stable thereafter. However, in the dystrophic rat the retinal DNA falls, reaching a level approximately 30% of that of the PVG after 10 weeks (Fig. 2.2). Accompanying histological studies revealed that the loss of RNA accompanies the degeneration of the photoreceptor cells and the loss of DNA accompanies the loss of the photoreceptor cell nuclei (the outer nuclei). The histological and biochemical findings were in such close agreement that the use of retinal RNA and DNA measurements suggested themselves as an accurate and rapid alternative method to histological screening for studying retinal degeneration (Dewar & Reading, 1974a and b).

Similar changes in nucleic acid metabolism have been observed in the mouse retina by Lolley (1973). In both the normal (DBA) sighted and dystrophic (C3H) mouse an increase in retinal RNA and DNA were seen immediately after birth, rising to a maximum at five days. In both the mouse and rat it was clear that the increase in retinal RNA concentration coincided with the maturational processes consequent to the opening of the eyes. The increase in DNA concentration during this period is attributable to increased mitotic division in the

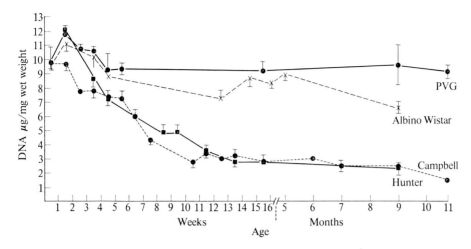

Fig. 2.2. The retinal DNA content of normal and dystrophic rats. PVG = pigmented sighted, Albino Wistar = albino sighted, Campbell = albino dystrophic, Hunter = pigmented dystrophic. Each point represents the mean ± S.E.M. of 4–26 animals. From Yates et al., 1974.

retina during this period. In both the sighted and dystrophic strains the retinal DNA and RNA content fell thereafter but the loss of RNA and DNA was considerably greater in the C3H mice. The loss of DNA in the sighted mouse retina corresponded to a loss of 1×10^6 cells, that in the C3H mouse to 6×10^6 cells per retina. In both the mouse and the rat the retinal RNA/DNA ratio remains constant in sighted animals but in animals suffering from retinal degeneration the RNA/DNA rises significantly. This suggests that the photoreceptor cells have a lower RNA/DNA ratio than the cells of the inner layers.

Thus in common with the defects in energy metabolism it would appear that the loss of retinal nucleic acids in the dystrophic animals is a consequence rather than a precipitant of the degeneration.

E. TOTAL PROTEIN AND AMINO ACID METABOLISM

The total protein content of the retina of the Campbell rat does not fall significantly until 7–8 weeks of age but thereafter it falls to approximately 40% of the protein content of the sighted retina (Yates et al., 1974). However, accurate determinations of protein content of an isolated dystrophic retina are difficult on account of the problem of particles of the protein-rich extralamellar material adhering to the tissue in a non-reproducible manner during dissection. Although the total protein content does not alter dramatically at an early age, the finding of Reading and Sorsby (1962) that the *in vitro* permeability to amino acid of the dystrophic retina alters before the age of 21 days prompted them to look for changes in protein metabolism in the affected retina. Consequently the uptake, transport and incorporation of ^{14}C glycine into total retinal protein was studied *in vitro* and *in vivo* (Reading & Sorsby, 1964). Experiments were carried out on litter mates of affected rats in comparison with normal animals at corresponding ages from 6 to 24 days after birth. Retinal tissue was incubated for 2-hour periods in a phosphate medium with added glucose and uniformly labelled ^{14}C glycine. *In vivo* each rat received a subcutaneous injection of 10 µCi of ^{14}C glycine in saline. Retinal protein was fractionated by tricholoracetic acid precipitation followed by removal of nucleic acid and lipids. The isolated protein fractions were then assayed for radioactivity.

In vitro glycine incorporation into retinal protein was lower in dystrophic than in normal retinae. At 6 to 8 days of age the differences were most pronounced ($p<0.04$) and became less with increasing age, until at 24 days, incorporation rates were equal in both normal and dystrophic animals (Fig. 2.3). No differences in amino acid uptake or in transport of amino acid into retinal tissue were found between normal and affected retinae at the early stages. Only

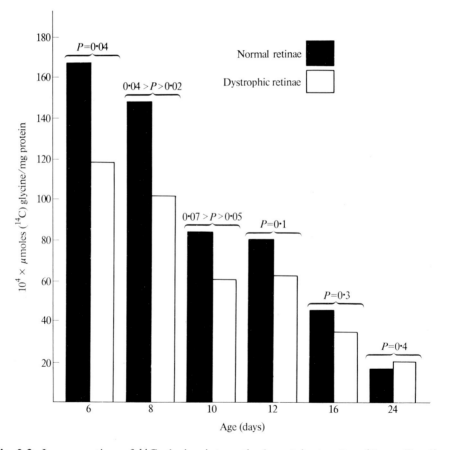

Fig. 2.3. Incorporation of ^{14}C glycine into retinal protein *in vitro*. (From Reading, 1965.)

at 24 days was the intracellular concentration of glycine less in dystrophic than in normal tissue. The *in vitro* results were confirmed by measuring the rate of retinal protein synthesis at various intervals after the subcutaneous injection of radioactive glycine into 8-day-old litter mate rats. The dystrophic animals showed a conspicuous decrease in the rate of protein synthesis and a reduction in the rate of protein synthesis and a reduction in the rate of protein breakdown, indicating a slower turnover of retinal protein. The rates of protein synthesis in the livers of both normal and affected rats were equal, so the differences observed in the retinae were unlikely to be due to strain differences. In addition, the rates of amino acid transport into the retinae were the same in normal and dystrophic animals.

The depression in protein synthesis and turnover was apparent some 6 to 8 days before histological changes were detectable. This suggested a discrepancy in the production of an essential protein, possibly one with structural as well as functional properties. The most obvious candidates for further investigation were proteins specifically concerned with the functioning of the photoreceptor cells and in particular with the metabolic changes occurring during the visual cycle. However, before a discussion of the abnormalities in the visual cycle found in the dystrophic retina, the features of the visual cycle in the normal retina will be presented briefly.

F. THE VISUAL CYCLE

The specific chemical macromolecules forming the basis of rod or scotopic vision in the vertebrate retina are the rhodopsins and porphyropsins. The former have a spectral sensitivity with maximum absorption at 500–505 nm whereas the latter show maximum absorption at 530 nm. We shall concern ourselves with rhodopsin since porphyropsin only occurs in freshwater fish and some amphibia. The visual pigments function by selectively absorbing parts of the visible spectrum and passing the absorbed energy into the photoreceptors. The dark adapted eye is most sensitive to light at wavelengths corresponding to the peak absorption for rhodopsin. Rhodopsin is a complex of a carotenoid (retinaldehyde), phospholipids (e.g. phosphatidylethanolamine) and a protein (opsin). Retinaldehyde is the aldehyde of vitamin A and is the essential light absorbing structure.

Rhodopsin is a red pigment which changes first to orange and then to yellow when exposed to light. This process of bleaching has been extensively investigated and it is now thought that part of the breakdown process takes place under the influence of light with possibly the later stages, and certainly the regeneration process, taking place in the dark. The sequence of reactions is shown in Fig. 2.4. The primary change produced by light is a stereoisomeric change in the carotenoid chromophore, conversion of the 11-cis isomer into all-*trans* retinaldehyde. Decomposition of metarhodopsin results in the formation of opsin and all-*trans* retinaldehyde as separate entities (Wald, 1960).

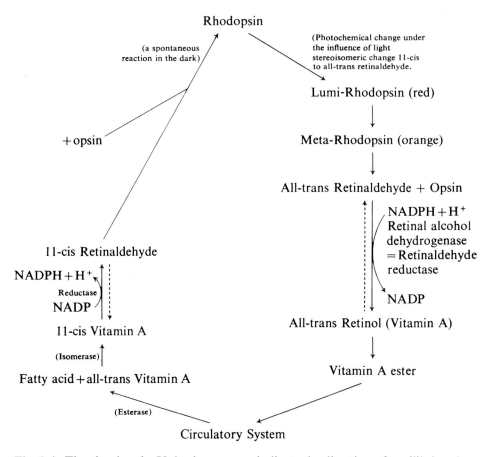

Fig. 2.4. The visual cycle. Unbroken arrows indicate the direction of equilibrium in a reaction.

In the formation of lumi-rhodopsin and meta-rhodopsin, conformational changes occur in the tertiary structure of opsin resulting in the 'unmasking' of two–SH groups and an acidic group per protein molecule. The nerve impulse is propagated at this metarhodopsin stage. Bonting (1969, 1975) has provided good experimental evidence for the chemical mechanisms involved. In the dark adapted state retinaldehyde forms a Schiff base with the amino groups of phosphatidylethanolamine:

$$R_1.CHO + H_2NR_2 = R_1.CH = N.R_2 + H_2O$$
$$\text{aldehyde} \quad \text{amine} \qquad \text{Schiff base}$$

When light falls on rhodopsin, after the 11-cis to all-*trans* retinaldehyde change, and during the formation of meta-rhodopsin, the all-*trans* retinaldehyde transfers to the ϵ-amino group of lysine ($NH_2(CH_2)_4 CH(NH_2) COOH$). This ϵ-amino group carries a positive charge and during its combination with retinaldehyde, this positive charge is lost. This opens up a channel for the passage of Na^+ and K^+ ions in the outer rod sac membrane, and so cationic fluxes occur that result in the propagation of the nerve impulse. Recent work by Hendricks et al. (1974) suggests that a release of calcium ions from the rod sac is also involved. In the regeneration process, the normal cationic gradient with high Na^+ concentration on the outside and high K^+ concentration on the inside of the membrane is re-established by the Na^+-K^+-ATPase (sodium pump) present in the rod sac membrane, the substrate ATP being supplied from the inner segment of the rod cells which are rich in mitochondria.

All-*trans* retinaldehyde is very rapidly removed by reduction to vitamin A alcohol (retinol). This reaction is catalysed by the enzyme retinaldehyde reductase (alcohol dehydrogenase) which is situated in the outer segment of the photoreceptor cell in close apposition to the visual pigment. This reduction is metabolically coupled to the oxidative dehydrogenases of the hexose monophosphate (HMP) shunt. In 1963 Futterman demonstrated that when light falls on the retina, reduction of the all-*trans* retinaldehyde to retinol is dependent on an adequate supply of reduced pyridine nucleotide NADPH. The enzymatic reduction involves the oxidation of NADPH to $NADP^+$, thus producing an increase in the ratio which, in turn, stimulates glucose $NADP^+$/NADPH oxidation by the HMP pathway which requires $NADP^+$ for its two dehydrogenases. The inter-connected pathways can be represented diagrammatically:

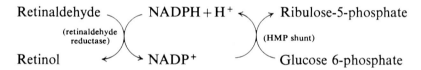

In the normal retina HMP shunt activity is low in dark adaptation with short bursts of activity when light falls on the retina. In the regeneration process the required stereoisomeric change in the vitamin A molecule from all-*trans* to 11-cis is mediated by a specific isomeric enzyme present in the pigment epithelium. We shall now go on to consider how these biochemical mechanisms associated with the visual cycle are altered in dystrophic retinae.

G. HEXOSE MONOPHOSPHATE SHUNT (HMP) PATHWAY

HMP shunt activity in developing normal and affected rat retinae was investigated *in vitro* by Reading (1964) by measuring the incorporation of ^{14}C into respiratory carbon dioxide produced by incubating excised retinal tissue with specifically 1-^{14}C or 6-^{14}C-labelled glucose substrates. The results showed a 40–50% increase in activity of the HMP shunt in affected retinae, but this difference was not statistically significant until 27–28 days. In normal undifferentiated retinae at 6–8 days of age HMP shunt activity is relatively high

Table 2.1. Activity of hexose monophosphate shunt in normal and dystrophic retinae

Age (days)	Normal	Dystrophic	% Increase in dystrophic
6–7	2·9 ($\pm$0·6)	4·1 ($\pm$0·8)	41·0 ($p=0\cdot3$–$0\cdot2$)
14	1·7 ($\pm$0·3)	2·5 ($\pm$0·5)	49·0 ($p=0\cdot2$–$0\cdot1$)
27–28	1·2 ($\pm$0·1)	1·8 ($\pm$0·2)	50·0 ($p=0\cdot05$–$0\cdot02$)

Results expressed as ratios of specific yields of $^{14}CO_2$ from 1-^{14}C glucose and 6-^{14}C glucose (SEM in parentheses). (From Reading, 1970.)

but this activity decreases as development proceeds (Table 2.1). These findings have been substantiated histochemically (Yates et al., 1975).

In 1965 Bonavita provided confirmatory evidence for increased HMP shunt activity in affected retinae by determining the specific activities of the two primary dehydrogenases of the pathway, namely: glucose-6-phosphate dehydrogenase (G6PD) and 6-phosphogluconate dehydrogenase (6PGD). Bonavita found decreases in the activities of both G6PD and 6PGD during development and differentiation in both normal and affected retinae. However, a significant difference was found between normal and affected animals in that G6PD specific activity reached a peak value in dystrophic retinae at 12 days of age (about 60% higher than normal), and this decreased to reach the values of normal retinae, both values then remaining about equal up to 120 days of age. A similar pattern was shown by 6PGD (i.e. a peak activity at about 12 days), though lower values were found in dystrophic retinae at all stages of development.

H. RETINAL ALCOHOL DEHYDROGENASE

This enzyme (RADH) is responsible for the reduction of retinaldehyde to retinol which, as we have already seen, is metabolically coupled to the HMP shunt in the retina. Since changes in the latter pathway had been found, it was thought that there might be a reflection of these manifested by early changes in RADH activity in dystrophic retinae. A study of RADH activity carried out by Reading and Sorsby (1966a) showed that the enzyme increased in activity in normal retinae until the animals were 1 month old, and then levelled off to reach a steady state of activity in the adult (Fig. 2.5). In dystrophic retinae, RADH enzyme activity developed similarly until the rats were about 2 weeks old, whereupon wide variations in activity were recorded. Subsequently an obvious decrease in activity was found from 3 weeks onwards, until retinae from 4-week-old dystrophic rats possessed only 40% RADH activity of corresponding normal retinae. Decrease in RADH activity in dystrophic retinae was therefore not obvious until after the third week of life. At 18 to 20 days of age, there was virtually no difference in enzymatic activity between normal and affected retinae. At this stage, the dystrophic animal shows substantial degenerative changes in the rod outer segments. It was concluded from these experiments that the decrease in RADH activity could be regarded as consequential to the primary biochemical lesion which causes cellular degeneration.

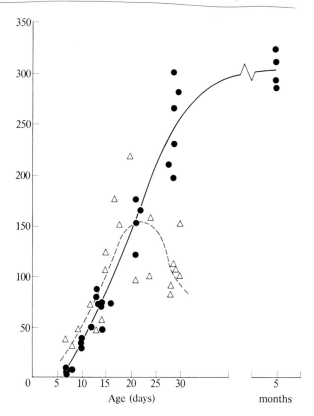

Fig. 2.5. Retinal alcohol dehydrogenase activities in developing normal and dystrophic rat retinae. ●—● normal retinae; △—△ dystrophic retinae. (From Reading & Sorsby (1966a).)

I. CHANGES IN ATPASE

The role of ATPase in Na^+ and K^+ transfer in cell membranes and the changes reported by Dowling and Sidman (1962) in the ERG at 22 days, coinciding with degeneration of rod inner segments and nuclei in the dystrophic rat, prompted Bonavita, Guaneri and Ponte (1966) to compare the development of activity of Na^+-K^+-activated ATPase in normal and dystrophic rats. They found that Mg^{2+}-activated and Na^+-K^+-activated ATPase activities increase during postnatal development of the retina in both normal and affected rats. However, a decline in Na^+-K^+-ATPase activity could be measured much earlier than 22 days after birth. These authors expressed their results in the form of the ratio Na^+-K^+/Mg^{2+} stimulated ATPase activity. In this way they found a pronounced decrease in the ratio by 12 days of age in dystrophic retinae. In normal retinae, this ratio remained constant over the period of 2 to 120 days of age.

J. ABNORMALITIES IN VISUAL PIGMENT METABOLISM

In their extensive investigation of retinal dystrophy in the rat, Dowling and Sidman (1962) demonstrated the existence of an overproduction of the visual pigment rhodopsin in the eye of the dystrophic rat. This was shown by electron microscopy and by rhodopsin analysis of the back portion of the eye which included retina, pigment epithelium plus choroid and sclera. The excess rhodopsin was in the form of swirling sheets or bundles of extracellular lamellae having a double membrane structure which, although disorganized, resembled the general appearance of normal outer segments. Dowling and Sidman (1962) described this as the first discernible abnormality in the dystrophic retina. By the age of 20 days, dystrophic rats possessed approximately twice as much rhodopsin per eye as corresponding normal rats. Rhodopsin content was found to increase rapidly in both affected and normal eyes until the thirtieth post-natal day. Subsequently the rhodopsin content of the dystrophic eye fell rapidly and by 40 days the level was half that of the normal (Fig. 2.6). The decrease in rhodopsin content, the onset of histological changes and changes in the ERG could be retarded by maintaining the rats in the dark from birth.

Dowling and Sidman stated that the excess visual pigment was qualitatively normal, judged by the wavelengths of the peaks of its absorption spectrum and those of the products of bleaching. Caution must be attached to any interpretation based on this observation, however, since the spectral characteristics simply show that the 'dystrophic visual pigment' contains the normal chromophore (11-cis retinal) and that the protein moiety (opsin) has a structure similar in molecular size and amino acid composition to that present in the normal unaffected eye. The protein moiety of the 'dystrophic pigment' could possess a relatively minor anomaly in one of its constituent peptide chains, viz. one or more amino acids 'out of place'. Such a situation accords with the genetic nature of the condition, and suggests how the visual pigment of the dystrophic animal could be excessively labile to the action of the light. As yet, there is no chemical evidence that the amino acid sequence in the opsin of dystrophic rats differs from that of normal opsin. However, studies of the early receptor potential (ERP) in retinitis pigmentosa patients suggest an abnormality in the visual pigment function that could be due to a slight alteration of chemical structure. These patients have a greater than normal ERP amplitude and a faster than normal ERP recovery rate (Berson & Goldstein, 1970; Berson, 1973).

Caravaggio and Bonting (1963) observed a similar phenomenon in the retina

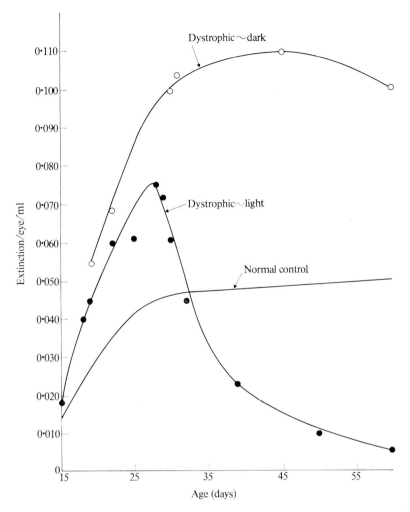

Fig. 2.6. Rhodopsin content of eyes from dystrophic rats raised under ordinary laboratory conditions of illumination (filled circles) and in darkness (open circles), compared with rhodopsin content of eyes from normal animals. (From Dowling & Sidman, 1962.)

of the dystrophic mouse. Further confirmation was provided by the work of Yamasaki and Muzuno (1971). However, in this latter case, although the dystrophic retina had an initial overproduction of rhodopsin, this was reduced at an early age. Rhodopsin and the electroretinogram were detectable earlier in dystrophic C3H mice than in normal mice of the T6 strain but by the tenth postnatal day the rhodopsin content of the former was significantly lower. After 12 days the ERG of the C3H mice diminished in proportion to the fall in rhodopsin content whereas in the normal mice the content of visual pigment continued to rise until the final adult level was attained after 21 days.

The protein patterns of C3H mice were compared with those from normal DBA/IJ mice during post-natal maturation by Farber and Lolley (1973). From 1 to 15 days of post-natal age the protein profiles of the C3H retina (assessed by polyacrylamide gel electrophoresis) remained very similar to those of the normally developing DBA retina despite the reduction in total protein content as a consequence of cell death. Between 5 and 20 days nearly 6 million cells die in the C3H retina (Lolley, 1973). However, the retinal protein profiles of the adults show distinctive differences. In the receptorless adult C3H retina three bands of protein were found to be missing two of which were identified as the visual pigment protein opsin and a cyclic AMP phosphodiesterase. The significance of the latter deficiency will be discussed in a later section of this chapter.

K. VITAMIN A METABOLISM

It has been known for a number of years (e.g. Osborne and Mendel, 1914; McCollum and Simmonds, 1917) that severe vitamin A deficiency causes histological degeneration of the retina and that rats raised on a vitamin A free diet (supplemented with vitamin A acid to maintain growth) develop a high degree of night blindness (Dowling & Gibbons, 1961). In rats fed a vitamin A diet the initial sign of deficiency is a decline in rhodopsin levels in the eye with a parallel rise in the log ERG threshold. Early in the deficiency these changes can be rapidly and completely reversed by the administration of vitamin A. With prolonged time on the diet the photoreceptor opsin levels decline and outer segments degenerate. Eventually the entire photoreceptor cell may be lost. In prolonged deficiency, recovery after vitamin A therapy is variable depending on the extent of visual cell degeneration. However, if the inner segments and nuclei remain intact, regeneration of new outer segments is possible. The histological picture of retinae from rats suffering from blindness resulting from long term vitamin A deficiency is strikingly similar to that of retinae from dystrophic rats. In both there is a degeneration of the photoreceptor cells while other retinal cells remained normal and healthy.

There is, however, a very important difference between nutritional and inherited blindness. Whereas in the former a loss of rhodopsin accompanies the increase in the ERG threshold, in the latter the rhodopsin level greatly increases (Dowling & Sidman, 1962; Dowling, 1964). Dowling (1964) concluded from these findings that the cause of the lesion in retinitis pigmentosa must be sought elsewhere than in a defect of vitamin A metabolism or rhodopsin synthesis. However, it is by no means proven that the same situation is present in man. In a study of rhodopsin density and visual threshold in human retinitis pigmentosa, Highman and Weale (1973) concluded that the results differed fundamentally from what Dowling and Sidman (1962) observed for the rat. They found no evidence of hyperdensity of the visual pigment in human retinitis pigmentosa cases. These findings are a salutary reminder of the dangers of arguing from one species to another. There is also evidence that the rat visual pigment is atypical among mammals (Lewin et al., 1970).

From a biochemical survey of retinitis pigmentosa patients in the Midlands area ranging in age from 9 to 65 years and covering a period of 11 years, Campbell and Tonks (1962) concluded that the most significant finding was a persistently low level of vitamin A in the blood, although the range of values was substantially the same in normal and affected persons (affected 38–162 international units per hundred ml: normal 69–167 international units per hundred ml). In retinitis pigmentosa patients the individual levels of vitamin A were below the normal mean in 91% of the adults and in all of the affected children. The low levels of vitamin A were found to persist in individual subjects over many years but they appeared in affected and non-affected relations in the same family. Campbell and Tonks also obtained evidence which suggested a defective absorption of vitamin A in these subjects. They suggested that low vitamin A levels were a positive factor in the development of retinitis pigmentosa and that the vitamin might circulate in a form which the retina could not use. Gouras et al. (1971) made a study of two patients with abetalipoproteinaemia and found that these patients developed a vitamin A deficiency when maintained on a normal diet. This deficiency was accompanied by severe impairment of vision including abnormalities of dark adaptation and the ERG which involved both rod and core receptor systems. Abetalipoproteinaemia involves disturbances of lipid and transport including an inability to form chylomicrons. Consequently intestinal absorption of vitamin A is impaired. In addition the absence of beta lipoprotein eliminates the main plasma fraction responsible for carrying carotenoids, the precursors of vitamin A. Massive oral doses of vitamin A reversed the visual abnormalities, cone function recovering more rapidly than rod function. The later states of abetalipoproteinaemia lead to retinitis pigmentosa and Gouras and his colleagues suggested that it is possible that some other forms of retinitis pigmentosa could also involve defects in vitamin A metabolism. However, there is evidence that argues against the suggestion. Both Wagreich et al. (1961) and Arden and his colleagues (personal communication, 1973) were unable to demonstrate the reduction in vitamin A blood levels in retinitis pigmentosa and a clinical study by Chatzinoff and his co-workers in 1968 failed to establish that vitamin A therapy had any beneficial effect on the course of the disease. In addition, it is now known that when there is a deficiency of vitamin A, the eye is, to some extent, protected. McLean (1970) showed that despite the ready exchange seen in vitamin A replete animals between the vitamin in the eye and the general body pool (Bridges & Yoshikami, 1969), the equilibrium is shifted markedly in favour of the eye when the vitamin is in short supply. When small doses of vitamin A were given to vitamin A deficient rats they were preferentially taken up by the eye over other tissues.

There are some indications that in humans affected by retinitis pigmentosa there is a defective transport mechanism for vitamin A. Vitamin A is absorbed in the upper part of the small intestine and reaches the liver where it is stored chiefly in the form of a palmitate ester (Moore, 1964). From the liver it is slowly released into the circulation as free retinol and then combines immediately with a specific transport protein (known as retinol-binding protein, RBP) to form a protein-retinol complex containing one molecule of retinol and one molecule of RBP (Kanai et al., 1968). Rahi (1972) considered that since low serum values for vitamin A are not a constant feature of retinitis pigmentosa (Krachmer et al., 1966) a quantitative or qualitative deficiency in RBP may be responsible for a vitamin A deficiency in the retina. In 1972 Rahi reported the results of an investigation of the RBP content of serum of patients with retinitis pigmentosa. Of the 51 patients examined, 82% showed a low level of RBP, the average being 76% of the mean normal adult value. However, this comparison of normal and affected subjects in this study is questionable since the control value was determined on pooled samples of serum from 20 healthy controls. Thus no estimate of the variation of the normal value was possible. If this finding is valid one could envisage an ineffective removal of retinol from the retina and pigment epithelium thus causing an increase in the local concentration of retinol. This interpretation is more probable than an actual deficiency of vitamin A occurring in the dystrophic retina. The results of Dowling and Sidman (1962) and Dowling (1964) suggest that such a deficiency is unlikely. On the other hand an inadequate removal of retinol from the retina could have serious consequences which will be discussed in a later section of this chapter.

L. VISUAL CYCLE DYSFUNCTION

In an effort to relate the abnormalities of visual pigment formation to visual cycle function, Reading (1966) compared distribution of retinaldehyde and retinol in normal and affected rat eyes under conditions of dark and light adaptation and found these to differ. Litter mates, 1–4 weeks old, were light or dark adapted before the eyes were enucleated and the retinae separated from the tissues of the back of the eyes (pigment epithelium, choroid and sclera). In the dark adaptation experiments, manipulations were carried out under the illumination of red dark room safe light. The tissues were assayed for both retinol and retinaldehyde, As expected, in normal animals dark and light adaptation produced an increase and a decrease respectively in retinaldehyde content.

In the case of prolonged bleaching, even when the retina had reached the stage of functional maturity at four weeks, the level of retinaldehyde was never reduced less than 11–12 micrograms per gram wet weight of tissue. On the other hand in affected animals, bleaching produced a conspicuous depletion of retinaldehyde in the retina. In addition, dark adaptation in affected animals resulted in the regeneration of less visual pigment than normal and this process decreased with increasing age. The most striking differences between normal and affected animals were seen in the retinol content of the back layers of the eye referred to as the pigment layers. This occurred at 2, 3 and 4 weeks of age and is illustrated in Fig. 2.7.

In the eye of the affected rat concentrations of retinol after bleaching were at least twice those in the normal rat at corresponding ages. Over the time course of these experiments, the wet weight of the retina and pigment layers from normal and affected animals were remarkably similar.

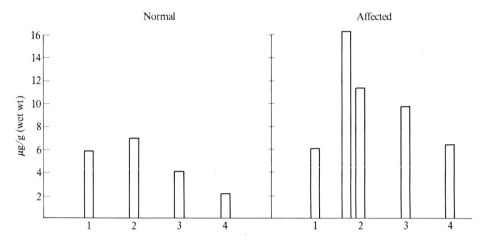

Fig. 2.7. Retinol (vitamin A alcohol) content of the 'pigment layers' from eyes of normal and dystrophic rats after light adaptation. (From Reading, 1970a.)

M. THE NATURE OF THE PRIMARY BIOCHEMICAL LESION

Reading (1970a) suggested that the sequence of biochemical events leading to the breakdown and disappearance of the neuroepithelium of the retina commenced with an excessive concentration of retinol arising in the pigment epithelium. This was a result of the action of light on an unusually labile type of visual pigment which is itself present in abnormally large amounts in the early stage of the condition. The build up of retinol in the pigment epithelium would tend to cause breakdown of the membranes of the lysosomes with which the pigment epithelium (on account of its phagocytotic function) is richly endowed. This would result in the release of acid proteases and hydrolases which account for the cellular digestion and subsequent disappearance of cellular debris.

This hypothesis was tested by Burden et al. (1971) who carried out a study of the activity of lysosomal enzymes in the retina and the structural integrity of lysosomes in the pigment epithelium of the normal and dystrophic rat eye using standard biochemical and cytochemical techniques. Experiments were carried out on the eyes of rats aged 1–12 weeks which covered the developmental period of the retina and the progressive stages of the degeneration. Measurement of total and free lysosomal enzyme activity in the retina was achieved by the use of acid protease as the lysosomal enzyme marker. Increases in both total and free

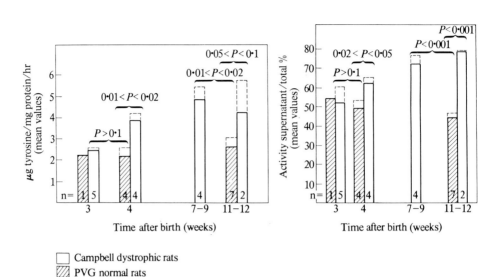

Fig. 2.8. Specific activity and percentage-free activity of acid protease in rat retina. n = number of experiments performed. (From Burden et al., 1971.)

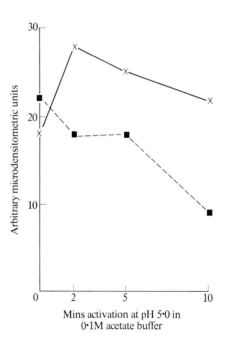

Fig. 2.9. Effect of preincubation (activation) on naphthylamidase in pigment epithelium. ×—×, 2-week-old normal rat; ---- 2-week-old dystrophic rat. (From Burden et al., 1971.)

activities above those found in the normal retina were first observed in the affected retina at four weeks. These differences became more apparent as degeneration progressed (Fig. 2.8).

The stability of the lysosomes in the pigment epithelium was studied by measuring the accessibility of the lysosomal enzymes, naphthylamidase and acid phosphatase, to their respective substrates. It was found that as early as 1–2 weeks of age the lysosomes in the affected pigment epithelium were less stable than those in the normal tissue (Fig. 2.9). This instability became more pronounced as the animals matured. These findings were corroborated by work of Ansell and Marshall (1974) who used an electron microscopic technique. Although these results are not in conflict with the hypothesis that the degeneration of the visual cells is initiated by the infiltration of lysosomal enzymes from the pigment epithelium, as we shall see later, they can be interpreted in an alternative manner.

Another assumption of the 'lysosome hypothesis' is that retinol labilizes lysosomes. Fell *et al.* had shown in 1962 that retinol labilizes lysosomes in a variety of tissues such as cartilage and liver. More recent *in vitro* work by Vento and Cacioppo (1973) and Dewar and Reading (1974a and b) on bovine and rat retina and pigment epithelium confirmed that retinol labilized lysosomes in these tissues, although the concentration of retinol (1·5 μg/mg original tissue) required for maximum (i.e. one hundred per cent) release of lysosomal enzymes was considerably greater than the retinol levels reported by Reading (1966) to exist *in vivo* (Fig. 2.10). However, these reported absolute *in vivo* levels are almost certainly an underestimate. The 'pigment layers' used in the assay consisted of pigment epithelium, choroid and sclera (i.e. the back of the eye minus the retina) and consequently the figures obtained must have been subject to considerable dilution due to the presence of excess tissue. In addition, the 'pigment layers' samples did not include the extralamellar material from the dystrophic rats.

The work of Young (1967, 1971) has shown that vertebrate rod photoreceptor cells continually renew photoreceptive outer segments (see Fig. 2.11). Protein is synthesized in the rod inner segments and then migrates into the bases of the rod outer segments where it is incorporated into the invaginating outer plasma membrane to form membranous outer segment discs (Young, 1969). This protein is primarily opsin. As new outer segment discs are formed they displace the previously synthesized discs outwards. When they reach the apex of the rod outer segment they are shed in groups of 10–30 and are ingested by the cells of the pigment epithelium to form inclusions called phagosomes which are subsequently degraded (Young, 1971). In the normal adult animal the rates of outer segment formation and degradation are equal but Kuwabara (1970) showed that if the eye is exposed to excessive light the equilibrium is altered so that phagosomes accumulate in abnormally high amounts. In the developing retina the rate of renewal and disposal are not equal and during the cytodifferentiation

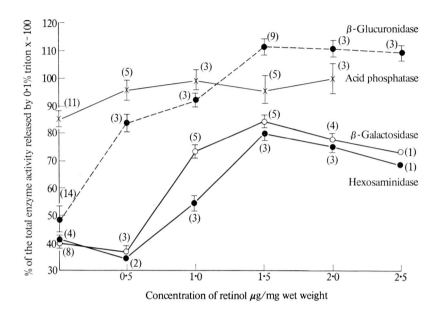

Fig. 2.10. The effect of retinol on the release of enzymes from retinal lysosomes *in vitro* during a 45 min incubation at 37°C. Points represent mean ±S.D. of the number of determinations shown in brackets. (From Dewar *et al.*, 1974.)

of the photoreceptor cells the rate of outer segment disc synthesis alters (Lavail, 1973).

Within the last few years attention has been focused on the possibility that an abnormality of the photoreceptor renewal-removal mechanism could account for some of the features of retinal dystrophy. Herron and his co-workers (1969) compared the rod outer segment rate in normal sighted Wistar rats and the dystrophic Campbell strain by an autoradiographic technique using ^{3}H methionine to label the protein. They found that the dystrophic rat had a normal rate of outer segment movement until the age of 18 days but thereafter the movement of lamellae slowed down. The pigment epithelium showed no ability to 'phagocytize' the rod outer segments and phagosomes were never seen within the pigment epithelial cells. The rate of segmental progress in the dystrophic retina was ultimately reduced to approximately one-sixth of the normal value. Similar findings were reported by Bok and Hall (1971) and Lavail *et al.* (1972). These findings reconcile a discrepancy between the early observation of Reading and Sorsby (1964) that protein synthesis in the affected retina is reduced in comparison to the normal retina and that of Dowling and Sidman (1962) that in the affected eye an excessive amount of rhodopsin-like protein was present from an early age. It is now clear that the accumulation of this protein is not due to an 'over-production' by the retina but a failure by the pigment epithelium to remove and digest it. This observation also explains the abnormal, high lipid

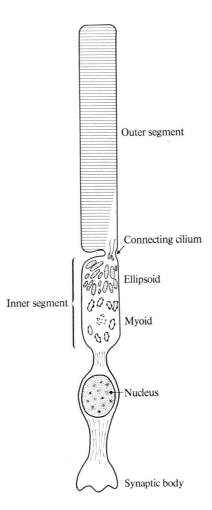

Fig. 2.11. Diagram showing the organization of a visual cell. The outer segment consists of a stack of disc-shaped membranes which contain visual pigment molecules. A connecting cilium joins the outer and inner segments. The ellipsoid portion of the inner segment is filled with mitochondria. The myoid portion is the region where most of the cell's protein, carbohydrate and (probably) phospholipid are produced. The Golgi complex is also located in the myoid. The nucleus contains the cell's genetic material and is the site of production of RNA. (From Young, 1973.)

phosphorus content of the adult dystrophic rat retina found by Swartz and Mitchell (1970). Lavail and his colleagues (1972) actually showed autoradiographically that the protein synthesis in the affected strain of rats was lowered in the rod cell layer. They also presented radioautographic and electron microscopic evidence that the extra lamellar material resulted in part from an increased protein synthesis in the pigment epithelium that the extralamellar material had a dual source.

Herron and his colleagues (1969, 1971) suggested that the inability of the pigment epithelium to phagocytize was due to a specific enzyme deficiency. An alternative possibility is that the spent rod discs are deficient in some way and hence the pigment epithelium will not engulf them (Ansell & Marshall, 1974). At present there is no evidence available to enable us to distinguish between these two possibilities.

The problem now arises as to how this failure of pigment epithelium phagocytosis in the dystrophic rat eventually results in the degeneration of the photoreceptor cell layer. Herron et al. (1969) were of the opinion that the death of the visual cells was primarily the result of poor nutritional supply to these cells due to the build up of extracellular lamellae inhibiting the metabolic flow and removing the visual cell from its nutritional source, the choriocapillaris. In view of the findings of Herron and his colleagues, it is possible to reinterpret the results of Burden et al. (1971). The fact that lysosomes in the dystrophic pigment epithelium were less stable than those in the normal tissue as early as one week of age may well contribute to the inability of the pigment epithelium to phagocytize efficiently. They also found that there was an increase in the total and 'free' lysosomal enzyme activity in the retina and this suggests that there is both an infiltration of lysosomal enzymes from the pigment epithelium to the retina and also a breakdown of retinal lysosomes. The breakdown of the retinal lysosomes would be accounted for if the accumulations of visual pigment in the rod outer sacs and extralamellar material breaks down to retinaldehyde resulting in high local concentrations of retinol. The increased activity of the HMP shunt in the dystrophic retina (Reading, 1964) would facilitate this conversion of retinaldehyde to retinol. The evidence that retinol can activate retinal lysosomes has already been quoted. As yet, however, there have been no estimates of free retinol in the dystrophic retina and extralamellar material so at present it is impossible to state unequivocally whether the labilization of retinal lysosomes by retinol observed *in vitro* also occurs *in vivo*.

Another important question to be answered is what proportion of the retinol in the outer segments is converted to retinyl esters. As we have seen in the normal eye after the breakdown of rhodopsin, the opsin produced is phagocytized by the pigment epithelium and the retinaldehyde is reduced to retinol in the rod outer segments. In 1964 Andrews and Futterman showed that the retinol is then esterified with fatty acids in the endoplasmic reticulum and then transferred to the pigment epithelium. It still remains to be determined whether the accumulated retinol is efficiently converted to retinyl esters in the dystrophic eye. This is important because whereas retinol has a labilizing action on lysosomes, retinyl esters do not. Unfortunately the Carr–Price reagent commonly used for the determination of retinol levels (e.g. Reading, 1966) cannot distinguish between retinol and its esters.

So far in this section we have concerned ourselves primarily with the rat and it is not clear how applicable these findings are to other species, let alone man. At present there is no evidence that there is a failure of phagocytosis in the pigment epithelium in C3H mice. Indeed indirect evidence by Farber and Lolley (1973) suggests that phagocytosis is normal. They found that although between 5 and 20 days 6×10^6 cells die in the C3H retina (i.e. 300 cells/min), little accumulation of protein breakdown products in the protein patterns of the retina was observed. This strongly suggests that the removal of dying cells is rapid enough to maintain a low level of breakdown products.

There is also no evidence that the phagocytic activity of the pigment epithelium in human retinitis pigmentosa cases is abnormal. There is one report that suggests that in these cases it is the rate of formation of rod outer segments that no longer matches the phagocytic activity of the pigment epithelium (Highman and Weale, 1973). It will be recalled that the findings of these workers differed fundamentally from those of Dowling and Sidman (1962). The apparent discrepancy between the experimental findings on rats and the data on the condition in humans reinforce the dangers of arguing from one species to another.

N. THE EFFECTS OF LIGHT DEPRIVATION AND ANTI-INFLAMMATORY DRUGS

It will be apparent from the preceding section that there is now sufficient information available on the mechanisms involved in hereditary retinal degeneration in the rat to enable tentative studies to be made into possible methods of retarding the progress of the degeneration. The use of massive doses of vitamin A would be unlikely to succeed in the rat and in fact would probably exacerbate the condition. The effectiveness in man is a matter of debate. Although the data obtained from rat work argue against its use, one has to keep in mind the findings of Highman and Weale, 1973 who failed to find an overproduction of rhodopsin in human retinitis pigmentosa cases.

Light deprivation has recently been suggested as a possible therapy for retinitis pigmentosa (Berson 1971, 1973) although the detrimental effects of light both on the normal and dystrophic retina have been known for some considerable time. As early as the beginning of the century it was observed that light had a detrimental effect on the dystrophic retina and suggestions were made that shielding the eyes from daylight may well help retard the degeneration (Johnson, 1901; Leber, 1916). In spite of their suggestions and a report that some cases of retinitis pigmentosa had received benefit from wearing yellow glasses (Collins, 1919) no long-term trials in patients were reported for over 50 years. However, since 1971 there has been a renewed interest in this mode of treatment and some clinical trials are in progress (Berson, 1971). Light exclusion can be produced almost completely by fitting an opaque flush fitting scleral lens to the eye. Berson (1971, 1973) has suggested that if light exclusion is begun in the earliest stages of the condition it may be possible to double a patient's visual lifetime by keeping one retina in reserve in darkness while the other retina functions in light.

There are good theoretical reasons for believing that this therapy will prove efficacious. Light deprivation would be expected to reduce the breakdown of visual pigment whether abnormally labile or not (see Fig. 2.6) and also to reduce the photoreceptor outer segment turnover (Kuwabara, 1970; Feeney, 1973). Thus light deprivation would both reduce the build up of unphagocytized rhodopsin and its breakdown. Light deprivation would also reduce the activity of the HMP shunt, thus reducing the rate of conversion of retinaldehyde to retinol.

Animal studies have confirmed the effectiveness of light deprivation in retarding retinal degeneration. The work of Dowling and Sidman (1962) has already been referred to and their findings that light deprivation retards retinal degeneration in the Campbell rat were confirmed by Dewar and Reading (1974a, b). A comparison of the retinal DNA content of Campbell rats kept under normal light/dark laboratory conditions for 8 weeks and of animals kept in complete darkness for this period, revealed that the latter had lost considerably less DNA (Table 2.2). This difference was mirrored by histological differences. Light deprivation also reduced the biochemical changes in the visual cortex which occur when the animal goes blind (Dewar & Reading, 1973, Dewar, 1975).

Further evidence for the important role of light in the degenerative process was obtained by Yates and co-workers (1974) who bred a new true breeding strain of pigmented dystrophic rats by crossing albino dystrophic Campbell rats with sighted pigmented Piebald Virol Glaxo (PVG) rats and selecting for pigmentation and blindness. In 1930, Lashley demonstrated that the presence of pigment in the eye reduces the amount of light impinging on the photoreceptor cells. Thus the presence of melanin in the dystrophic eye should retard the retinal degeneration. A full-scale comparative study of the histology and biochemistry of the albino dystrophic Campbell strain and the pigmented dystrophic strain (known as 'Hunter') revealed that this was indeed the case. The onset and progression of the lesion in the pigmented strain was delayed by about one week compared with the albino dystrophic rats (Table 2.2).

It is apparent from Table 2.2 that dark rearing by no means abolishes the degeneration in the Campbell rat but merely retards it. In addition, if rats of the Hunter strain were dark reared in the same manner for 8 weeks, the degree of improvement (as assessed by the retinal DNA level) was negligible, although a slight improvement was seen histologically (Dewar & Reading, 1974b). This suggested that there may be other mechanisms involved which are light independent.

In view of the evidence that lysosomal activation by retinol may play a significant role in the degeneration of the retina, attempts have been made to stabilize retinal lysosomes against the action of retinol. There is evidence that anti-inflammatory agents have a stabilizing effect on lysosomes and can counteract the labilizing effects of histamine in inflammation (Chayen et al., 1972) and

Table 2.2. The retinal nucleic acid content of 8-week-old normal, dystrophic and light-deprived dystrophic rats

	RNA μg/mg wet weight	DNA μg/mg wet weight	RNA/DNA
PVG (normal sighted)	2·23 ± 0·07 (24)	9·33 ± 0·32 (24)	0·24 ± 0·008 (24)
Campbell (albino dystrophic)	2·14 ± 0·10 (18)	2·67 ± 0·14 (18)	0·80 ± 0·04 (18)
Campbell—light deprived	2·15 ± 0·12 (20)	4·81 ± 0·23 (20)	0·45 ± 0·02 (20)
	N.S.	$t = 7·42$ $P = 0·0001$	$t = 7·21$ $P = 0·0001$
Hunter (pigmented dystrophic)	1·95 ± 0·16 (14)	4·33 ± 0·29 (14)	0·46 ± 0·03 (14)
Hunter—light deprived	2·28 ± 0·16 (19)	4·55 ± 0·29 (14)	0·50 ± 0·03 (19)
	N.S.	N.S.	N.S.

Values represent mean ± S.E.M. n shown in brackets.
(From Dewar & Reading, 1974b.)

this led Dewar and his co-workers (Dewar, 1975; Dewar et al., 1975a) to examine the effect of acetylsalicylic acid on retinal lysosome stability *in vitro*.

It was found that in concentrations of 0·25–0·50 mM, acetylsalicylic acid significantly reduced the release of β-glucuronidase, acid phosphatase, hexosaminidase and β-galactosidase from retinal lysosomes but had no effect on the activity of these enzymes after release. These concentrations of acetylsalicylic acid also dramatically reduced the release of the enzymes produced by 1·5 µg/mg retinol. As can be seen in Fig. 2.10, this concentration of retinol normally causes 100% release. However, if 0·25–0·50 mM acetylsalicylic acid was present in the medium the release was reduced to as little as 13–43% depending on the enzyme (Fig. 2.12). In common with its effect on liver lysosomes (Tanaka &Lizuka, 1968) acetylsalicylic acid labilized retinal lysosomes at high concentrations.

The fact that aspirin stabilizes retinal lysosomes against the labilizing effect of retinol *in vitro* suggested that long-term administration of aspirin may retard the progress of retinal degeneration *in vivo*. However, this is proving difficult to demonstrate in practice. The *in vitro* results indicated that the desired stabilizing effect of aspirin occurs over a comparatively narrow concentration range and to maintain the correct concentration of aspirin in the rat retina *in vivo* is difficult in view of the known pharmacokinetic properties of aspirin (Rowland & Riegelmann, 1968). Administration of acetylsalicylic acid (solubilized with N-methylglucamine) by injection and in the drinking water for an eight week period has been shown to have a slight effect on retinal DNA levels in Campbell rats (Dewar & Reading, 1974b). In a later study 75 Campbell rats maintained on 100 mg/kg/day aspirin from the age of 10 days until eight weeks had a mean retinal DNA level ($\pm$ S.D.) of $3\cdot76 \pm 0\cdot81$ µg/mg compared with a control value of $3\cdot17 \pm 1\cdot24$ (34 rats) (Dewar A.J. and Barron G. unpublished). The RNA DNA values were $0\cdot69 \pm 0\cdot65$ and $0\cdot60 \pm 0\cdot16$ respectively. Although these differences were statistically significant their functional significance (if any) remains to be assessed. Certainly the few histological studies made so far fail to show any dramatic effects of long-term aspirin administration. A combination of aspirin administration and dark rearing also failed to produce any greater beneficial effect than dark rearing by itself.

In view of the dramatic effects of aspirin in stabilizing retinal lysosomes against retinol *in vitro*, the *in vivo* results are disappointing. The failure may stem from the difficulties in maintaining the correct concentration *in vivo* or may indicate that *in vivo* high levels of retinol do not exist in the dystrophic retina owing to its conversion to retinyl esters.

There is, however, a small piece of purely circumstantial evidence which suggests it may be worth while persevering with the testing of anti-inflammatory drugs *in vivo*. In 1964 a clinical study by Powell and Field revealed that diabetic patients suffering from rheumatoid arthritis and requiring continuing large doses of salicylates had a significantly reduced incidence of retinopathy compared with the general diabetic population.

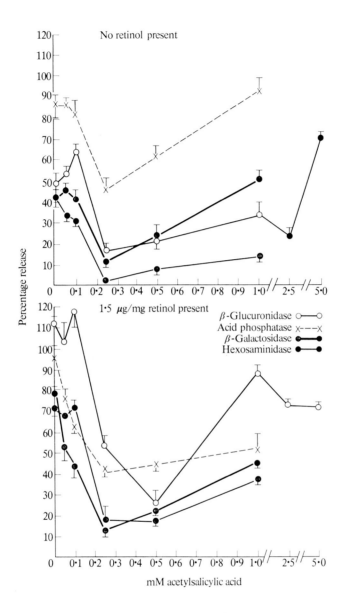

Fig. 2.12. The effect of acetylsalicylic acid on the release of lysosomal enzymes from retinal lysosomes both in the presence and absence of 1·5 μg/mg retinol *in vitro*. Incubations were carried out at 37°C for 45 min under nitrogen. The percentage release of the lysosomal enzyme in each case estimated by comparing the amount of enzyme activity released into the supernatant fraction with the release produced by the presence of 0·1% Triton X-100 in the incubation medium. Values are mean ±S.D. of between 3 to 14 determinations. The reduction in release of enzymes produced by 0·25–0·50 mM acetylsalicylic acid is statistically significant in all cases. (Data from Dewar *et al.*, 1974.)

The above refers to attempts to combat premature retinol-induced release of lysosomal enzymes in the dystrophic retina. There also remains the problem of how to correct defects in the phagolysosomal system in the pigment epithelium—that is assuming that the failure of pigment epithelium observed in the dystrophic rat also occurs in man—at present a questionable assumption.

As yet very little is known about the lysosomal enzymes of the pigment epithelium (Feeney, 1973). Acid phosphatases, aryl sulphatase and acid lipases have been shown to be present using histochemical and subcellular fractionation techniques. However, there is far less N-acetyl-glucosaminidase (the enzyme responsible for cleaving the sugar chain of rhodopsin) and the proteases required for the hydrolysis of effete rhodopsin molecules have not been characterized at all.

Miyaura (1970) and Miyaura *et al.* (1970) have used animals treated with sodium iodate to devise a new therapy for retinitis pigmentosa. Using the ERG as a measure, it was found that vitamin B_{12} prevented the evolution of experimental retinitis pigmentosa by $NaIO_3$ in the rat. Vitamin B_{12} also improved the regeneration of visual purple in frogs. On the basis of these findings, Miyaura (1970) administered vitamin B_{12} to 20 cases of retinitis pigmentosa. The response of the patients to vitamin B_{12} was good on visual acuity, visual fields and dark adaptation. The effectiveness of B_{12} (which was given in very large doses of over 1,000/day) was especially marked in children.

O. OTHER ENZYME DEFECTS IN RETINAL DYSTROPHY

Light, as we have seen, has a key role in the degeneration of the dystrophic retina. However, complete light exclusion only retards but does not halt the degenerative process. There also appears to be a limit to the improvement produced by light deprivation (Dewar & Reading, 1974b). It thus appears probable that there are also light independent factors contributing to the overall degeneration of the retina. For a number of years it has been apparent that the underlying genetic abnormality is a multiple gene defect involving a number of different proteins. The possible abnormality in the visual pigment and in the enzymes involved in pigment epithelium phagocytosis have already been mentioned. Evidence has been accumulating for the existence of a number of other defective enzyme systems in the dystrophic retina.

1. LACTATE DEHYDROGENASE

In 1959 Futterman and Konishita fractionated extracts of bovine retinae by zone electrophoresis on starch paste and assayed the isolated fractions for lactic dehydrogenase (LDH) activity. They found five discrete fractions, all with the capability of converting pyruvate to lactate, but having different affinities for the pyridine nucleotide coenzymes, NADH and NADPH which depended on substrate (pyruvate) concentration.

Whereas Bonavita et al. (1963) reported only four distinct fractions from the rat retina, Graymore (1964a, b) using cellulose acetate strips as the supporting medium for electrophoresis, was able to demonstrate five fractions. These five fractions consisted of two major and discrete fractions corresponding to the 'M' (muscle) and 'H' (heart) isoenzymes. The remaining three fractions were hybrids of these. It is considered that the 'M' isoenzyme is concerned predominantly with anaerobic glycolysis (i.e. lactate formation) whereas the 'H' isoenzyme is associated with highly oxygenated tissues and is inhibited by high concentrations of pyruvate. Tissues with predominantly 'H' type LDH tend to oxidize pyruvate aerobically via the Krebs cycle rather than convert it to lactate, so the retina is characterized by having a predominantly 'M' type pattern.

The dystrophic rat retina has been shown to have an anomalous LDH isoenzyme pattern (Bonavita et al., 1963; Graymore, 1964a, b). They demonstrated that the 'M' type LDH was deficient from birth. It was suggested that the 'M' type LDH had an important role in the biochemical changes occurring during retinal differentiation. Consequently a lack of this isoenzyme would be expected to have grave consequences on retinal development and function.

Sweasey and his co-workers (1971) determined the total LDH activity and isoenzyme patterns in the retinal tissues of normal sheep and sheep suffering from Bright Blindness. This is a condition characterized by progressive degeneration of the neuroepithelium of the retina. The histopathology of Bright Blindness has similarities to that of the hereditary retinal degeneration in rats (Graymore, 1970). In normal sheep the total LDH activity increased only slightly with age but there were pronounced changes in the LDH isoenzyme pattern, notably an increase in the ratio of H-type to M-type monomer. Total LDH activity was less in the retinae of blind sheep and alterations in isoenzyme pattern were also observed, again an increase in the ratio of 'H' to 'M' LDH. Whereas the H/M ratio rose from 1·39 to 2·31 during normal development, in the blind sheep this ratio rose to 3·48. The increase in the ratio was due to a reduction in the level of 'M' type LDH.

2. PHOSPHODIESTERASE

3' 5' cyclic adenosine monophosphate cyclic (AMP) influences protein synthesis at both the transcriptional and translational levels (Chambers & Zubay, 1969) and has been implicated in the processes of differentiation of the retina (Chader, 1971). There is increasing evidence that cyclic AMP and other cyclic nucleotides such as cyclic GMP may have an important role in photoreceptor function. Adenyl cyclase is present in photoreceptors and its specific activity has been reported to be one of the highest in any known tissue (Miller et al., 1971). This enzyme is inactivated in the outer segments of the photoreceptor cells in proportion to the degree of bleaching of rhodopsin and electrophysiological experiments on the photoreceptors of the compound eye of Limulus indicated that cyclic AMP mimicked the effect of illumination (Miller et al., 1971). The outer segments of frog rods contain a light inhibited adenylate cyclase (Bitensky et al., 1971). These observations have led to the suggestion that adenyl cyclase and cyclic AMP may have a function in the translation of light energy into a nerve impulse. Pannbacker and co-workers (1972) suggested that cyclic GMP was as likely an intermediate in this process as cyclic AMP. This suggestion was prompted by their observation that bovine rod outer segments contained high activities of guanylate cyclase as well as a protein kinase stimulated by both cyclic AMP and cyclic GMP.

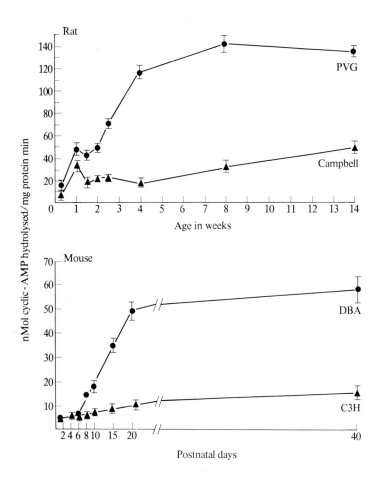

Fig. 2.13. cAMP–PDE specific activity in normal and dystrophic mice and rats. Each point represents the mean ±S.E.M. Number of determinations of rat cAMP–PDE is 4–10. In the case of mice 5 or more determinations were used. (Data from Schmidt and Lolley, 1973, Dewar et al., 1

For the inhibition of adenyl or guanylate cyclase to significantly alter the concentration of cyclic nucleotides within the short time in which visual excitation occurs, the receptor region must also be rich in phosphodiesterase, the enzyme that destroys cyclic nucleotides.

The existence of high phosphodiesterase activity in purified outer segments of bovine rods was demonstrated by Pannbacker *et al.* (1972). This enzyme was particulate and unaffected by light. It hydrolysed cyclic GMP more rapidly than cyclic AMP at low substrate concentrations. It seems probable therefore that retinal phosphodiesterase is involved in the control of cyclic nucleotide concentration during visual excitation or adaptation. The fact that inhibitors of phosphodiesterase such as dibutyryl cyclic AMP, caffeine, theophylline and papaverine degrees the amplitude of receptor potential in frogs provides further evidence for this view. The effectiveness of these compounds in depressing receptor amplitude parallel their effectiveness as inhibitors of phosphodiesterase from rod outer segments. Papaverine is the most effective followed by theophylline and caffeine. A study of a number of PDE inhibitors on bovine retina undertaken by Chader *et al.* (1974) convincingly demonstrated that even small changes in PDE activity may be reflected in relatively large changes in nucleotide concentration.

An abnormality in cyclic AMP metabolism in the C3H mouse retina occurring before the onset of photoreceptor cell degeneration has been demonstrated by Schmidt and Lolley (1973). (Figs 2.13, 2.14). In the retina of normal DBA mice the specific activity of cyclic AMP phosphodiesterase increases eightfold between the 6th and 20th postnatal day. Before 6 days kinetic analysis indicates PDE with a single Michaelis constant (Km) value for cyclic AMP (low Km-PDE). However, after 6 days a second PDE with a high Km for cyclic AMP (high Km-PDE) appears. The appearance and increasing activity of the high Km-PDE accompanies the differentiation and growth of the photoreceptor outer segments. Experiments using microchemical techniques indicated that the low Km-PDE is associated primarily with the inner layers of the retina. In the C3H mouse the postnatal increase in cAMP-PDE is substantially lower because the high Km-PDE fails to appear after the 7th postnatal day. The dystrophic retina also metabolizes cyclic GMP at a slower rate. The high Km-PDE possessed greater affinity for cyclic GMP than cyclic AMP. Polyacrylamide gel electrophoresis of adult C3H mouse retinal protein revealed that the C3H retina was deficient in three bands of protein, one of which was identified as opsin and one as a cyclic nucleotide phosphodiesterase (Farber & Lolley, 1973).

Recently evidence has been obtained that the high Km-PDE is also missing from the retina of the dystrophic rat (Dewar *et al.*, 1975b). A significant difference in the specific activities of cyclic AMP-PDE in the normal and dystrophic rat is found between 9 and 12 days, i.e. before the onset of photoreceptor cell degeneration. Kinetic analysis revealed that as in the mouse the high Km-PDE fails to appear during photoreceptor cell development. However, in both

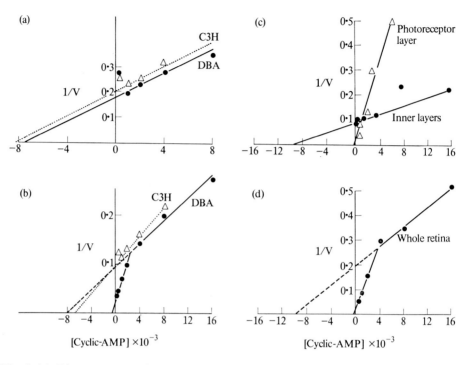

Fig. 2.14. Lineweaver–Burk plots of cAMP–PDE activity in homogenates of DBA (●—●) and C3H (△– – – –△) mouse retinae: (A) 2 postnatal days and (B) 12 postnatal days, (C) freeze-dried microdissected photoreceptor layer (△) and bipolar-plus-ganglion-cell layer (●) and (D) freeze-dried samples of whole retina (●) from adult DBA mice. The enzyme activity (nanomoles of cyclic-AMP hydrolysed per milligram of protein per minute), was determined in duplicate samples at each level of substrate. An extrapolation of the broken Lineweaver–Burk plot (– – – –) shows a second apparent Michaelis constant (low K_m-PDE) for cyclic-AMP in the whole DBA retina (B) and (D). (From Schmidt and Lolley, 1973.)

mouse and rat there is no difference between the brain PDE in the normal and dystrophic strains.

The defect in the high K_m-PDE in the dystrophic retina would be expected to have repercussions on the level of cyclic AMP. In the mouse, adenyl cyclase activity increases normally for the first 7 days of age and thereafter it becomes greater than normal. The level of cyclic AMP becomes significantly abnormal after 10 days of age. In a 20-day-old dystrophic mouse the retinal cyclic AMP level is of the order of 90 pmoles/mg protein as compared to the 20 pmoles/mg protein found in the normal retina (Lolley et al., 1974). On the basis of this finding it would be tempting to produce a unified hypothesis of retinal degeneration by postulating that the high levels of retinal cyclic AMP are responsible for the inability of the pigment epithelium to phagocytize since there is evidence that cyclic AMP controls this process and inhibits it at high concentrations (Weissman et al., 1972; Zurier et al., 1973).

However, further work by Lolley and his colleagues suggest that the situation is unlikely to be as simple as this. Quantitative histochemical and biochemical studies in mice (Lolley et al., 1974) revealed that cyclic AMP is concentrated in the inner layers of the retina where it is associated with neuronal function and the content of cyclic AMP and the activity of adenylate cyclase are both low in normal photoreceptor cells. These findings appear to conflict with the earlier report of Miller and co-workers (1971). Thus the elevated adenyl cyclase and cyclic AMP levels in the dystrophic mouse retina are located primarily in the inner layers. Lolley et al. (1974) concluded that the elevated levels of cyclic AMP and adenylate cyclase in the dystrophic mouse were probably not involved in the aetiology of the degeneration of the photoreceptors. Another finding which cast serious doubt as to whether cyclic AMP is available to serve as substrate for the PDE enzyme of the photoreceptor cells *in vivo* was that in the immature retina of C3H mice, although the photoreceptor cells are deficient in cyclic nucleotide PDE activity, the cyclic AMP concentration does not increase.

In spite of these observations it is not unreasonable to suppose that the lack of the high Km phosphodiesterase in the dystrophic retina will have some repercussions on photoreceptor cell function. The high concentrations of guanosine nucleotides in the retina and the greater affinity of the high Km-PDE enzyme for cyclic GMP prompted Farber and Lolley (1974) to investigate the distribution of cyclic GMP in the normal and dystrophic mouse retina. They found that in the normal retina cyclic GMP increases post natally nearly doubling between day 6 and day 12 to a value of approximately 70 pmoles/retina. In the dystrophic retina the content of cyclic GMP is normal for 6 days and then it increases, reaching a maximum value of approximately 180 pmole/retina at days 14 to 16. Subsequently the cyclic GMP decreases to stabilize at a value of approximately 60 pmole/retina by 22 days. The early accumulation of cyclic GMP can be attributed to the inability of the photoreceptor cells to hydrolyse it due to their lack of the high Km PDE enzyme. Indeed the rise in cyclic GMP content in the dystrophic retina at 6 days occurs at the time when the deficiency in this enzyme is first detectable. The decrease in retinal cyclic GMP content after 16 days is attributable to a depletion in the photoreceptor cell population.

Histochemical studies revealed that approximately 60 per cent of the retinal cyclic GMP content was located in the photoreceptor cell layer. The photoreceptor layer in the degenerating dystrophic retina contained four to five times more cyclic GMP per milligram of protein than the photoreceptor cell layer of the normal retina.

Farber and Lolley (1974) consider that the imbalance in cyclic GMP in the dystrophic mouse retina which occurs before histological signs of degeneration is responsible for disturbances in the metabolism and function of these cells. The elevated cyclic GMP levels in the photoreceptors may produce degenerative changes by affecting the homeostatic balance.

As yet the imbalance in cyclic GMP metabolism in the photoreceptor cells of the dystrophic mouse have not been demonstrated in other species. However, since the dystrophic rat also lacks the high Km-PDE, it is probable that elevated cyclic GMP levels also occur in the photoreceptor cells of dystrophic rats.

The finding by Lolley et al. (1974) that the cyclic AMP metabolism of the *inner* layers of mouse retina become abnormal at an early stage of photoreceptor degeneration is interesting in view of the later work by Blanks et al. (1974a, b) who studied the effect of photoreceptor degeneration on synaptogenesis in dystrophic mice. Their findings suggest that the inherited abnormality may alter the synaptic contacts between photoreceptor cell and bipolar cells and thus influence the maturation of cells within the inner layers.

P. CONCLUSIONS TO BE REACHED FROM THE STUDIES OF RETINAL DEGENERATION IN ANIMALS

It is probable that inherited retinal degeneration is a defect involving a number of different enzyme systems. 'M' type LDH and a high Km cyclic nucleotide phosphodiesterase are missing in dystrophic animals and this is likely to interfere with normal photoreceptor development and function. The pigment epithelium of dystrophic rats has lost the ability to phagocytize spent photoreceptor cell outer segments. This may be due either to an inherited defect in the enzymes or membrane structures involved in phagocytosis in the pigment epithelium or to some change in properties in the outer segments which interferes with the ability of the pigment epithelium to engulf them. The failure to remove spent outer segments results in a build up of visual pigment between the retina and pigment epithelium. This accumulation could result in retinal degeneration either by interfering with the nutritional supply of the retina or by an accumulation of its breakdown product, retinol, labilizing retinal and pigment epithelium lysosomes. Light deprivation retards degeneration by slowing down both the production of rod outer segments and the breakdown of rhodopsin but is unlikely to ameliorate any of the effects of the lack of 'M' LDH and the high Km-PDE.

Q. THE CORTICAL EFFECTS OF RETINAL DYSTROPHY

In spite of some findings to the contrary (e.g. Metzger et al., 1966) there is sufficient evidence to suggest that visual stimulation can affect the visual cortex both structurally and biochemically. The first exposure of rats to light results in changes in both synapse number and dimensions (Cragg, 1967) and visual deprivation leads to a loss of dendritic spines in certain layers (Fifkova, 1968; Valverde, 1971). It has also been shown by Gyllensten and his colleagues (1965) that in the visual cortex of visually deprived mice there is a reduced nuclear diameter and a decreased volume of cell cytoplasm in addition to a reduced capillary supply.

The structural effects of the first exposure to light are accompanied by a number of biochemical changes which are found only in the visual cortex; a transient increase in the incorporation of labelled lysine into protein (Rose, 1967), a transient increase in the incorporation of orotic acid into RNA (Dewar et al., 1973a) and a transient increase in the rate of formation of uridine tri-phosphate (UTP) from uridine (Dewar, 1974). Visual stimulation of mature rabbits has been reported to increase the incorporation of amino acids into occipital cortex protein (Talwar, 1966) and stimulation of rats by means of 'enriched environments' produces increases in cortical weight, protein and cholinesterase content which are most pronounced in the visual cortex (Bennett, et al., 1964).

In view of these findings one would expect to find significant changes in the structure and biochemistry of the visual cortex in dystrophic animals. Although comparatively little work has been undertaken to investigate this, the evidence so far available suggests that this is indeed the case.

The visual cortex of adult dystrophic rats of the Campbell strain has a lower RNA/DNA ratio than that of a corresponding sighted rat of the PVG strain (Dewar & Reading, 1973). The former also exhibits a reduced rate of incorporation of intraventricularly administered ^{14}C orotic acid into RNA. This difference is not seen in any other cortical area nor is it seen in any other organ so it is unlikely to be a reflection of a strain difference in gross metabolic rate. Investigations of the visual cortex acid soluble precursor pool and its nucleotide components eliminated the possibility that the reduced labelling of visual cortex RNA in the blind rats was merely a reflection of reduced precursor availability or a difference in the rate of formation of UTP from orotic acid. The apparent reduction in the rate of RNA synthesis in the blind visual cortex was accompanied by a fall in the magnesium activated RNA polymerase activity of the tissue. There was no evidence of an increased ribonuclease activity in the blind visual cortex. The reduction in the RNA synthesis is of the order of 30%.

Studies at the subcellular level (Dewar et al., 1973b) have provided evidence that the transport of newly synthesised RNA from nucleus to cytoplasm is significantly retarded in the Campbell visual cortex and that the difference in the labelling of RNA is neuronal and glial nuclear preparations were less pronounced. Evidence from sucrose gradient analysis and (^3H) methyl-L-methionine labelling suggested that the blind visual cortex has a reduced rate of ribosomal RNA production.

These differences in RNA metabolism are accompanied by changes in protein metabolism. In the blind visual cortex the incorporation of intraperitoneally administered L-(1-^{14}C)-leucine into both nuclear and cytoplasmic protein is reduced although the rate of transport of the amino acid to the visual cortex is unchanged. Of the nuclear proteins, only the two fractions associated with the nucleolus (the site of ribosomal RNA production) are affected, the rates of labelling of the acid chromatin and histone fractions are not reduced (Dewar & Winterburn, 1973).

The bulk of these studies were carried out on 3-month-old rats but the metabolic differences between the blind and sighted animals were also apparent at the age of 8 weeks. However, no differences were discernible before 4 weeks (Dewar, 1975). It will be remembered that the Campbell rat does not become completely blind until 3–4 weeks of age.

A number of factors may contribute to the observed alterations in RNA and protein metabolism in the visual cortex of dystrophic rats. A direct relationship between synaptic stimulation and neuronal RNA metabolism has been demonstrated by Gisiger (1971) in the mammalian sympathetic ganglion so it is possible that the reduction in afferent impulses reaching the visual cortex has a direct effect in reducing the rate of RNA and protein metabolism.

A circadian rhythm in the RNA polymerase activity and nuclear RNA content exists in the cortex of rats exposed to alternating 12 hour periods of light and dark (Merritt & Sulkowski, 1971), the difference between the values at the nadir and zenith of the rhythm is of the order of 40%. Since the dystrophic rats are initially sighted it is probable that they exhibit similar circadian rhythms to the sighted rats. As yet, it is not known how the circadian rhythms of the Campbell rats are affected by their going blind. However, it is unlikely that any alterations would be restricted to the visual cortex.

The most probable explanation of these biochemical findings are that they are a reflection of morphological changes resulting from light deprivation. It is known that the incorporation of RNA precursors into RNA is approximately four times greater in neuronal than in glial nuclei (Løvtrup–Rein, 1970) and that this difference is largely attributable to the greater production of ribosomal RNA by the former. Similarly the incorporation of amino acids into protein is 2·5–6 times greater in neurones than in glia (Blomstrand & Hamberger, 1969). The data could be explained by the blind visual cortex becoming more 'glia-like' in its overall biochemical properties. This would occur if there was a reduction in the number of neurones and an increase in the number of glia.

Confirmatory evidence for this view has been obtained by McCormick (1973) who undertook an extensive quantitative histological study of the visual cortices of the Campbell and PVG rats. Up to 30 days of age no difference was discernible between the cortices of normal and dystrophic rats but after 30 days a significant difference in mean cell size was apparent. This change was diffuse and not restricted to any particular cortical layer. The thickness of the cortex was unaltered.

The alteration in mean cell size in the dystrophic visual cortex was due to an increase in the number of cells of dimensions less than 79 μM^2 and a decrease in the incidence of large cells greater than 147 μm^2. Electron microscopy identified the former as microglia and the latter as large neurones. There was no change in the number of cells of intermediate size (79–147 μm^2). The reduction in mean cell size in the dystrophic visual cortex was 30% at 70 days of age.

Structural changes have also been reported in the visual cortex of the dystrophic mouse (Gyllensten & Lindberg, 1964). There was a decrease in nuclear diameter and internuclear material in layers 3 and 4 but not in layers 5 and 6.

R. BEHAVIOURAL STUDIES IN DYSTROPHIC ANIMALS

Before proceeding to a discussion of the use of retinotoxic drugs some recent behavioural studies of dystrophic animals merit attention since they may prove to have important repercussions in the interpretation of the biochemical data.

In dystrophic mice by 20 days the electroretinogram is abolished and the photoreceptor cells have almost completely disappeared (Karli, 1952; Noell, 1958). However, in spite of this the mice show open-field behavioural responses which are believed to be visually mediated (Nagy & McKay, 1972). These mice have also been shown to be capable of discriminating light intensity (Karli, 1954; Ross et al., 1966; Nagy & Misanin, 1970) but Bonaventure and Karli (1961) demonstrated that the threshold was raised by 5 log units.

In 1972, Anderson and O'Steen and Bennett et al. reported that rats suffering from light-induced retinal degeneration were still able to discriminate patterns and light and dark. However, extreme durations of exposure caused these rats to lose eventually their ability to learn to discriminate between light and dark (Bennett et al., 1973). These findings suggested to Lavail et al. (1974) that animals affected with inherited retinal degeneration might lose visual function at a more advanced stage of the disease than was previously thought. To investigate this Lavail and his colleagues studied affected animals at relatively advanced ages (6 months to 2 years) and correlated behaviour with light and electron microscopic findings in the same animals.

Lavail et al. (1974) found that the dystrophic rats suppressed lever-pressing behaviour in response to onset of light at ages from 6 months to more than 2 years. These rats continued to show undiminished behavioural responses to light while their pupils were dilated with atropine, thus demonstrating that they do not receive their visual cues by means of a direct mechanosensory pupillary response to light. Electron microscopic examination of the retina revealed the existence of some surviving photoreceptor cells which, although lacking outer segments, did make synaptic contact with bipolar and/or horizontal cell processes. Lavail and his colleagues were of the opinion that the persistence of some photoreceptor cells and their synaptic connections with other retinal elements may explain the light mediated responses of the dystrophic rats. However, they did suggest other possibilities worthy of investigation. In 1972 Lavail et al. showed that the pigment epithelial cells appear to produce a rhodopsin-like protein at early stages of the disease. If this protein persists in the older rats the pigment epithelial cells may be responsible for transducing the light stimulus. A similar mechanism has been suggested for the dystrophic mouse (Karli et al., 1965). In addition, the possibility of some direct photosensory pathway from the iris cannot be ruled out.

III Retinotoxic Substances

A. INTRODUCTION

A number of substances including drugs of therapeutic use cause specific retinotoxic effects.

Their effects are characterized by a degeneration of the neuroepithelium of the retina, directly or indirectly involving the pigment epithelium, mimicking some of the hereditary clinical conditions classed as tapeto-retinal degenerations.

The important substances, or groups of drugs are:

1. sodium iodate
2. sodium iodoacetate
3. the diabetogenic agent dithizone (diphenylthiocarbazone)
4. some schistosomicidal drugs; in particular 1:5 di-(p-aminophenoxy) pentane (M and B 968A)
5. sodium fluoride
6. chloroquine and some of the phenothiazines, a group of drugs causing damage to the neuroepithelium of the retina but differing from the above in that they do so only after long-term therapy.

Sorsby and Harding (1962a) showed that sodium fluoride produces an inconsistent reponse, but when administered with small amounts of an oxidizing agent such as permanganate produces marked retinal toxicity.

All the substances mentioned with the exception of chloroquine and the phenothiazines produce degeneration experimentally in rabbit retina when given intravenously. M and B 968A shows retinotoxic effects in cats, monkeys and man after oral administration (Raison & Stanton, 1955, Edge et al., 1956). They all produce retinal changes which follow almost identical ophthalmoscopic and histological patterns.

At first sight, the retinotoxic substances appear to have little in common either chemically or pharmacologically. However, they all produce marked degeneration of the visual cells of the mammalian retina.

The object of this research was to find substances which mimicked the progressive hereditary retinal degeneration seen particularly in human retinitis pigmentosa, in the hope that if specific toxic compounds were found, a detailed study of their pharmacology and biochemistry would elucidate their mode of action and throw some light on the aetiology of the human degeneration condition.

B. INVOLVEMENT OF THIOL GROUPS

It was found that cysteine and certain aliphatic thiol compounds protected the retina against the toxic effects of sodium iodate suggesting that the retinotoxic effect is determined by an alteration in -SH (thiol) levels in the retina (Sorsby &

Harding, 1962b), cysteine being a -SH (thiol) group containing amino acid. Reading and Sorsby (1966b) using a potentiometric method for the determination of -SH groups, investigated the possibility of the existence of a common factor present in the retinotoxic substances. 'Total' -SH represents all the thiol groups present in the tissue, whereas the 'free' or soluble -SH fraction represents the thiol groups present in small soluble peptides (e.g. glutathione) or amino acids like cysteine or methionine. It is possible to estimate differentially these classes since they have different reactivities towards a specific thiol group reactant such as p-chloromercuribenzoate. The 'soluble' thiol groups react much more readily than the insoluble or 'bound' thiol groups. So if a tissue is allowed to react with a specific thiol-group reagent for a short time (say 10–15 min) only the 'soluble' -SH groups will react whereas if the tissue is allowed to react for 1 hour or more all thiol groups will react ('bound' and 'soluble') (Calcutt & Doxey, 1959).

A consistent rise in 'total' -SH values of rabbit retina, following intravenous injection, was produced by all the compounds (chloroquine and phenothiazine groups were not tested) and was the first pharmacological or biochemical effect found to be common to these substances. This required careful interpretation. A rise in 'total'- SH values is consistent with the idea of specific protein denaturation occurring in the retina; '-SH' groups previously inaccessible to chemical reaction are 'unmasked' by the denaturing process. This does not amount to a destruction nor a decomposition of the protein, but simply means that the tertiary structure of the native protein is altered—usually by uncoiling globular protein. These experiments led to a further detailed investigation of the action of di-(p-aminophenoxy)-pentane.

C. DI-(p-AMINOPHENOXY)-PENTANE (M AND B 968A)

This drug is of especial interest, since it has a potent schistosomicidal activity in addition to its marked toxicity towards the retina in humans as well as lower mammals. Studies on di-(p-aminophenoxy)-pentane took into account knowledge of the role of thiol groups in the retina which have been implicated in the process of visual excitation (Wald, 1955). The action of light in the bleaching of rhodopsin in the vertebrate retina is associated with 'unmasking' of -SH groups: two per molecule of visual pigment. Thiol groups are not involved in the direct chemical combination between 11-cis retinaldehyde and the visual pigment protein opsin, but they are involved in the conformation of the protein moiety by means of hydrogen bonding. Another observation associated with this drug was made by Arden and Fojas (1962). They examined the effects of a series of diaminodiphenoxypentanes on the electro-retinogram (ERG) of the rabbit and found that these primary amines affected the amplitude and form of the dark-adapted ERG and reduced the rate of dark adaptation. They argued that the results accorded with an alteration in the rate of regeneration of visual pigment.

Previous biochemical studies by H.W. Reading and M. Davies (unpublished) has shown no marked or specific effect on enzymes of carbohydrate metabolism in mammalian retina by the diaminodiphenoxyalkanes and in addition Goodwin et al. (1957) had found that these drugs prevented the resynthesis of visual pigment in amphibia. The effect of 1:5 di-(p-aminophenoxy)-pentane on enzymatic pathways associated with the visual cycle was investigated extensively by Reading (1970b). These consisted of the hexose monophosphate shunt (HMP) of direct oxidation of glucose, retinal alcohol dehydrogenase (retinaldehyde reductase) and effects on the stability of the visual pigment. As previously outlined the HMP shunt pathway was implicated in the normal functioning of the visual cycle in mammalian retina by the researches of Futterman (1963). Light falling on the retina stimulates the HMP shunt through its metabolic coupling to retinal alcohol dehydrogenase.

HMP shunt activity was evaluated by determination of the proportions of respiratory carbon dioxide produced by incubation of retina with specifically carbon 1 or carbon 6 labelled glucose.

Although there was an initial indication of a slight stimulation of HMP activity 7 hours after intravenous injection, the general tendency was towards an overall decrease in HMP shunt activity. Values, however, showed a high degree of variability.

Determinations of retinal alcohol dehydrogenase were made 1 hour to 7 days after injection of the diaminodiphenoxypentane. Intravenous injection of the drug produced a marked and rapid effect on enzyme activity. One hour after injection, enzyme activity was inhibited about 42% with a return to normal levels after 24 hours (Table 2.3). Determinations of the visual pigment content of rabbit retina following intravenous injection of 5 mg/kg of the diaminodiphenoxy-

Table 2.3. The effect of 1:5-di (p-aminophenoxy)-pentane dihydrochloride (5 mg/Kg i.v.) on retinaldehyde reductase in rabbit retina

Time after injection	Mg Retinaldehyde produced/hour per gram wet weight of retina
Control value (no drug administered)	291 ± 21
1 hour	169 ± 7
4 hours	228 ± 25
24 hours	321 ± 22
48 hours	301 ± 39
7 days	277 ± 14

Values represent mean $\pm$ SD of six determinations. (Data from Reading, 1970b.)

pentane were made on digitonin extracts of retinal tissue after light or dark adaptation.

Under conditions of dark adaptation, injection of the drug produced a reduction in visual pigment content of the retina, even when the injection itself was carried out in the dark, i.e. under illumination from a red darkroom safelight. This reduction amounted to about 15% of the normal visual pigment content of the retina. In addition, when the drug was administered to the dark adapted animal, and this was followed by light adaptation, the visual pigment content of the retina was reduced to well below the value found by similar light adaptation in control animals.

In light adapted animals, the drug did not produce any further decrease in visual pigment in the retina. However, the drug caused impairment of the retina's ability to resynthesize visual pigment during dark adaptation. In this case, inhibition amounted to about 42%. Inhibition of the coupled enzyme systems, i.e. retinal alcohol dehydrogenase (retinaldehyde reductase) and HMP shunt confirmed that the drug affects the normal functioning of the visual cycle and the reduction of retinaldehyde released by the action of light on the retina. The observed effects on the HMP shunt are probably sequential to the primary effect on retinal alcohol dehydrogenase; the dehydrogenase being inhibited there will be a consequent drop in the metabolically coupled HMP shunt activity.

The most interesting aspects of the drug's effects on the retina are those concerning its action on the stability of the visual pigment, especially during dark adaptation. No such effect was observed after the retina was in the fully light adapted condition, i.e. when the visual pigment was fully bleached. Injection of the drug did not reduce the amount of residual visual pigment even after 24 hours continuous illumination.

However, the drug did affect the capacity of the retina to re-synthesize visual pigment after bleaching. One explanation of the varied effects and suggesting a basic mode of action could be that the diaminodiphenoxypentane, and possibly, the other retinotoxic substances affect the visual pigment itself causing conformational changes in the protein moiety of rhodopsin. It has become evident from the researches of Professor Bonting and his associates at Nigmegan University (Bonting, 1969, 1974) that the conformation and structure of the protein part of rhodopsin is of paramount importance to its functioning as a light sensitive structure and propagation of the nerve impulse.

This recent knowledge, which has been discussed earlier in the chapter, throws some light on the mode of action of the retinotoxic compounds, especially the diaminodiphenoxypentanes. It would appear that their action in 'unmasking' -SH groups in retinal tissue is undoubtedly a measure of protein denaturation. This type of denaturation is of a relatively mild form and is often reversible. It entails alteration of the tertiary structure and shape of the protein molecule. Since a similar process occurs in rhodopsin under the action of light, viz: the 'unmasking' of two -SH groups per protein molecule, it is not difficult to see why these substances affect the visual cycle in the retina. In particular, diaminodiphenoxypentane affects the stability of rhodopsin, even in the dark, and increases retinal total -SH group content (Reading & Sorsby, 1966b) and inhibits retinal alcohol dehydrogenase (Reading, 1970b). Alcohol dehydrogenase is well known to be dependent on intact -SH groups for its activity and is particularly susceptible to inhibition by thiol-group inhibitors. Retinal alcohol dehydrogenase is situated in the distal parts of the visual cells of cattle and rat retinae so that both visual pigment and alcohol dehydrogenase are in close apposition.

D. RETINAL SENSITIVITY TO TOXIC AGENTS

The specific sensitivity of the retina to these agents is most likely due to thiol (-SH) groups being of paramount importance to the structural and functional integrity of the visual pigment and also that alcohol dehydrogenase is structurally and functionally associated with the visual mechanism. Excessive denaturation of visual pigment will increase its breakdown with a similar sequence of events to that occurring in the 'retinitis pigmentosa' rat with the final stages involving release of lysosomal proteases digesting the cellular material of the neuroepithelium.

E. SUBSTITUTED PHENOTHIAZINES AND THE QUINOLINE DERIVATIVES (CHLOROQUINE)

Chlorpromazine induces skin and ocular (lens and corneal) pigmentation in 20–40% of patients treated with more than 300 mg of the drug. On the other hand, piperidylchlorophenothiazine (Sandoz, N.P. 207) shows true retinotoxic properties. Experimentally in cats and rabbits, NP207 produces a pigmentary retinopathy (Meier-Ruge & Cerletti, 1966). In addition, many clinicians have described reversible deposition of chloroquine in cornea and lens together with irreversible retinal damage in patients treated over long periods with the drug (Hobbs et al., 1959).

Meier-Ruge and Cerletti (1966) carried out careful cytochemical and autoradiographic investigations of these effects and showed that the phenothiazine (NP207) accumulates in melanin-bearing structures of the eye and other melanin-containing tissues and that it affects the metabolic status of the rod-cell ellipsoids, so that an inhibition of oxidative phosphorylation and anaerobic glycolysis takes place. This means that there is a serious decrease in the normal energy-supplying processes for the rod cell. The metabolic effects are likely to be due to a blocking action on flavine nucleotides. Chloroquine was shown by the same workers to attack the pigment epithelium and they concluded that chloroquine retinopathy is due to an insufficiency of the specific permeability of the pigment epithelium.

In this way, it can be seen that these classes of drugs act in a different manner to the ones mentioned previously. The main differences lie in the fact that both the quinolines and the phenothiazines are stored in tissues of the eye and were it not for the property they would probably not show a specific effect on eye tissues. Their overall effect appears to be on energy metabolism and in fact these compounds in low concentration actually stabilize membranes decreasing permeability and so prevent access to substrates etc., whereas the diaminodiphenoxyalkanes would appear to be acting as denaturants of lipoprotein membranes.

IV Biochemical aspects of diabetic retinopathy

A. INTRODUCTION

Diabetic retinopathy is widely regarded as an outcome of generalized vascular disorder peculiar to diabetes. Whether the morphological changes in this class of retinopathy is regarded as an ocular expression of a generalized small vessel disease, or as a vascular disease of the eye conditioned by the special metabolic characteristics of the diabetic retina affects the biochemical approach to the problem.

Biochemical investigations have been carried out on retinal tissue from two sources: tissue obtained from animals with spontaneous or induced diabetes (diabetes being induced usually by alloxan) and retinae in which retinopathies which closely resemble or are identical with diabetic retinopathy have been induced by specific agents. In a very small number of instances human diabetic retinae have been investigated.

B. ACTION OF INSULIN IN DIABETES

The exact mode of action of insulin is not known in molecular terms. The early idea that it controlled the activity of enzymes concerned in glucose utilization is not now accepted. It appears that insulin is concerned in regulating the synthesis of a membrane-bound carrier transport system for glucose and other metabolites, by enhancing the rate of translation of specific messenger-RNA at the ribosomes. The major defect in insulin deficiency is a greatly decreased affinity for glucose in muscles and liver so that much higher levels of glucose are required in the blood to effect penetration. Most of the metabolic effects of insulin deficiency appear to be compensatory in order to allow the body to maximise its blood glucose concentration and thus overcome the defect in utilization. Thus the combined effect of the decreased conversion of glucose to fatty acids, decreased protein synthesis and gluconeogenesis from amino acids are responses which lead to excess glucose production. The above general observations on the metabolic consequences of insulin deficiency in diabetes lead to the specificity of retinal metabolism since differences between retinal and somatic metabolism could be responsible for the changes seen in diabetic retinopathy. The retina shares with brain the capacity for requiring glucose as its major nutrient and source of energy and it is tempting to speculate that changes in carbohydrate metabolism brought about by insulin lack are concerned in the lesion in the diabetic retina.

C. RETINAL BIOCHEMISTRY AND DIABETES

Biochemical experimental work on the retina must of necessity nearly always be carried out with excised isolated tissue incubated in a nutrient medium. Most work in which diabetic and non-diabetic retinae have been compared has been done on retinae obtained from the laboratory rat or dog.

There are some important divisions between the experimental studies. These can be subdivided as follows:

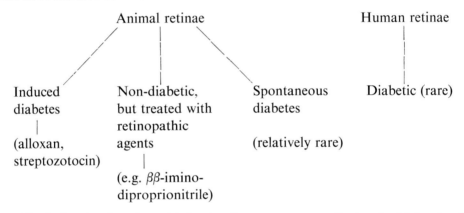

The isolated retinae has a high rate of oxygen consumption and a high rate of glycolysis (Graymore, 1970). Pyruvate is oxidized to carbon oxide and water by the citric acid or Krebs tricarboxylic acid cycle, thus providing the major part of the energy made available by the oxidation of glucose. Intermediates of the Krebs cycle are interconverted to amino acids, mainly glutamic, aspartatic, glutamine and γ-amino butyric acid. In fact there is good evidence that a very large proportion of the glucose oxidized by the retina has to pass through the amino acid stage before being completely oxidized (Catanzaro et al., 1962). The amino acids are incorporated into protein and possibly utilized as transmitter agents. This is especially the case with glutamic and γ-amino butyric acids.

In addition, like brain, retina has the capacity to oxidize glucose by the direct oxidation or hexose monophosphate shunt pathway. This may account for 10% of the total glucose metabolized. In neo-natal retina this figure is even higher (Graymore, 1970).

D. EFFECT OF GLUCOSE CONCENTRATION AND INSULIN IN NORMAL RETINA

Keen and Chlouverakis (1965) showed that increasing glucose concentration in the media of incubated dog retina did not alter uptake into the tissue. Rat retina does, however, show some dependence on glucose concentration. The addition of insulin to the medium in a range of concentrations always failed to affect glucose uptake of normal dog or rat retina. It appears, therefore, that the retina is an insulin-insensitive tissue.

E. DIABETIC RETINA—ANIMALS SPONTANEOUSLY DIABETIC OR DIABETIC BY INDUCTION

No striking changes have been observed in the catabolism of glucose in diabetic retinae. Graymore and Towlson (1970) found no change in glycolysis in retina from the diabetic rat; they also found no changes in the pathways of glucose oxidation, whilst Keen and Chlouverakis (1965) found no change in uptake of glucose or lactic acid output in diabetic rat or dog retinae.

An important difference in glucose metabolism between normal and diabetic rat retina is the accumulation of glycogen in the diabetic retina; both *in vivo* and *in vitro* (Kurimoto & Newell, 1963). This is interesting since Kuwabara (1965) showed that injury to the retina by toxic agents caused accumulation of glycogen mainly in the Muller fibres. Certainly the alloxan-diabetic rat retina synthesizes more glycogen than normal but it is not known where this is localized. The control mechanism for this does not appear to be an increase in free fatty acid concentration which would tend to depress glycolysis, rather it appears to be a consequence of the overall increase in blood glucose concentration. In the human there seems to be no direct evidence that glycogen accumulates in the diabetic retina before pathological changes occur.

F. FATTY ACID METABOLISM IN THE DIABETIC RETINA

Keen and Chlouverakis (1965) investigated the effect of increased concentration of non-esterified fatty acids (NEFA) on isolated rat retina. They acted on an observation which they made when testing the effect of incubating normal isolated retinae with plasma from a diabetic patient. They found that this reduced glucose uptake. The effect seemed to be caused by the presence in the plasma of a non-dialysable factor which was present in plasma from ketotic patients but absent from normal plasma. About the time that Keen and Chlouverakis made this observation, Randle and his colleagues (1963) showed that in certain tissues raised concentration of NEFA and ketone bodies inhibit glucose breakdown. Keen and Chlouverakis (1965), however, found that NEFA levels did not affect glucose uptake so they concluded that the 'Randle' effect did not operate in retina. In addition, insulin did not affect the uptake of NEFA. The only positive finding was that alloxan diabetic retina has a higher oxidative capacity for NEFA than normal as judged by production of carbon dioxide. This may be an adaptive sequel to the raised level of NEFA in diabetes.

It is therefore undecided that a raised concentration of NEFA in diabetes has any effect on the development of diabetic retinopathy. It must be pointed out, however, that oxidation of fatty acids is accompanied by reduction of the co-enzyme nicotinamide adenine dinucleotide ($NAD^+ \rightarrow NADH$). In this context Heath et al. (1962) found an increase in NADH in the retina of alloxan diabetic rats. In normal tissues there is more oxidized than reduced co-enzyme ($NAD^+ > NADH$). Such a change in the ratio of oxidized to reduced co-enzyme would be expected to have a very marked effect on the course of oxidative reactions in metabolism. For instance, such changes alter the proportion of alcoholic to acidic metabolites produced from the biogenic amines (Eccleston et al., 1969).

In particular in the retina a change in the ratio of oxidized to reduced nicotinamide nucleotide co-enzymes would also affect the production of vitamin A alcohol, the final product of bleaching in the visual cycle. This occurs since alcohol dehydrogenase present in the outer rod cell sacs of the retina requires NADPH as a specific co-enzyme. One of Heath's findings was that the decrease in NAD/NADH is accompanied by a decrease in NADPH. In these circumstances less retinaldehyde can be converted to alcohol and vitamin A ester so that less escapes from the retina. This compound cause structural damage or changes in the vasculature. This suggestion has never been fully investigated in the diabetic retina.

In the same context, one of the few clinical observations which may have some biochemical significance, and which does not appear to have received further attention, was that made by Powell and Field (1964). They reported that the incidence of retinopathy was much below expectation in diabetics with rheumatoid arthritis and who had been treated over long periods with salicylates. Other workers (Eshmann et al., 1963) have reported regression of retinopathy in diabetics treated with p-aminosalicylic acid preparations. The retinal vascular changes which involve excessive fragility and permeability also include premature pericyte degeneration as well as degenerative changes in the endothelial lining (Garner, 1970). Such changes could be caused by the release of lysosomal enzymes. This would be in line with changes in membrane integrity, possibly associated with changes in fatty acid metabolism. In other words, the vascular destruction could be due to partial digestion by lysosomal enzymes. In this case salicylates would act by stabilizing lysosomal membranes; a property which they are known to possess (Tanaka & Lizuka, 1968).

G. MUCOPOLYSACCHARIDES OF THE DIABETIC RETINA

Retinae from alloxan-diabetic rats show an increase in PAS-staining material, particularly of the vessels (Heath, 1970) which led to the belief that mucopolysaccharide synthesis may be increased in the diabetic retina. Where mucopolysaccharide analyses have been carried out, it appears that the major component is chondroitin sulphate, whilst histological examination has shown that the main formation of sulphated mucopolysaccharide was in the ground substance between the visual cells. On the other hand, Paterson and Heath (1965) reported a significant fall in chondroitin sulphate in diabetic retinae.

In fact, there is no evidence that increase in PAS-staining is due to increased mucopolysaccharide. PAS-staining is not specific and will react with reducing compounds attached to proteins such as collagenous protein of basement membrane.

H. CHANGES IN BASEMENT MEMBRANE OF KIDNEY, LENS AND RETINA IN DIABETES

It is noteworthy that the basement membrane of the retinal vessels, glomerulus and lens capsule have been shown to be antigenically similar in the normal rat (Roberts, 1957). Anti-rat glomerulus serum coupled with fluorescein reacts in tissue sections with lens capsule and small blood vessels of the retina. All these structures are affected in chronic diabetes so studies have been carried out in the hope of finding some common target mechanism. The basement membranes of these tissues contain a prominent glycoprotein component: collagen linked to carbohydrate molecules (Patterson & Heath, 1967b). In the diabetic kidney the basement membrane thickening is associated with an increase in glycoprotein fraction.

A factor common to all collagen is the presence of the amino acid hydroxyproline as part of the protein structure. Collagen is the only protein in the body to contain this amino acid which it does in fairly large proportions.

The studies of Heath and his colleagues (Heath et al., 1967); Paterson and Heath, (1967b) stimulated great interest in the nature of the chemical structure of collagen in diabetes. They found the glomerulus basement membrane lens capsule and retinal vascular tree of the diabetic all contain more hydroxyproline than normal. The increase in hydroxyproline has been found in alloxan-diabetic animals as well as in retinal vessels of human diabetics. It appears, therefore, that thickening of the basement membrane or capillaries in the diabetic is due to increased formation of collagenous glycoprotein.

Paterson and Heath (1967a, b) found that administration of a lathyrogen (iminodiproprionitrile, IDPN) to rats, produced retinopathy and caused a great increase in hydroxyproline content of the lens capsule.

Such substances as IDPN cause cross linkages in the molecular structure of collagen molecules and it is possible that these changes result in increased permeability of the basement membrane. This would explain the increase in 'leakiness' of the retinal capillaries in diabetes. It does not, however, explain the marked catastrophic effects of the blood vessels in the retina, in contrast to general vascular capillaries. In discussing experimentally-induced retinopathy, it must be stressed that although diabetes can be induced experimentally in animals with a number of agents, the pathological picture characteristic of human diabetic retinopathy does not always develop. This could be due to the experimental condition being different to the natural disease. Some dogs with spontaneous diabetes show similar retinal lesions to humans. This may simply reflect the duration of the disease, since the human lesions develop after many years and artificially induced diabetes is a relatively acute state. Other factors such as hormonal imbalance or dietary abnormalities may play a part since in the experimental animal diabetic cataract invariably occurs without retinal complications. For this reason, experiments using animals rendered diabetic have not yielded a great deal of clinically useful information. Improved histological techniques, including flat retinal preparations and treatment of the retina with trypsin, have shown microscopic changes in the vascular system which appear to be proceeding towards full retinal changes (Heath, 1970). The paramount question is how to relate experimentally induced retinopathy to the condition in humans in order to elucidate the aetiology of diabetic retinopathy.

Poor control of diabetes appears to contribute to retinopathy in man, but Heath (1970) has questioned whether this is a contributory factor in experimental animals since he has maintained alloxan-diabetic rats for long periods with the minimum of control and with severe hyperglycaemia without exacerbation of retinal changes. Such conditions invariably induce diabetic caterats which would indicate that some factor other than poorly controlled diabetes must be operative and to contribute to diabetic retinopathy.

The irrational nature of experimental retinopathy reflects the situation in the human condition which adds further weight to the argument for extra-causal factors. Episodic alterations in metabolism or diet may be responsible or exacerbate the condition of retinopathy. In experimental animals this is often not achieved since the animals are normally kept under careful environmental conditions. However, attempts to simulate the human situation by feeding high-fat or high-carbohydrate diets have not produced severe retinopathy as expected. Administration of corticotrophin or corticosteroids to diabetic animals have not produced conclusive evidence of the development of lesions. More positive results have been obtained by physical means, in which the retinal vessels are occluded by injection of inert foreign bodies of extremely small size.

In the case of IDPN retinopathy, although this chemical may produce lesions of small vessel embolisation, the manner in which this toxic nitrile acts is not known and virtually no information is available on the specific metabolic processes involved. It is therefore not yet possible to reproduce the complete syndrome in animals but it appears that the most successful course would be to superimpose the unknown causal factor or factors on the diabetic state. In fact, Heath has suggested that future research should be directed to exposing diabetic animals to more metabolic fluctuation.

I. PLATELET ABNORMALITIES

Another important factor which must be considered concerns recent work by Dr. Heath and his colleagues at University College Hospital, London (Heath et al., 1971). The idea for the investigation of platelet abnormality arose from the susceptibility of diabetic patients to cardio-vascular complications. Heath and his colleagues examined two groups of diabetics considered to be either prone or resistant to retinal vascular complications. The resistant group had had diabetes for at least seventeen years but showed minimal or no ophthalmoscopic signs of retinopathy. The susceptible group consisted of patients with severe retinopathy. Both groups were compared with age-matched normal healthy subjects.

No significant differences in platelet adhesiveness were found between 28 control subjects and a group of 22 diabetics without retinopathy and another group of 22 diabetics with retinopathy. In order to examine the platelet abnormalities more fully, ADP-activated aggregation was investigated on the same groups. These results showed that on a group basis, platelets from patients with serious or minimal retinopathy did not aggregate in the presence of ADP more rapidly or to a greater extent than normal healthy subjects. However, the platelets of the sub-group of patients with actively deteriorating retinopathy showed greater sensitivity towards ADP. Their platelets aggregated at lower concentrations of ADP. In addition this group showed a decreased rate of disaggregation of platelet aggregates. If this situation occurred *in vivo*, capillary occlusion may take place under these circumstances. The change in aggregation towards ADP was not due to any differences in the activity of ADP splitting enzyme systems in the blood.

J. DRUG THERAPY OF DIABETIC RETINOPATHY

Apart from the photo-coagulation techniques aimed at arresting the vascular lesions in the retina, various attempts have been made at drug therapy to arrest degeneration in diabetic retinopathy. As already mentioned, there exists the chance observation that patients on long-term salicylate therapy have shown less tendency towards retinopathy, but other drugs have received attention. Preparations of rutin have been tried sporadically (Donegan & Thomas, 1948) but the general consensus of opinion seems to be that such drugs have an equivocal effect on the progress of the degeneration. Most investigators have concluded that the administration of rutin derivatives to patients with diabetic retinopathy does not materially affect the progress of the condition. There does seem to be some evidence that this therapy will prevent further loss of visual acuity associated with decreasing capillary fragility rather than improve reduced visual acuity. In recent years there has been a great deal of attention in treatment of exudative retinal lesions with preparations of α-tocopherol (vitamin E). In effect, such treatment has proved just as equivocal as previous ones. Japanese workers have been particularly active in this field. In America De Hof and Ozazewski (1954) recently published the results of a trial they made with α-tocopherol. They concluded that given in 300 and 600 mg doses every day for protracted periods there was no demonstrable effect on the progression of retinopathy. However, these observations were made on only 12 patients, but there are other reports where various claims have been made although it is impossible to assess the overall value of the use of vitamin E.

V The present position: an overview

The research described in this chapter characterizes to some extent the contribution of the biochemist to the field of neuro-ophthalmology over the past twenty years or so. This contribution is considerable since it involves basic research into normal retinal and optic tract function. A number of scientific centres are currently working on the mechanism of the visual cycle and the propagation of the nerve impulse. Much of this work involves comparative biological studies, especially that concerned with the nature and characterization of light-sensitive pigments.

However, when we examine the field with respect to the cause of retinal dysfunction and blindness, it would appear that less and less attention is being paid to this important aspect. The clinical advantages from such research are not always readily apparent and much painstaking and protracted work must be done before substantial advances can be made.

The most important problem requiring careful and detailed investigation is retinopathy associated with diabetes mellitus. In Great Britain, for instance, this is responsible for 7% of all new registrations of blindness each year. It is, therefore, the most important single systemic cause of blindness. Apart from the studies on platelet adhesives, very little is known of the basic biochemical changes associated with diabetes which can be associated with breakdown of retinal vasculature. In the field of inherited retinal disease which represents about 10 to 14% of registrations, although the prognostic indications are relatively depressing, biochemical research has produced some encouraging information. It is possible to question on a sound basis the wisdom of the administration of prolonged and massive vitamin A therapy in young retinitis pigmentosa patients or to people genetically at risk to this condition. There is also a scientific basis of support for the palliative treatment of reducing patients to exposure to light and there is tentative support for some approach to therapeutic measures involving the use of anti-inflammatory drugs. It remains to be seen whether new drugs of this type will be forthcoming and whether they will form a basis of a mixed therapy. These conditions are interesting in relative importance as causes of blindness, now the incidence of bacterial and viral infections are diminishing.

References

ANDERSON K.V. & O'STEEN W.K. (1972) Black-white and pattern discrimination in rats without photoreceptors. *Exp. Neurol.* **34**, 446–454.

ANDREWS J.S. & FUTTERMAN S. (1964) Metabolism of the retina. V. The role of microsomes in vitamin A esterification in the visual cycle. *J. Biol. Chem.* **239**, 4073–4080.

ANSELL P.L. & MARSHALL J. (1974) The distribution of extracellular acid phosphatase in the retinae of retinitis pigmentosa rats. *Exp. Eye Res.* **19**, 273–281.

ARDEN G.B. & FOJAS M.R. (1962) The mode of action of diaminodiphenoxyalkanes and related compounds in the retina. *Vision Res.* **2**, 163–174.

BARNETT K.C. (1966) Primary tapeto-retinal degeneration in dogs. In Graham Jones, (Ed.) *Aspects of Comparative Ophthalmology*, pp. 77–87, Pergamon Press, Oxford.

BENNETT E.L., KRECH D. & ROSENZWEIG M.R. (1964) Reliability and regional specificity of cerebral effects of environmental complexity and training. *J. Comp. Physiol.* **57**, 440–441.

BENNETT M.H., DYER R.F. & DUNN J.D. (1972) Light-induced retinal degeneration—effect upon light-dark discrimination. *Exp. Neurol.* **38**, 80–89.

BERSON E.L. (1971) Light deprivation for early retinitis pigmentosa—a hypothesis. *Arch. Ophthal.* **85**, 521–529.

BERSON E.L. (1973) Experimental and therapeutic aspects of photic damage to the retina. *Invest. Ophthal.* **12**, 35–44.

BERSON E.L. & GOLDSTEIN E.B. (1970) Recovery of the human early receptor potential during dark adaptation in hereditary retinal disease. *Vision Res.* **10**, 219–226.

BITENSKY M.W., GORMAN R.E. & MILLER W.H. (1971) Adenyl cyclase as a link between photon capture and changes in membrane permeability of frog photoreceptors. *Proc. Natn. Acad. Sci. U.S.* **68**, 561–562.

BLANKS J.C., ADINOLFI A.M. & LOLLEY R.N. (1974a) Synaptogenesis in the photoreceptor terminal of the mouse retina. *J. Comp. Neurol.* **156**, 81–94.

BLANKS J.C., ADINOLFI A.M. & LOLLEY R.N. (1974b) Photoreceptor degeneration and synaptogenesis in retinal degenerative (rd) mice. *J. Comp. Neurol.* **156**, 95–106.

BLOMSTRAND C. & HAMBERGER A. (1969) Protein turnover in cell enriched fractions from rabbit brain. *J. Neurochem.* **16**, 1401–1407.

BOK D. & HALL M.O. (1971) The role of the pigment epithelium in the etiology of inherited retinal dystrophy in the rat. *J. Cell. Biol.* **49**, 664–682.

BONAVENTURE N. & KARLI P. (1961) Sensibilite visuelle spectrale chez des souris a retine entierement depourvue de cellules visuelles photoreceptrices. *C.R. Soc. Biol.* **155**, 613–623.

BONAVITA V. (1965) Molecular and kinetic properties of NAD- and NAPD-linked dehydrogenases in the developing retina. In '*Biochemistry of the Retina*', pp. 5–13, London Academic Press.

BONAVITA V., GUANERI R. & PONTE F. (1966). ATPase in the normal and dystrophic developing retina of the rat. *Experientia*, **22**, 720–721.

BONAVITA V., PONTE F. & AMORE G. (1963) Neurochemical studies on the inherited retinal degeneration of the rat. I. Lactic dehydrogenase in the developing retina. *Vision Res.* **3**, 277–280.

BONTING S.L. (1969) Recent insights in the photoreceptor mechanism. In *Proceedings of the 2nd International Meeting of the International Society for Neurochemistry, Milan*, p. 31. Milan, Tamburini Editore.

BONTING S.L. (1975) Photoreceptor membrane properties and visual excitation. *Biochemical Society Transactions*, **2**, 1228–1229.

BOURNE M.C., CAMPBELL D.A. & TANSLEY K. (1938) Hereditary degeneration of the rat retina. *Brit. J. Ophthalmol.* **22**, 613–623.

BRIDGES C.D.B. & YOSHIKAMI S. (1969) Uptake of tritiated retinaldehyde by the visual pigment of dark-adapted rats. *Nature (London)*, **221**, 275–276.

BROTHERTON J. (1962) Studies on the metabolism of the rat retina with special reference to retinitis pigmentosa. 2. Amino acid content as shown by chromatography. *Exp. Eye Res.* **1**, 2246–2252.

BRUCKNER R. (1951) Spaltlampemikroskopie und Ophthalmosckopie am Auge von Ratte und Maus. *Documenta Ophthalmologica*, **5–6**, 452–544.

BURDEN E.M., YATES C.M., READING H.W., BITENSKY L. & CHAYEN J. (1971) Investigation into the structural integrity of lysosomes in the normal and dystrophic rat retina. *Exp. Eye Res.* **12**, 159–165.

CALCUTT G. & DOXEY D. (1959) The measurement of tissue-SH. *Expt. Cell Res.* **17**, 542–543.

CAMPBELL D.A. & TONKS E.L. (1962) Biochemical findings in human retinitis pigmentosa with particular relation to vitamin A deficiency. *Brit. J. Ophthalmol.* **46**, 151–164.

CARAVAGGIO L.L. & BONTING S.L. (1963) The rhodopsin cycle in the developing vertebrate retina, II. Correlative study in normal mice and in mice with hereditary retinal degeneration. *Exp. Eye Res.* **2**, 12–19.

CATANZARO R., CHAIN E.B., POCCHARI F. & READING H.W. (1962) Metabolism of glucose and pyruvate in rat retina. *Proc. Roy. Soc. B.* **156**, 139–143.

CHADER G.J. (1971) Hormonal effects on the neural retina: induction of glutamine synthetase by cyclic $3'5'$ AMP. *Biochem. Biophys. Res. Commun.* **43**, 1102–1105.

CHADER G.J., HERZ L. & FLETCHER R.T. (1974) Cyclic nucleotide hydrolysis: some possible natural regulators in retina and rod outer segments. *J. Neurochem.* **23**, 873–874.

CHAMBERS D.A. & ZUBAY G. (1969) The stimulatory effect of cyclic adenosine $3'5'$ monophosphate on DNA-directed synthesis of β-galactosidase in a cell-free system. *Proc. Natn. Acad. Sci. U.S.A.*, **63**, 118–122.

CHATZINOFF A., NELSON E., STAHL N. (1968) Eleven-CIS vitamin A in the treatment of retinitis pigmentosa. A negative study. *Arch. Ophthalmol.* **80**, 417–419.

CHAYEN J., BITENSKY L. & UBHI G.S. (1972) The experimental modification of lysosomal dysfunction by anti-inflammatory drugs acting *in vitro*. *Beitr. Path. Bd.* **147**, 6–20.

COGAN D.G. (1950) Symposium: primary chorioretinal aberrations with night blindness. *Pathology Transactions of the American Academy of Ophthalmology and Otolaryngology*, **4**, 629–661.

COLLINS E.T. (1919) Abiotrophy of the retinal neuropithelium of 'retinis pigmentosa'. *Trans. Ophthal. Soc. U.K.* **39**, 165–195.

CRAGG, B. G. (1967). Changes in synaptic numbers and dimensions following first exposure of rats to light. *Nature (London)*, **215**, 251–253.

DAWSON G. & NEWELL F.W. (1974) Arachidonic acid and retinal pigmentary degeneration. *Lancet* **i**, 1119.

DE HOFF J.B. & OZAZEWSKI J. (1954) Alpha-tocopherol to treat diabetic retinopathy. *Am. J. Ophthal.* **37**, 581–582.

DEWAR A.J. (1974) Changes in incorporation of (^{14}C) uridine into uridine nucleotides in rat visual cortex during first exposure to light. *Exp. Neurol.* **45**, 134–140.

DEWAR A.J. (1975) Biochemical studies on the dystrophic rat retina and brain. *Biochemical Society Transactions* **2**, 1233–1238.

DEWAR A.J., BARRON G. & READING H.W. (1975a) The effect of retinol and acetylsalicylic acid on the release of lysosomal enzymes from rat retina *in vitro*. *Exp. Eye Res.* **20**, 63–72.

DEWAR A.J., BARRON G. & RICHMOND J. (1975b) Cyclic AMP phosphodiesterase activity in the dystrophic rat retina. *Biochemical Society Transactions*, **3**, 265–268.

DEWAR A.J. & READING H.W. (1973) A comparison of RNA metabolism in the visual cortex of sighted rats and rats with retinal degeneration. *Exp. Neurol.* **40**, 216–231.

DEWAR A.J. & READING H.W. (1974a) Stabilization and labilization of lysosomes in rat retina. *Biochemical Society Transactions* **2**, 645–647.

DEWAR A.J. & READING H.W. (1974b) The role of retinol in, and the action of anti-inflammatory drugs on, hereditary retinal degeneration. In *Impairment of Cellular Function during Ageing and Development* in vivo *and* in vitro. Cristofalo V.J. & Holeckova E., New York, Plenum Press, pp. 281–295.

DEWAR A.J., READING H.W. & WINTERBURN A.K. (1973a) RNA metabolism in the cortex of newly weaned rats following first exposure to light. *Life Sciences* **13**, 565–573.

DEWAR A.J., READING H.W. & WINTERBURN A.K. (1973b) RNA metabolism in subcellular fractions from rat cerebral cortex and from the visual cortex of rats with retinal degeneration. *Exp. Neurol.* **41**, 133–149.

DEWAR A.J. & WINTERBURN A.K. (1973) Metabolism of nuclear and cytoplasmic proteins in the visual cortex of sighted rats and rats with retinal degeneration. *Exp. Neurol.* **41**, 584–598.

DONEGAN J.M. & THOMAS W.A. (1948) Capillary fragility and cutaneous lymphatic flow in relation to systemic and retinal vascular manifestations: Rutin therapy. *Am. J. Ophthal.* **31**, 671–684.

DOWLING J.E. (1964). Nutritional and inherited blindness in the rat. *Exp. Eye Res.* **3**, 348–356.

DOWLING J.E. & GIBBONS I.R. (1961). The effect of vitamin A deficiency on the fine structure of the retina. In *The Structure of the Eye*, pp. 85–99. New York, Academic Press.

DOWLING J.E. & Sidman R.L. (1962) Inherited retinal dystrophy in the rat. *J. Cell Biol.* **14**, 73–109.

ECCLESTON D., READING H.W. & RITCHIE I.M. (1969) 5-hydroxytryptamine metabolism in brain and liver slices and the effects of ethanol. *J. Neurochem.* **16**, 274–276.

EDGE N.D., MASON D.F.J., WIEN R. & ASHTON N. (1956). Pharmacological effects of certain diaminodiphenoxyalkanes. *Nature (London)* **178**, 806–807.

ESMANN V., JENSON H.J. & LUNDPACK K. (1963) Disappearance of waxy exudates in diabetic retinopathy during administration of p-amino salicylate (PAS). *Acta. med. Scand.* **174**, 99.

FARBER D.B. & LOLLEY R.N. (1973) Proteins in the degenerative retina of C3H mice: deficiency of a cyclic nucleotide phosphodiesterase and opsin. *J. Neurochem.* **21**, 817–828.

FARBER D.B. & LOLLEY R.N. (1974) Cyclic guanosine monophosphate: elevation in degenerating photoreceptor cells of the C3H mouse retina. *Science* **186**, 449–451.

FEENEY L. (1973) The phagolysosomal system of the pigment epithelium. A key to retinal disease. *Invest. Ophthalmol.* **12**, 635–638.

FELL H.B., DINGLE J.T. & WEBB M. (1962) Studies on the mode of action of excess of vitamin A. 4: the specificity of the effect on embryonic chick-limb cartilage in culture and on isolated rat liver lysosomes. *Biochem. J.* **83** 63–69.

FIFKOVA E. (1968) Changes in the visual cortex of rats after unilateral deprivation. *Nature (London)*, **220**, 379–380.

FUJIWARA H. (1969) Biochemical studies on retinitis pigmentosa. Report 4. Study on the lipid metabolism of retinitis pigmentosa. *Folia Ophthal. Jap.* **20/4**, 354–371.

FUTTERMAN S. (1963) Metabolism of the retina III. The role of reduced triphosophopyridine nucleotide in the visual cycle. *J. Biol. Chem.* **238**, 1145–1150.

FUTTERMAN S. & KINOSHITA J.H. (1959) Metabolism of the retina, II. Heterogeneity and properties of the lactic dehydrogenase of cattle retina. *J. Biol. Chem.* **234**, 3174–3178.

GARNER A. (1970) Pathology of diabetic retinopathy. *Brit. Med. Bull.* **26**, 3174–3178.

GISIGER V. (1971) Triggering of RNA synthesis by acetylcholine stimulation of the postsynaptic membrane in a mammalian sympathetic ganglion. *Brain Res.* **33**, 139–146.

GOODWIN L.G., RICHARDS W.H.G. & UDALL V. (1957) The toxicity of diaminodiphenoxylakanes. *Brit. J. Pharmacol.* **12**, 468–474.

GOURAS P., CARR R.E. & GUNKEL RlD. (1971) Retinitis pigmentosa in abetalipoproteinaemia. Effects of vitamin A. *Invest. Ophthal.* **10**, 794–793.

GRAYMORE C.N. (1964a) Possible significance of the iosenzyme of lactic dehydrogenase in the retina of the rat. *Nature (London)*, **201**, 615–616.

GRAYMORE C.N. (1964b) Metabolism of the developing retina, 7. Lactic dehydrogenase isoenzyme in the normal and degenerating retina. A preliminary communication. *Exp. Eye Res.* **3**, 5–8.

GRAYMORE C.N. (1970) Biochemistry of the retina. In Graymore C.N. (Ed.) *Biochemistry of the Eye*, pp. 645–735. London, Academic Press.

GRAYMORE C.N., TANSLEY K. & KERLY M. (1959) Metabolism of the developing retina, 2. The effect of an inherited retinal degeneration on the development of glycolysis in the rat retina. *Biochem. J.* **72**, 459–461.

GRAYMORE C.N. & TOWLSON M. (1970) Biochemistry of the retina. In Graymore C.N. (Ed.) Biochemistry of the Eye, pp. 645–735. London, Academic Press.

GYLLENSTEN L. & LINDBERG J. (1964) Development of the visual cortex in mice with inherited retinal dystrophy. *J. Comp. Neurol.* **122**, 79–82.

GYLLENSTEN L., MALMFORS J. & NORRLIN M.C. (1965) Effect of visual deprivation on the optic centres of growing and adult mice. *J. Comp. Neurol.* **124**, 149–160.

HEATH H. (1970) Experimentally induced retinopathies in relation to the problem of diabetes. *Brit. Med. Bull.* **26**, 151–155.

HEATH H., BRIDGEN W.D., CANEVER J.V., POLLOCK J., HUNTER P.R., KEESEY J. & BLOOM A. (1971) Platelet adhesiveness and aggregation in relation to diabetic retinopathy. *Diabetologic*, **7**, 308–315.

HEATH H., PATERSON R.A. & HART J.C.D. (1967) Changes in the hydroxyproline, hexosamine and sialic acid of the diabetic human and BB'-imminodipropionitrile-treated rat retinal vascular systems. *Diabetologia*, **3**, 515–518.

HEATH H., RUTTER A.C. & BECK T.C. (1962) Reduced and oxidised pyridine nucleotides in the retinae from alloxan-diabetic rats. *Vision Res.* **2**, 333–342.

HENDRIKS T., DAEMEN F.J.M. & BONTING S.L. (1974), in press.

HERRON W.L., RIEGEL B.W., MYERS O.E. & RUBIN M.I. (1969) Retinal dystrophy in the rat—a pigment epithelial disease. *Invest. Ophthal.* **8**, 595–604.

HERRON W.L., RIEGEL B.W. & RUBIN M.L. (1971) Outer segment production and removal in the degenerating retina of the dystrophic rat. *Invest. Ophthal* **10**, 54–63.

HIGHMAN V.N. & WEALE R.A. (1973) Rhodopsin density and visual threshold in retinitis pigmentosa. *Amer. J. Ophthal.* **May 1973**, 822–832.

HOBBS H.E., SORSBY A. & FREEDMAN A. (1959) Retinopathy following chloroquine therapy. *Lancet* **ii**, 478–480.

JOHNSON G.L. (1901) Contributions to the comparative anatomy of the mammalian eye, chiefly based on ophthalmoscopic examination. *Phil. Trans. Roy. Soc. London* **194**:1, 82–92.

KANAI M., RAZ. A. & GOODMAN DE W.S. (1968) Retinol-binding protein: the transport of protein for vitamin A in human plasma. *J. Clin. Invest.* **47**, 2025–2044.

KARLI P. (1952) Retines sans cellules visuelles. Recherches morphologiques, physiologiques et physiopathologiques chez les rongeurs. *Arch. d.Antat. d'Hist. d'Embryol.* **35**, 1–76.

KARLI P. (1954) Etude de la valeur fonctionelle d'une depourve de cellules visuelles photoreceptrices. *Arch. Sci. Physiol.* **8**, 305–328.

KARLI P., STOECKEL M.E. & PORTE A. (1965) Degenerescence des cellules visuelles photoreceptrices et persistance d'une sensibilite de la retine a la stimulation photique. Observations au microscope electronique. *Z. Zellforsch.* **65**, 238–252.

KEELER C.E. (1927) Rodless retina: an ophthalmic mutation in the house mouse. *Mus musculus. J. Exp. Zool.* **46**, 355–407.

KEEN H. & CHLOUVERAKIS C. (1965) Metabolic factors in diabetic retinopathy. In Graymore C.N. (Ed.) *Biochemistry of the Retina*, pp. 123–138. London, Academic Press.

KRACHMER J.H., SMITH J.I. & TOCCI P.M. (1966) Laboratory studies in retinitis pigmentosa. *Arch. Ophthal.* **75**, 661–4.

KURIMOTO S. & NEWELL F.W. (1963) Localisation of phosphorylase in the alloxan-diabetic rat retina. *Invest. Ophthal.* **2**, 24–31.

KUWABARA T. (1965) Some aspects of retinal metabolism revealed by histochemistry. In Graymore C.N. (Ed.) *Biochemistry of the Retina*, pp. 93–98. London, Academic Press.

KUWABARA T. (1970) Retinal recovery from exposure to light. *Am. J. Ophthalmol.* **70**, 187–198.

LASANSKY A. & DE ROBERTIS E. (1960) Submicroscopic analysis of the genetic dystrophy of visual cells in C3H mice. *J. Biophy. and Biochem. Cytol.* **7**, 679–683.

LASHLEY K.S. (1930) The mechanism of vision III. The comparative visual activity of pigmented and albino rats. *J. Genet. Psychol.* **37**, 481–484.

LAVAIL M.M. (1973) Kinetics of rod outer segment renewal in the developing mouse retina. *J. Cell Biol.* **58**, 650–661.

LAVAIL M.M., SIDMAN R.L. & O'NEILL D. (1972) Photoreceptor pigment epithelial cell relationships in rats with inherited retinal degeneration. Radioautographic and electron microscope evidence for a dual source of extra lamellar material. *J. Cel. Biol.* **53**, 185–209.

LAVAIL M.M., SIDMAN M., RAUSIN R. & SIDMAN R.L. (1974) Discrimination of light intensity by rats with inherited retinal degeneration. A behavioural and cytological study. *Vision Res.* **14**, 693–702.

LEBER T. (1916) Die Pigment degeneration der Netzhaut und mit ihr Verwandte Erkrankungen in Graefe Saemisch Heso: in *Handbuch der Gesamten Augenheilkunde* ed 2 Englemann W., Leipzig **7**, 1076–1225.

LEWIN D.R., THOMPSON J.N., PITT G.A. & HOWELL J.M. (1970) Blindness resulting from vitamin A deficiency in albino and pigmented guinea pigs and rats. *Int. J. Vit. Res.* **40**, 270–275.

LOLLEY R.N. (1972) Changes in glucose and energy metabolism *in vivo* in developing retinae from visually competent (DBA/IJ) and mutant (C3H/HeJ) mice. *J. Neurochem.* **19**, 175–185.

LOLLEY R.N. (1973) RNA and DNA in developing retinae: comparison of a normal with the degenerating retinae of C3H mice. *J. Neurochem.* **20**, 175–182.

LOLLEY R.N., SCHMIDT S.Y. & FARBER D.B. (1974) Alterations in cyclic AMP metabolism with photoreceptor cell degeneration in the C3H mouse. *J. Neurochem.* **22**, 701–707.

LOVTRUP-REIN H. (1970) Synthesis of nuclear RNA in nerve and glial cells. *J. Neurochem.* **17**, 853–863.

LUCAS D.R. (1954) Retinal dystrophy in the Irish setter I. Histology. *J. Exp. Zool,* **126**, 537–547.

LUCAS D.R., ATTFIELD M. & DAVEY J.B. (1955) Retinal dystrophy in the rat. *J. Path. and Bacteriol.* **70**, 469–474.

McCOLLUM E.V. & SIMMONDS N. (1917). A biological analysis of pellegra producing diets. II. The minimum requirements of the two unidentified dietary factors for maintenance as contrasted with growth. *J. Biol. Chem.* **32**, 181–191.

McCORMICK S. (1973) *Studies on the visual system of the rat.* Ph.D. Thesis. University of Edinburgh.

McLEAN C. (1970) *Studies on the distribution of vitamin A in the rat.* Ph.D. Thesis. University of Liverpool.

MEIER-RUGE VON W. & CERLETTI A. (1966) Zur experimentallen Pathologie der Phenothiazin—Retinopathie. *Ophthalmologica* **151**, 512–533.

MERRITT J.H. & SULKOWSKI T.S. (1971) Rhythmicity of RNA polymerase and RNA levels in nuclei of rat cerebral cortex. *J. Neurochem.* **17**, 1327–1328.

METZGER P.H., CEUNOD M., GRYNBAUM A. & WAELSCH H. (1966) Visual stimulation: lack of effect on protein synthesis in monkey brain. *Life Sciences* **5**, 1115–1120.

MILLER W.H., GORMAN R.E., & BITENSKY M.W. (1971) Cyclic adenosine monophosphate: function in photoreceptors. *Science* **174**, 295–297.

MIYAURA K. (1970) Experimental studies upon the influence of vitamin B_{12} on the retina. Report II. The influences of vitamin B_{12} on the evolution of experimental retinitis pigmentosa by $NaIO_3$. *Acta. Soc. Ophthal. Jap.* **2413**, 260–267.

MIYAURA K., KOZAKI M., IWAI H., OURA T., IOSHITI G., VEMURA I., MATSUSHITE K. & TANI Y. (1970) Clinical effects of VB_{12} cobamide on retinitis pigmentosa. *Jap. J. Clin. Ophthal.* **24**, 31–37.

MOORE T. (1964) Systemic action of vitamin A. *Exp. Eye Res.* **3**, 305–315.

NAGY Z.M. & MCKAY C.S. (1972) Development of adult-like open field behaviours in young retinal degenerate C3H mice. *Develop. Psychobiol.* **5**, 249–258.

NAGY Z.M. & MISANIA J.R. (1970) Visual perception in the retinal degenerate C3H mouse. *J. comp. Physiol. Psychol.* **72**, 306–310.

NOELL W.K. (1958) Differentiation metabolic organization and viability of the visual cell. *A.M.A. Arch. Ophthal.* **60**, 702–733.

OSBORNE T.B. & MENDEL L.B. (1914) The influence of cod liver oil and some other fats on growth. *J. Biol. Chem.* **17**, 401–408.

PANNBACKER R.G., FLEISCHMANN & REED D.W. (1972) Cyclic nucleotide phosphodiesterase. High activity in a mammalian photoreceptor. *Science (Wash. D.C.)* **175**, 757–758.

PATERSON R.A. & HEATH H. (1965) The glycosaminoglycans of the normal and alloxan-diabetic rat retina and aorta. In Graymore C.N. (Ed.) *Biochemistry of the Retina,* pp. 143–148. London Academic Press.

PATERSON R.A. & HEATH H. (1967a) The effect of BB'iminodiproprionitrile treatment and alloxan diabetes on the glycosaminoglycans of the rat retina and aorta. *Biochem. biophys. Acta* **148**, 207–214.

PATERSON R.A. & HEATH H. (1967b) Chemical changes in human diabetes, cataractous and $\beta\beta'$iminodiproprionitrile treated rat lens capsule. *Exp. Eye Res.* **6**, 233–238.

POWELL E. & FIELD R.A. (1964) Diabetic retinopathy and rheumatoid arthritis. *Lancet* **i**, 17.

RAHI A.H.S. (1972) Retinol-binding protein (RBP) and pigmentary dystrophy of the retina. *Brit. J. Ophthal.* **36**, 647–651.

RAISON C.G. & STANTON O.D. (1955) The schistosomicidal activity of symmetrical diaminodiphenoxyalkanes. *Brit. J. Pharmacol.* **10**, 191–199.

RANDLE P.J., GARLAND P.B., HALES C.N. & NEWSHOLME E.A. (1963) The glucose fatty acid cycle; its role in insulin sensitivity and the metabolic disturbances of diabetes millitus. *Lancet* **i**, 785–789.

READING H.W. (1964) Activity of the hexose monophosphate shunt in the normal and dystrophic retina. *Nature (London)* **203**, 491–492.

READING H.W. (1965) Protein synthesis and the hexosemonophosphate shunt in the developing normal and dystrophic retina. In Graymore C.N. (Ed.) *Biochemistry of the Retina* pp. 51–72. London, Academic Press.

READING H.W. (1966) Retinal and retinol metabolism in hereditary degeneration of the retina. *Biochem. J.* **100**, 34P–35P.

READING H.W. (1970a) Biochemistry of retinal dystrophy. *J. Med. Genetics*, **7**, 277–284.

READING H.W. (1970b) Effects of a retinotoxic phenoxyalkane on the visual cycle in rabbit retinae. *Biochem. Pharmacol.* **19**, 1307–1313.

READING H.W. & SORSBY A. (1962) Metabolism of the dystrophic retina. I. Comparative studies on the glucose metabolism of the developing rat retina, normal and dystrophic. *Vision Res.* **2**, 315–325.

READING H.W. & SORSBY A. (1964) Metabolism in the dystrophic retina. II. Amino acid transport and protein synthesis in the developing rat retina, normal and dystrophic. *Vision Res.* **4**, 209–220.

READING H.W. & SORSBY A. (1966a) Alcohol-dehydrogenase activity of the retina in the normal and dystrophic rat. *Biochem. J.* **99**, 3c–5c.

READING H.W. & SORSBY A. (1966b) Retinal toxicity and tissue-SH levels. *Biochem. Pharmacol.* **15**, 1389–1393.

ROBERTS D. ST. C. (1957) Studies on the antigenic structure of the eye using the fluorescent antibody technique. *Brit. J. Ophthal.* **41**, 338–347.

ROSE S.P.R. (1967) Changes in incorporation of ^{3}H lysine into protein in rat visual cortex following first exposure to light. *Nature (London)* **215**, 253–255.

ROSS S., NAGY Z.M., KESSLER C. & SCOTT J.P. (1966) Effects of illumination on wall-learning behaviour and activity in three inbred mouse strains. *J. Comp. Physiol. Psychol.* **62**, 338–340.

ROWLAND M. & RIEGELMAN S. (1968) Pharmacokinetics of acetylsalicylic acid and salicylic acid after intravenous administration in man. *J. Pharmaceutical Sci.* **57**, 1313–1319.

SCHMIDT S.Y. & LOLLEY R.N. (1973) Cyclic-nucleotide phosphodiesterase. An early defect in inherited retinal degeneration of C3H mice. *J. Cell Biol.* **57**, 117–123.

SORSBY A. (Ed.) (1972). *Modern Ophthalmology*. 2nd edition. London, Butterworths.

SORSBY A. & HARDING R. (1962a) Experimental degeneration of the retina. VIII. Dithizone retinopathy: its independence of the diabetogenic effect. *Vision Res.* **2**, 149–155.

SORSBY A & HARDING R. (1962b) Experimental degeneration of the retina. VII. The protective action of thiol donors against the retinotoxic effect of sodium iodate. *Vision Res.* **2**, 139–148.

SORSBY A., KOLLER P.C., ATTFIELD M., DAVEY J.B. & LUCAS D.R. (1954) Retinal dystrophy in the mouse: histolofical and genetic aspect. *J. Exp. Zool.* **125**, 171–197.

SWARTZ J.G. & MITCHELL J.E. (1970) Biosynthesis of retinal phospholipids: incorporation of radioactivity from labelled. *J. Lipid Res.* **11**, 544–550.

SWEASEY D., PATTERSON S.P. & TERLECKI S. (1971) Lactate dehydrogenase (LDH) isoenzymes in the retina of the sheep and changes associated with progressive retinal degeneration (Bright Blindness). *Exp. Eye Res.* **12**, 60–69.

TALWAR G.P., CHOPRA S.P., GOEL B.K. & D'MONTE B. (1966) Correlation of the functional activity of the brain with metabolic parameters. III. Protein metabolism of the occipital cortex in relation to light stimulus. *J. Neurochem.* **13**, 109–116.

TANAKA K. & LIZUKA Y. (1968) Suppression of enzyme release from isolated rat liver lysosomes by non-steroidal anti-inflammatory drugs. *Biochem. Pharmacol.* **17** 2023–2032.

VALVERDE F. (1971) Rate and extent of recovery from dark rearing in the visual cortex of the mouse. *Brain Res.* **33**, 1–11.

VENTO R. & CACIOPPO F. (1973) The effect of retinol on the lysosomal enzymes of bovine retina and pigment epithelium. *Exp. Eye Res.* **15**, 43–49.

WAGREICH H., LASKY M.A. & ELKAN B. (1961) Some biochemical studies in retinitis pigmentosa. *Clin. Chem.* **7**, 143–148.

WALD G. (1955) The photoreceptor process in vision. *Am. J. Ophthal.* **40**, 18–41.

WALD G. (1960) The visual function of the vitamins. In *Vitamins and Hormones*, vol. XVIII, pp. 417–430 New York, Academic Press.

WALKERS P.T. (1959) Anaerobic glycolysis in rats affected with retinitis pigmentosa. Its significance in relation to various forms of primary pigmentary degeneration of the retina. *Brit. J. Ophthal.* **43**, 686–696.

WEISS H. & KOSMATH B. (1971) Therapeutische Anwendung von Phosphatiden bei retinalen Erkrankungen. Verlaufige Mitteilung unber das Verhalten der Shescharge bei Maculopathien und Retinitis Pigmentosa. *Klin. Mondsbl. Augenheilt.*, **158**/2, 278–285.

WEISSMANN G., ZURIER R.B. & HOFFSTEIN S. (1972) Leukocytic proteases and the immunologic release of lysosomal enzymes. *Amer. J. Path.* **68**, 539–563.

YAMASAKI I. & MIZUNO K. (1971) Rhodopsin and ERG of hereditary dystrophic mice and experimental retinitis pigmentosa. *Jap. J. Ophthal.* **14**, 151–158.

YATES C.M., DEWAR A.J., WILSON H., WINTERBURN A.K. & READING H.W. (1974) Histological and biochemical studies on the retina of a new strain of dystrophic rat. *Exp. Eye Res.* **18**, 119–133.

YATES C.M., READING H.W., BITENSKY L. & CHAYEN J. (1975) Activity of the hexose monophosphate shunt in the outer segments of normal and dystrophic rat retinae. In preparation.

YOUNG R.W. (1967) The renewal of photoreceptor cell outer segments. *J. Cell. Biol.* **33**, 61–72.

YOUNG R.W. (1969) A difference between rods and cores in the renewal of outer segment protein. *Invest. Ophthal.* **8**, 222–231.

YOUNG R.W. (1971) The renewal of rod and core outer segments in the rhesus monkey. *J. Cell Biol.* **49**, 303–318.

ZURIER R.B., HOFFSTEIN S. & WEISSMANN G. (1973) Mechanisms of lysosomal enzyme release from human leukoctyes. I. Effect of cyclic nucleotides and colchicine. *J. Cell Biol.* **58**, 27–41.

3 Biochemistry and basic mechanisms in epilepsy

H. F. Bradford

P. R. Dodd

I Introduction, 116

    A. The overall picture of the clinical pathology, 116

II Experimental models of epilepsy, 118

    A. The criteria for clinical relevance, 118
    B. Agents producing focal epilepsy, 119
        1. Cobalt, 119
        2. Aluminium oxide, 120
        3. Foci produced by freeze-lesions, 121
        4. Tungstic acid, 122
        5. Penicillin and other topically applied agents, 122
    C. Models of reflex epilepsy, 123
        1. The photosensitive baboon, 123
    D. Electrical stimulation as a model, 124

III The concept of a focus, 124

    A. Organization, 124
    B. Common features of experimental foci, 125
        1. Electrical features, 125
        2. Structural features, 126
        3. Lowered threshold to chemical convulsants, 126
        4. Biochemical changes common to different foci, 126
    C. The mechanism of spread and temporal limitation, 127
        1. Cations and epilepsy, 131

IV Acetylcholine and monoamine involvement, 132

    A. Acetylcholine, 132
    B. Monoamines, 133

V Amino acid involvement, 135

    A. Amino acids and neurotransmission, 135
    B. The amino acid content of epileptic foci, 136
    C. Amino acid changes in whole brain, 137
    D. Amino acid changes: metabolic implications, 137
    E. Glutamate and GABA, excitation and inhibition, 142
        1. GABA and its enzymes, 143
        2. Glutamate and its enzymes, 146
    F. Glutamate and spreading depression, 150
    G. Taurine and epilepsy, 151

VI Folate and epilepsy, 153

VII Summary and speculation, 154

I Introduction

A. THE OVERALL PICTURE OF THE CLINICAL PATHOLOGY

The fundamental processes which lead to the generation of epileptic seizures remain obscure despite the considerable advances of the past 30 years both in the range and the effectiveness of anticonvulsant medication, and in our understanding of the neurophysiological basis of excitation in the CNS. The improved armoury of anticonvulsants has provided a new impetus and has allowed a great improvement in the possibilities for effective clinical management of many forms of epilepsy. However, further fundamental advance, leading to provision of a really satisfactory therapy, requires a detailed knowledge of the basic biochemical and physiological mechanisms which lead from the underlying histological or biochemical pathology to the persistent and episodic neuronal hyperactivity which is characteristic of epilepsy.

A large proportion of epilepsy in the overall population (30–40%, Sutherland & Tait, 1969; 25%, Currie et al., 1971), appears to arise in the temporal lobes, and overall probably about 80% of these cases show a detectable histopathology. These structural abnormalities tend to occur in the medial region of the lobe and are commonly associated with the hippocampus. Ammons horn sclerosis (termed: mesial temporal sclerosis) has an incidence of about 50–60% (Corsellis, 1970; Brown, 1973). The temporal lobes appear to be particularly susceptible to influences causing seizures, including many of extra cranial origin such as hypoxia and hypoglycaemia. It has been argued that the blood supply to the structures in the medial temporal lobes and their anatomical siting renders them particularly vulnerable to hypoxic or to mechanical damage. Therefore, birth trauma (i.e. hypoxia, or local ischaemia of the hippocampus and adjacent structures due to bilateral head deformation), or febrile convulsion (i.e. convulsive hypoxia) in the young may result in histological changes in the medial temporal grey matter which in later life lead to epileptogenicity (Earle et al., 1953; Ounsted et al., 1966; Falconer, 1971). Whether the sclerosis results from episodes at birth or from episodes occurring post-natally, remains, however, a contentious issue. The opposite case has also been argued, namely, that the observed sclerosis and other changes appear only after onset of the epileptic condition, being caused by prolonged seizures rather than generating them (for reviews see Corsellis, 1970; Meldrum, 1975). In a current reappraisal based on a re-examination of the clinical literature and on experimental work involving induction of seizures in baboons, cats and rats, Meldrum (1975) concludes that both situations may obtain, namely, that temporal lobe sclerosis may be produced by prolonged convulsion, and that sclerotic lesions experimentally induced in the hippocampal region can give rise to temporal lobe seizures. In addition, temporal lobe sclerosis exists in some cases of epilepsy without apparently being the initiating focus of activity. Local and systemic factors likely to be causative in the formation of the sclerosis are hypoxia, hyperpyrexia and brain oedema, probably due to astrocytic swelling causing an insufficiency in the local vascular supply. It still remains to be demonstrated, however, that the form of sclerosis resulting from sustained seizure in humans (e.g. febrile convulsions in children, or status epilepticus) can produce a focus subsequently generating temporal lobe epilepsy, as proposed by Falconer et al., 1964, and Ounsted et al., 1966.

Removal of temporal lobe foci localized by EEG and by electrical stimulation and recording during surgery leads to improvement in 40–60% of patients in a highly selected group who were mostly refractory to anticonvulsant therapy and showed a unilateral focus. (Falconer & Taylor, 1968; Rasmussen, 1969, 1974.) About 80% of the resected tissue samples which included the spiking focus and the local pathological tissue (and would typically include the uncus amygdala and interior hippocampus), showed some histological abnormality, 47% showing mesial temporal lobe sclerosis, the others showing glial malformation and hamartoma. However, 20% were without demonstrable lesion.

It appears that other forms of epilepsy are associated in a more clear-cut fashion with specific organic changes such as tumours, hamartoma and cortical scarring through head injury or infection (Currie et al., 1971). Brain scarring in missile wounds (e.g. shrapnel) and other forms of head injury leads to a chronic epileptic condition in 20–30% of cases (Falconer, 1971; Corsellis, 1970; Newcombe, 1969). Surgical removal of these foci is only about half as effective as removing temporal lobe sclerosis, and this raises the question of whether surgery is also effective through interruption of specific pathways (Bates, 1962).

The absence of a firm linkage to specific patterns of neuropathology, such as mesial temporal lobe sclerosis, and the absence of any detectable histological lesion in non-focal, generalized or centrencephalic epilepsy indicates the difficulty of isolating any biochemical factors involved in generating the condition. It is striking, for instance, that only particular kinds of scarring can lead to the development of spiking foci, and this does not include the scars produced by surgery itself. However, the kinds of changed tissue organization found in active scars might well give a lead to the underlying pathophysiology or pathobiochemistry. Gliosis, with a changed ratio of neurones to glial cells, could critically alter metabolic inter-relations between these cell types, and also the composition of the extracellular fluid. Partially damaged but surviving neurones could leak excitant substances or show a hyperexcitability to their usual volley of input signals. Alternatively, active scars could simply provide a region of high permeability in the blood-brain barrier, allowing local entry of excitant substances from the blood.

In constructing a philosophy for an experimental approach, a useful distinction may be drawn between the processes which initiate and maintain the local hyperexcitability and those involved in the spread of the hyperactivity from the focus, with recruitment of many normal neurones and precipitation of seizures. It is the initiating mechanisms which form the point of central interest to the fundamental scientist, and although they may arise systemically (due for instance, to hypocalcaemia), it is those whose origin is intracranial which are likely to cause the majority of epileptic conditions. These mechanisms may be diffuse or focal in nature as judged by EEG, although focal discharge often leads on to a major seizure, and improved methods of investigation are uncovering an increasing range of possible focal lesions in what might previously have been classified as idiopathic epilepsy.

It is the purpose of this review to discuss the biochemical findings in experimental epilepsy in relation to possible biochemical mechanisms involved in generating and controlling seizures. Biochemical events may well be fundamentally involved in the local trigger mechanisms initiating neuronal hyperactivity as well as in those controlling its duration. Anticonvulsant action will be mediated by biochemical factors which may act by damping the initiating mechanisms or by preventing seizure propagation, and an analysis of anticonvulsant action provides another biochemical approach to elucidating the basic mechanisms of the seizures themselves.

II Experimental models of epilepsy

A. THE CRITERIA FOR CLINICAL RELEVANCE

A wide and bizarre range of animal preparations and treatments are in use in laboratories across the world for their propensity to produce seizures, both focal and general, and although it is desirable to establish criteria for assessing their merit as models yielding information of value in understanding clinical epilepsy, clear-cut criteria are not easily established. It is for instance desirable that seizures should be recurrent, episodic and associated with an abnormal repetitive high-voltage EEG pattern. They should be controllable with clinically effective anticonvulsants, and the animals should, where possible, be free of other, complicating, neurological disorders and show no obviously abnormal behaviour between epileptic episodes. Since neither clinical nor experimental models always conform to these criteria, their rigid application is unproductive. In spite of this, quite a wide spectrum of non-conforming models continues to provide information of considerable value, sometimes just in simple ways by indicating how systemic factors can affect convulsive thresholds. Attempts have been made to produce experimental representations of focal, temporal lobe, and generalized epilepsy, and even of petit-mal (Marcus, 1972). Some of the agents employed are effective only in certain animal groups which may exclude primates. Indeed it might be argued that only primates can have the degree of complexity of brain structure necessary for a realistic model of human epilepsy, but this is a narrow view, and most mammalian brains, and even invertebrate nervous systems, can provide experimental models generating valuable insights into the processes controlling the balance between excitation and inhibition and an understanding of the aberrations which might lead to the loss of control of neuronal discharge.

B. AGENTS PRODUCING FOCAL EPILEPSY

An impressively broad spectrum of agents is employed to produce localized tissue lesions in the cerebral cortex of laboratory animal species ranging from frogs to monkeys. The most widely used substances are cobalt metal, aluminium oxide, tungstic acid gel, penicillin, ouabain, tetanus toxin and strychnine. Foci produced mechanically by local cortical freezing with ethylchloride spray, cold metal rods or chips of solid carbon dioxide have also been popular techniques. Some workers have produced lesions surgically, or by electrolysis (Green et al., 1957). The condition induced by these substances may be acute or chronic, may remain localized or become general. Thus it may be manifest as a spiking-focus in the EEG producing specific motor response such as contra-lateral limb jerk, or activity may spread from the focus and precipitate a full grand-mal seizure. Chemical agents commonly precipitate major seizures with little or no delay, depending on the route of administration. The full range of the agents in use and technical details for their application will be found in either Jaspers et al., 1969, or Pupura et al., 1972, and only a brief account of the most widely used techniques will be given here.

1. COBALT METAL

This is applied as the fine metal powder either sprinkled on the surface of the sensori-motor cortex (Dawson & Holmes, 1966) or inserted into the tissue of the cortex trapped in small rods of gelatin (Fischer et al., 1967). Motor effects such as contralateral limb jerks, appear days or weeks after implantation and last for 2–6 weeks depending on the method of application. Generalized epileptic episodes, although they do occur, are rare and the epileptogenic activity does not appear to spread beyond the focus. Foci which were electrically active for 1–2 months have been established in amygdala and hippocampus (Mutani, 1967, 1967a).

An epileptogenic EEG pattern is detectable in minutes or hours after application and persists for the duration of the focus (Jaspers & Koyama, 1973; Hommes & Obbens, 1971). Even when sprinkled on to the undamaged surface of the brain, cobalt causes a massive degree of local cortical necrosis, abcess formation and chronic inflammation (Payan, 1971; Emson & Joseph, 1975). However, since a large selection of other metals closely related to cobalt in the periodic table are equally effective in causing these tissue changes but do not induce an epileptic focus, it must be assumed that some special interaction occurs between the tissue and cobalt (Chusid & Kopeloff, 1962). The relatively long periods involved in the onset, duration and cessation of the epileptic condition (Fig. 3.1) allow the biochemical factors associated with the transitions from stage to stage to be followed with a greater chance of cause being separated from effect. It is possible to follow their progression by quantitating the motor-effects, such as contra-lateral limb jerks, from motor cortex implantations

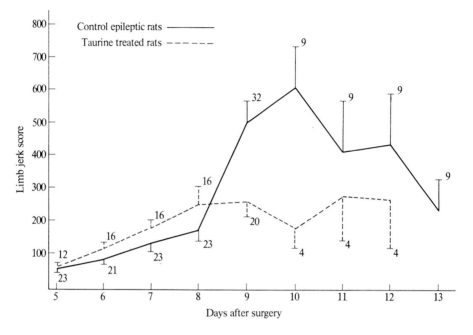

Fig. 3.1. Effect of taurine on frequency of limb-jerks in rats suffering from cobalt-induced epilepsy. Taurine was given orally by stomach tube (600 mg/kg) after each trial on days 6–11. Numbers of animals are given together with the mean ±SD of limb jerk score (from Wheler et al., 1975a).

(Dawson & Holmes, 1966; Wheler et al., 1974; Fig. 3.1), or by continuous recording of the EEG at the focus through chronically implanted electrodes (Dow et al., 1962, 1972; Hommes & Obbens, 1972). The effects of anticonvulsant drugs on the established EEG pattern or limb-jerk frequency may then be examined.

2. ALUMINIUM OXIDE

This agent differs in its actions from cobalt by causing a chronic focus which readily produces generalized epileptic episodes as well as purely focal manifestations. The foci, once established, may last for years in the monkey (Ward, 1969), and the agent appears to be mainly effective in primates, dogs and cats, variable or no effects being reported for small laboratory mammals. Cortical and subcortical foci are readily induced in rhesus monkeys and display their presence within 1–8 months after implantation as spontaneous seizures. Several intracortical injections increase the severity of the epilepsy and reduce the length of the period before onset. The area of the implantation determines the nature of the responses which develop and these may vary considerably. Thus a common procedure is multiple subcortical injections into the sensory and motor face and hand areas of monkeys. Seizures develop within 1–2 months, with epileptogenic EEG patterns occurring much earlier as in the case of cobalt and other experimental foci. Although some brain areas such as thalamus, hypothalamus and caudate nucleus, appear to be refractory to the treatment, alumina injections into the temporal lobe and amygdaloid complex produced what has been described as 'behavioural', or 'psychomotor fits' as well as grand-mal seizures (Gastaut et al., 1959). Monkeys with active cortical foci induced by this agent respond well to anticonvulsant drugs which are effective in humans with seizures of cortical onset. Once the condition develops, it may last indefinitely with spontaneous recurrence of seizures (Ward, 1969, 1972, Nie & Ettlinger, 1974).

3. FOCI PRODUCED BY FREEZE-LESIONS

Local freezing of small areas of cortex with solid carbon dioxide, cooled metal rods or ethylchloride spray produces sub-chronic epileptogenic foci which are detectable as EEG abnormality within minutes or hours after freezing and are presumed to arise from mechanical damage caused in ice-crystal formation. The condition lasts for days rather than hours or weeks. Thus, the activity reaches a maximum after about 12 hours in the rat and thereafter declines. The spontaneous seizures which accompany the electrical activity show the same pattern of development and decline (Lewin, 1972). Focal discharge may still, however, be evoked several weeks after spontaneous activity has ceased by systemic injections of chemical convulsants such as pentylenetrazole, or by photic stimulation of occipital lesions. 'Mirror-foci' readily develop in the contralateral homotropic cortex after 2 or 3 days and may become independent of the primary focus within days or weeks depending on the species, and may even become more active than the original focus (Morrell, 1960). These mirror foci are not restricted to freeze-lesions and occur in a wide range of experimental epilepsy. They appear to be generated in regions receiving hyperactivity from the initial focus through synaptic connection. For instance, they appear to reach the opposite hemisphere by transmission through the fibres of the corpus callosum. After a certain period of synaptic bombardment they commonly become independent of the primary focus (Wilder, 1972). Other secondary foci may equally well develop in subcortical structures which make adequate synaptic contact with the primary focus (Wilder, 1972), and in monkeys such sub-cortical pathways provide a route for establishing cortical mirror foci (Nie et al., 1974).

They offer the possibility of an active focus free of the damage or contamination caused by the applied agent, provided the agent does not also reach the region of the mirror focus by moving through the connecting fibres. Whether or not mirror foci occur in man does not yet appear to be established, and the full significance of these foci for clinical epilepsy therefore remains uncertain, though surgical removal of a primary focus would clearly be of limited value once a secondary focus had been established (Nie & Ettlinger, 1974), and untreated foci might produce progressive epilepsy.

4. TUNGSTIC ACID

The acid is injected sub-cortically in the form of a gel and seizures follow in 2 or 3 hours preceded by an electrically active focus. Once established the condition lasts for about 12 hours (Ward, 1969).

5. PENICILLIN AND OTHER TOPICALLY APPLIED AGENTS

It is curious that the penicillin molecule should be a potent convulsive agent readily producing acute foci when applied to the cortical surface at 1–2 mM concentrations. The precise mechanism of action remains obscure but the focus which is established usually lasts for only a few hours and shows the usual abnormal electrical activity including the paroxysmal depolarization shifts (PDS) which are a common electrophysiological feature of most cortical epileptogenic foci (see below). Strychnine has a similar time course and displays similar electrical properties. The basis for its action is most likely to be related to its ability to block the glycine-mediated post-synaptic inhibition in spinal motoneurons, though whether it exerts similar effect in the cortex remains controversial (Ajmone–Marsan, 1969). Ouabain, the potent sodium, potassium and amino acid transport inhibitor, is also, not surprisingly, an effective topical convulsant agent which probably works through its ability to paralyse the sodium pump thereby allowing intracellular sodium levels to rise with ensuing membrane depolarization and increased excitability. It would be cells positioned towards the periphery of the treated area, and not fully paralysed by the agent, which would remain most hyperexcitable, since total stoppage of the sodium-pump would reduce membrane potential well beyond the critical firing-level to an excessively low level (cathodal block) rendering the cells inexcitable. Tower (1969) has suggested that ouabain may also be effective through an ability to mobilize membrane-bound calcium ions, thereby rendering the membranes 'unstable', presumably again through increased sodium influx.

The sites of action of many agents generating epileptic foci are known, even if precise details of the mechanisms leading to seizure in these are not known. This knowledge allows examination of the possible involvement of these systems (e.g. transmitter synthesis, ion transport) in the control of excitability in the normal brain and in seizure generation in these specific circumstances, but it is likely to be of little relevance in investigating the nature of the trigger mechanisms in human epilepsy, except where some derangement in the specific system is observed in the clinical condition presenting epilepsy. Thus seizures arising during vitamin B6 deficiency may be mimicked by anti pyridoxal agents which have been shown to reduce profoundly the production of the inhibitory agent GABA, the substance which is likely to mediate inhibition at many cortical and sub-cortical synapses. This precise biochemical lesion is not, however, likely to underlie most other forms of epilepsy and the use of hydrazide and semi-carbazide antipyridoxal agents will not assist the investigation of most clinical epilepsy even though they might help to define the role of GABA in controlling overall excitability in the normal brain. The fact that a wide spectrum of agents leads to similar patterns of abnormal activity (e.g. PDS, see p. 125) in an experimental focus is interesting and important, but in the final analysis it is the precise events leading to this abnormal activity which are the more fundamental in determining the *aetiology of the disease*.

C. MODELS OF REFLEX EPILEPSY

The apparent importance of genetic susceptibility in the development of many forms of epilepsy (Falconer, 1971; Lindsay, 1971) highlights the possible value of animal models showing genetic susceptibility to specific stimuli. These include predisposition to audio and photic stimuli, and usually result in the precipitation of generalized seizures. (Collins, 1972; Naquet & Meldrum, 1972.) These varieties of epilepsy are often referred to as Reflex Epilepsy because of their requirement for a sensory stimulation to set them off, and it is not surprising that the visual and auditory systems with their massive sensory inputs to the brain should be the neuronal inputs most readily precipitating epileptic seizures. However, unlike photogenic seizures, audiogenic seizures do not appear to occur in man (Naquet, 1969). Models of reflex epilepsy have proved valuable in testing anticonvulsants and photic stimuli are useful for detecting low thresholds to seizure in humans, and in terms of fundamental advance of direct clinical relevance, the photosensitive baboon is providing one of the most useful and interesting models currently available.

1. THE PHOTOSENSITIVE BABOON

The particular animal which is employed for epilepsy studies is the Senegalese baboon, *Papio papio*, one particular area of Senegal providing the group of baboons with the greatest incidence of true photosensitivity (60–70%, Naquet & Meldrum, 1972). Using stroboscopic light stimulation on the closed eyelids of baboons strapped in a chair and usually asleep, it is possible to study both the natural photo epilepsy (manifested electrically and as myoclonus or generalized epilepsy) and the effects of convulsants and anticonvulsants on the threshold for the photically-induced myclonic responses. (Meldrum *et al.*, 1974, 1975, 1975a).

D. ELECTRICAL STIMULATION AS A MODEL

Electrical stimulation has been applied directly to the brain surface to stimulate activity since last century when David Ferrier employed surface stimulation to localize brain function (Ferrier, 1876). In the past 40 years electrical stimulation has been employed as the main test system in the development of most of the anticonvulsant drugs in current use. In these experiments the ability of the drug to prevent full tonic extension of the hind limb during maximal electroshock seizure (MES) in a range of experimental animals is used as an index of its anticonvulsant action and is thought to be related to the ability of the drug to block generalized seizure spread (Spinks & Waring, 1963). Though extremely successful in this respect, the technique reveals little of the *mechanism* of seizure generation since it presumably works by a direct depolarizing effect on the cerebral tissue. Thus, it may be pictured as bypassing the initiating mechanisms which operate in human and experimental epilepsy and which comprise the essential pathophysiology. Thus, although a short delay may occur between mild electrical stimulation and precipitation of seizure this is too short for the model to provide a useful experimental tool for studying cause *as distinct* from effect in epilepsy.

A more useful model employing electrical stimulation has been developed by Goddard *et al.*, 1971. This involves daily electrical cortical stimulation with subthreshold currents over periods of weeks. The result is the development of a chronically active epileptogenic focus which allows study of both ictal and interictal periods. Because of the slow development of the focus and the absence of any gross histopathology, the model is known as the 'kindling effect' and is becoming widely used in experimental epilepsy.

III The concept of a focus

A. ORGANIZATION

The tissue contained within a region designated an epileptic focus will contain both normal and abnormal histological components. The abnormal tissue structure may be obviously necrotized and contain inflamed areas of degenerating tissue, as in cobalt foci (Payan, 1917), or local gliosis and scar formation as in electrolytic and surgical lesions (Green *et al.*, 1957). In human neuropathology the foci include sclerosis of specific regions (e.g. medial temporal lobe sclerosis) or local cellular changes related to such events as neoplasia, hamartoma, infection and meningeal adhesion. These widely ranging tissue changes undoubtedly disrupt local tissue organization, alter local vascular supply and change the nature and dimensions of local extracellular spaces and cellular relationships. The degree of cellular damage sustained will vary from cell death with replacement by scar tissue and glial growth, to partial impairment following from changed tissue organization and vascular supply. In many foci there would be a gradient of decreasing impairment from the centre to the periphery, and this is in accord with the finding that neuronal hyperactivity is usually associated with the region at the periphery of a focus (Koyama, 1972; Prince & Futamachi, 1970; Schmidt et al., 1959). In some cases neurones would simply be absent from the central region and in others too damaged to be able to sustain hyperactivity. Where neuronal organization allows it, therefore, foci are likely to consist of a central inactive region surrounded by an annulus of hyperactive cells at the boundary juxtaposed with the normal cells (Spinks & Waring, 1963).

B. COMMON FEATURES OF EXPERIMENTAL FOCI

Although it seems unlikely that a single kind of pathophysiology with an associated specific neurochemistry underlies all forms of epilepsy, the existence of a number of specific features in a wide range of experimental foci does indicate some degree of convergence in their mechanism.

1. ELECTRICAL FEATURES

Firstly, there is the epileptiform EEG, which is always present, In addition neurones in many acute and chronic experimental foci (e.g. penicillin, strychnine, cobalt and alumina) display a characteristic large depolarization shift called the paroxysmal depolarization shift (PDS). These recurrent shifts of depolarization take the neuronal membrane through the critical firing level causing increased spike discharge, and subsequently gross depolarization with inactivity due to cathodal block. This is followed by a period of hyperpolarization before onset of the next PDS. The cause of the PDS is unknown, but all evidence points to their being synaptically generated and graded in nature, appearing to be greatly enhanced post synaptic potentials of an excitatory nature. Recordings from neurones in these foci (between the PDS events) do not indicate any intrinsic hyperexcitability, suggesting that the properties of the neuronal membrane itself are not altered (Ajmone–Marsan, 1969; Prince, 1969), and that the PDS arises as the result of abnormal increase in *synaptic* activity. Such excitatory activity could result from increased excitatory synaptic input, decreased inhibitory synaptic input, or direct stimulation of the synaptic receptors on the membrane of the cells showing the PDS by agonists appearing locally in the interstitial fluid.

2. STRUCTURAL FEATURES

In addition to the overall structural changes described above, some foci show characteristic structural changes in neurones which have been visualized with Golgi–Cox stains. These involve loss of dendritic spines and smoothing out of the dendritic surface, reduced dendritic aborization and the appearance of varicosities. This implies a substantial loss in the number of dendritic synapses, and therefore a reduction in excitatory input to the neurones (partial deafferentiation). Such changes have been observed in alumina foci in monkeys (Westrum et al., 1965; Ward, 1969) and in human hippocampal sclerosis (Brown, 1973; Scheibel & Scheibel, 1973). In the case of the alumina focus these changes are most marked in the region of the primary focus which is the seat of the epileptiform EEG. Arguing from evidence accruing from deafferentiation of spinal and brainstem neurones, Ward (1969) contends that partial deafferentiation in the focus could readily lead to the augmented spontaneous neuronal activity which is observed, and increased synaptic receptor sensitivity may be part of the mechanism (Sharpless, 1969).

3. LOWERED THRESHOLD TO CHEMICAL CONVULSANTS

Convulsive and anticonvulsive influences interact in an additive fashion in both normal and epileptic animals. Most experimental foci are activated by chemical convulsants at doses well below the level precipitating generalized seizures, e.g. pentylene tetrazole, folate, glutamate or potassium activation of alumina (Kopeloff, 1954; Kopeloff & Chusid, 1967) or cobalt foci (Zuckerman & Glaser, 1970). The motor activity evoked under these conditions takes the form which is characteristic for the focus concerned, and the parallel increase in electrical discharge remains focal (Bradford & Dodd, 1975; Hommes & Obbens, 1972). In the same way, one or other anticonvulsant controls the motor effects of these epilepsies and greatly raises the threshold for the convulsants.

Photogenic and audiogenic models show lowered or raised sensory stimulation thresholds in the presence of convulsants or anticonvulsants respectively (Meldrum et al., 1975, 1975a), and many of these models provide useful test systems for anticonvulsants where their effects may be quantitated. What makes the foci susceptible to the activating effects of these convulsant agents is not clear but it is possible that there is a breakdown of the local blood-brain barrier in the focal area, allowing a more rapid and effective build-up of convulsant concentration in that region. Certain kinds of scar may develop this increased permeability to agents in the blood and become of central importance in controlling epileptic conditions in man.

4. BIOCHEMICAL CHANGES COMMON TO DIFFERENT FOCI

From the wealth of conflicting data on biochemical factors which are variously reported to be either raised, lowered, or to show no change in experimental foci, certain compositional changes do consistently emerge. The most significant of these are changes in cation and amino acid content. Sodium could be expected to show an increase and potassium a decrease in level in tissue samples from central regions of foci which are necrotized and show a greatly increased CSF space. The relatively normal cells of the primary electrically-active focus and its associated mirror focus should show a smaller change in the same direction since sustained excitation taxes the ability of the sodium-pump to extrude sodium and recover potassium. Fig. 3.2, shows that the expected changes are found in cobalt-induced epilepsy, and it must be stressed that changes *result* from focal hyperactivity and are not likely to be its cause. Among the many amino acids present only six appear to change in the epileptogenic region of the focus. Glutamate, aspartate, GABA and taurine commonly diminish in amount, and glycine and serine commonly increase (Fig. 3.3, Table 3.1). These changes occur in foci induced by agents as disparate as penicillin (Gottesfeld and Elazar, 1972) and cobalt metal (Koyama, 1972) as well as in human temporal lobe epilepsy (Tower, 1960; Van Gelder *et al.*, 1972) and their possible significance is discussed below.

C. THE MECHANISM OF SPREAD AND TEMPORAL LIMITATION

There is at present little understanding of the nature of the events which cause a 'quiescent' spiking focus to interact with its surrounding normal tissue and initiate the development of a major seizure, or become sufficiently active to initiate a detectable motor or sensory event. Ward (1969) regards an understanding of this change from interictal to ictal state as critical for an understanding of epilepsy. What prevents some foci (e.g. Co focus) and not others (e.g. Alumina focus) from spreading into generalized seizures poses another central enigma. However, once the change is under way, seizures may spread at rates which differ from 3–4 mm/min as seen in the motor march of Jacksonian epilepsy, to the instantaneous spread seen in major clinical seizures and many chemically induced convulsions. From this variation it is clear that several modes of propagation are possible (Jaspers, 1969). These appear to be: (1) by diffusion of chemical substances released at the advancing front of activity and having an excitatory effect e.g. K^+ or the glutamate ion (Prince *et al.*, 1973); (2) by conduction horizontally in cortical layers via interneurones and other trans-cortical pathways; (3) projection from the focus to thalamus and other subcortical structures or brain stem nuclei, and subsequent radiation from these sites across the whole cortex of each hemisphere by thalamo-cortical or other ascending pathways; (4) from hemisphere to hemisphere via the corpus callosum. Some forms of focal epilepsy such as the Co-focus in rat may show limited spread to adjacent motor areas of the cortex (e.g. from fore to hind-limb and trunk) but rarely spread to become generalized.

The factors imposing a temporal limitation on seizures and determining the

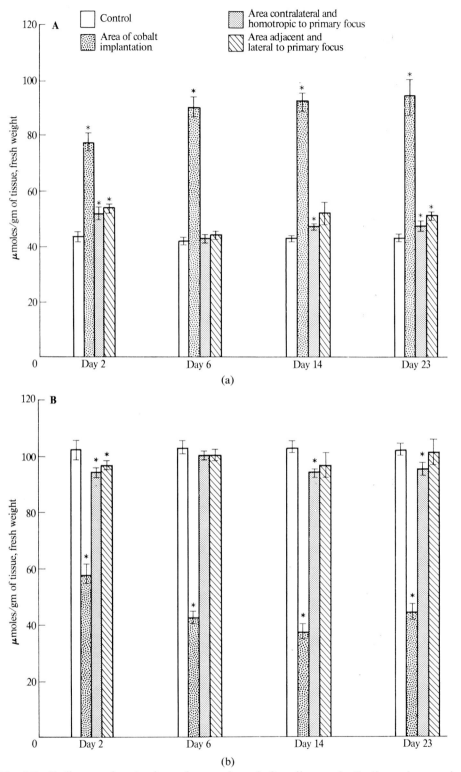

Fig. 3.2. Sodium and potassium changes in cobalt epilepsy. A. Sodium changes in cerebral cortex from control and cobalt-treated rats on different days following cobalt implantation. *Different from control ($P \leq 0.05$). B. Potassium changes in same areas ($P \leq 0.05$). (From Hunt & Craig, 1973.)

Table 3.1. Amino acid changes in various experimental primary cortical epileptogenic foci

Focus	Animal	% change from control value								Reference
		GABA	Glutamate	Glutamine	Aspartate	Taurine	Serine	Glycine	Alanine	
Cobalt	Mouse	18↓	12↓	a	a	15↓	60↑	32↑	69↑	Van Gelder, 1972
Cobalt	Cat	50↓	28↓	54↑	20→	a	57↑	49↑	—	Van Gelder, 1972
Cobalt	Cat	75↓	50↓	a	32→	60↑	43↑	53↑	29↑	Koyama, 1972
Cobalt	Rat	50↓	50↓	a	60→	60→	—	15↑	—	Emson & Joseph, 1974
Cobalt	Rat	—	50↓	—	77→	40↓	—	—	225↑	Craig & Hartmann, 1973
Freeze-lesion	Cat	a	50↓	50↓	—	—	—	—	—	Berl et al., 1959
Penicillin	Cat	33↓	50↓	—	—	—	—	—	—	Gottsfeld & Elazar, 1972

Values are those from maximum convulsive stage.
a = value unchanged or less than 10% changed from control.

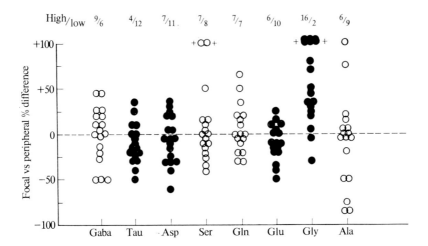

Fig. 3.3. A comparison between focal epileptogenic nervous tissue and tissue of surrounding cortex with respect to amino acid content. Data refer to the percentage by which the concentration of an amino acid in the focus differs from its matching peripheral sample (focal–peripheral)/(peripheral) × 100. The results are therefore independent from possible erroneous estimation of concentrations in normal human cortex. Dark circles call attention to amino acids having a consistently differing concentration in the focal area as compared to surrounding regions (over 50%). Ratios indicate the number of focal samples in which the level of a particular amino acid was either above or below the concentration found in the matching peripheral sample. (From Van Gelder et al 1972.)

length of the period between fits (interictal period) are not simple. Neuronal exhaustion is commonly invoked as an explanation (Williams, 1970) and this presumably refers to the effects of the massive metabolic response which accompanies paroxysmal neuronal discharge. These are geared to the sodium and potassium movements which occur. The increased intracellular sodium stimulates the sodium pump which requires substantial quantities of energy for its operation. As a consequence carbohydrate and energy metabolism are greatly stimulated and this implies augmentation of glucose uptake, glycolysis, turnover of intermediary metabolites and accelerated respiration, in attempts to maintain adequate levels of ATP and phosphocreatine. These none the less show appreciable decrease in their pool size (Table 3.2) (see Tower, 1960; Sokoloff, 1969). In addition, potassium which is released will no doubt depolarize surrounding glial cells (Kuffler, 1967), and the sodium influx into these cells will trigger a similar pattern of metabolic response. The systemic effects of the acidosis developing from increased lactate and carbon dioxide production are other consequences of the increased metabolism which also help to bring a seizure to an end. These changes are commonly featured in any description of the biochemistry of epilepsy and are perhaps the most dramatic and readily detectable

Table 3.2. Cerebral metabolism of cat brain *in vivo* during seizures (from Tower, 1960)

	Control	During seizure
Cerebral blood flow	—	↑
Cerebral oxygen consumption (μM)§	112*	268*
Cerebral lactic acid level (M/g)	2·2	5·5
Cerebral creatine-P level (M/g)	2·4	1·2
Cerebral ATP level (μM/g)	1·3	0·5
Cerebral ADP level (μM/g)	1·1	1·7
Cerebral inorganic P level (μM/g)	4·7	6·7

* Perfused brain *in situ*. § Per 100 g brain per minute

biochemical events. However, they clearly occur as the *result* of seizures rather than being involved in their initial cause, though they clearly influence the length of the interictal period and may have some influence in seizure spread. In fact, the changes in ATP and phosphocreatine may be prevented from occurring by hyperventilation to prevent tissue hypoxia; under these conditions seizures are *not* prevented (Collins *et al.*, 1970; & Plum, 1968). The postictal depression which follows major seizures is likely to be due to several factors all closely related. The rising intracellular sodium levels will first make the cells hyperexcitable as the membrane potential falls to the critical firing level, and then gross depolarization will prevent action potential generation (cathodal block). The sodium pump will subsequently extrude the sodium and repolarize the cell membrane, but only as ATP and phosphocreatine become available through increased mitochondrial activity. Any excessive depletion of transmitter stores caused by the paroxysmal firing will require refurbishment and this process will also influence the length of the fit-free period. However, since the true course for these events involves minutes or hours rather than longer periods, these considerations probably apply most appropriately to the condition of status epilepticus where the brief respite from the seizures is short. The more common, longer, interictal periods involve other factors which could include a positive inhibition from the thalamus via the thalamocortical pathways (Jaspers, 1969).

1. CATIONS AND EPILEPSY

The important role of sodium, potassium and chloride ions in all excitatory events creates a centre of interest when considering basic mechanisms of seizure initiation. The ability of ouabain to create a temporary epileptogenic focus is probably due to partial paralysis of the Na^+-K^+-ATPase activity which constitutes the sodium-pump (see above). It follows that partial inhibition of this enzyme in scar tissue or chronic experimental epilepsy could lead to a similar but more chronic condition than that produced by ouabain. Aluminium foci, for instance, appear to show lowered Na-K-ATPase activity (Harmony et al., 1968). However, freeze-foci and their mirror foci appear to show increased Na-K-ATPase activity (Lewin & McCrimmond, 1967, 1968), though the same foci yielded synaptosomes showing a reduced ability to concentrate potassium, and the extent of their incapacity correlated with the degree of activity of the focus, inactive foci being normal in this respect (Escueta & Reilly, 1971; Escueta & Appel, 1969, 1970, 1970a). This work indicates that in freeze lesions, at least, some primary deficiency in ion transport might be present in the presynaptic membrane which could have a direct effect on neuronal excitability. It is notable therefore that the important anticonvulsant Phenytoin appears to accelerate the activity of the sodium-pump (Woodbury & Kemp, 1970).

Calcium involvement in excitability is through its role in transmitter release and in control of membrane stability (Shanes, 1959). Neural tissues *in vitro* show increased respiration and glycolysis in calcium-free media and this is interpreted as being due to a change in membrane structure leading to increased sodium influx, and therefore to sodium pump stimulation. Hypocalcaemia is also known greatly to increase susceptibility to seizure, perhaps through related mechanisms (Corriol, et al., 1969).

IV Involvement of acetylcholine and monoamines

A. ACETYLCHOLINE

In looking for neurochemical mechanisms in seizure initiation, it is natural to look for abnormal function among the transmitter compounds likely to be mediating excitation and inhibition in the normal tissue, particularly in the cerebral cortex. For acetylcholine, the overall picture which emerges from the literature is a more likely involvement in the propagation of seizures, though participation in their initiation cannot yet be dismissed. The acetylcholine mediated pathways in the mammalian cerebral cortex appear to be both excitatory and inhibitory, with a preponderance of the latter. Brodman's layers II, III and IV mostly show inhibition on application of ACh, and deeper layers show excitation. The cholinergic innervation of the cortex forms part of a diffuse projection system from brain stem and reticular formation, and this includes major pathways for cortical arousal (see Phillis, 1970, and Krnjevik, 1974, for reviews).

When both acetylcholine (0·1%) and anticholinesterases are topically applied to cerebral cortex (intact or undercut), epileptic bursts of very rhythmic activity interspersed with silence are observed (Ferguson & Jasper, 1971; Ferguson & Cornblath, 1974). The waves are interpreted as being due to periodic inhibition from the deeper cholinergic layers gating the otherwise continuous excitatory drive upon the pyramidal cells by the prolonged depolarization. The effect is readily blocked by atropine or scopolamine indicating a direct involvement of ACh receptors. In addition, ACh is released to superfusion cups containing anticholinesterases following sensory stimulation (e.g. via reticular formation, Jaspers & Koyama, 1969) provided the cortex is not undercut. The amounts released would be sufficient to initiate an epileptiform discharge if it were to accumulate in the normal situation (Jaspers, 1969). Organophosphorus cholineresterase inhibitors have long been known to be potent agents in initiating convulsions, and parallel acetylcholine accumulation can be demonstrated in the tissue under these circumstances (Stone, 1957), though eserine and atropine at doses affecting cerebral cholinergic systems seem to be without effect on the threshold or pattern of light-induced seizures in the baboon (Meldrum *et al.*, 1970). Equally it has long been known that the levels of ACh in the cerebral cortex are greatly reduced during electroshock (Richter & Crossland, 1949) and this is paralleled by large changes in synaptic vesicle content (Essman, 1973). Thus interference with the cholinergic systems of the cerebral cortex, particularly by amplifying the effects of its transmitter which may be mediating arousal, can lead to paroxysmal neuronal firing within the cortex. These same pathways may, however, be central in the propagation of experimental and clinical seizures *caused* by other mechanisms. There is as yet no large body of evidence of changed levels of acetylcholine, or its synthetic or degradative enzymes in a range of clinical or experimental foci. Such evidence would strongly implicate cholinergic mechanisms in the initiation of seizures. Recently, Joseph and Emson (1974) have reported a sharp fall in cholineacetylase in both primary and secondary cobalt foci of rats, and Tower and Elliott (1952) long ago described small elevations in cholinesterase activity beyond normal levels in human foci (mostly temporal lobe) excised by surgery. In the latter study total ACh levels were not different, nor were rates of synthesis of ACh in slices prepared from the foci and incubated *in vitro*, though the ability to form bound ACh and to release it in response to raised potassium was diminished in both cases. The authors speculated that a chronic tendency to release ACh rather than bind or store it could lead to a hyperactive state in the focus. In addition, it has been shown that chronic primary and secondary foci established in the hippocampi of cats are greatly activated by small quantities of Ach injected directly into the foci (Guerrero–Figueroa *et al.*, 1964).

B. MONOAMINES

The overall picture of the evidence for monoamines again suggests a role secondary to the onset of seizures, or, at most an indirect effect through an increase in susceptibility, rather than a direct participation in a trigger mechanism. Thus, monoamine oxidase inhibitors, and other agents likely to increase synaptic availability of monoamines, greatly raise the threshold of experimental animals to convulsants as different as metrazole and hyperbaric oxygen (Chen et al., 1954; Schlesinger et al., 1968; De La Torre & Mullan, 1970; Haggendal, 1968, 1968a). Periods of maximum seizure risk within the lifespan of genetic strains of mice and rabbits correlated with periods of decreased brain levels of noradrenaline and serotonin (Schlesinger et al., 1965; Netthaus, 1970). In contrast to their depleting action on acetylcholine, both electroconvulsive shock and metrazole increase the pool size and the turnover of brain serotonin, noradrenaline and dopamine in experimental animals (Kety et al., 1967; Essman, 1973), and this could well be the basis for its therapeutic effect in treating mental depression. These increases may continue for days after cessation of the electroshock, and an increase in the activity of the synthetic enzymes such as tryosine hydroxylase, has also been detected (Musacchio et al., 1969). In addition, there appears to be a tendency for drugs which increase the brain levels of biogenic amines to decrease seizure susceptibility. This has been found after treatment with monoamine oxidase inhibitors (Prockop et al., 1959), pyrogallol and imipramine (Lehman, 1967). Essman (1973) finds a clear correlation between the length of the period of post-ictal motor depression and the noradrenaline content of mouse brain following electroshock treatment. One might therefore expect that seizures would be self-limiting, in their frequency, due to the elevation of brain levels of monoamines.

Whole brain levels of noradrenaline, dopamine and serotonin have been measured in rats with focal cobalt epilepsy and found not to differ significantly from those of sham-operated controls, even when the threshold for metrazole-induced seizure had fallen to 50% of the control dose. Turnover of these mono-amines was also indistinguishable between control and cobalt-epileptic animals (Colasanti & Craig 1973). Thus, although concentration of these transmitter substances in the local area constituting the spiking focus needs to be examined before unequivocal statements can be made, the data available indicates little direct involvement of monoamines in the initiation of seizures, though the opposite view is proposed by Bowen et al. (1975), who found noradrenergic varicosities sprouting in freeze-lesion foci a few hours after they were found.

Although reserpine treatment had little effect on the threshold or response of the photosensitive baboon to stroboscopic light stimulation, both serotonin, and LSD blocked the responses for some hours. Audiogenic seizures in rodents and action-myoclonus in man are similarly affected by treatment with these agents, and the mode of action is interpreted as being via afferent pathways and their interaction with cortical activity. This is based on electrophysiological evidence of inhibition produced by the drugs on afferent relay systems such as the lateral geniculate nucleus (Meldrum & Balzano, 1972).

V Involvement of amino acids

A. AMINO ACIDS AND NEUROTRANSMISSION

Currently a great deal of interest is centred on the possible involvement of certain amino acids in the mechanisms of epilepsy. This is partly the result of an awakening in neurophysiology to the paradoxical fact that a number of simple and ubiquitous amino acids which are involved in a wide range of metabolic pathways in the CNS are likely also to function as major synaptic transmitters in the brain and spinal cord. At present a list of these compounds would include glutamate and aspartate as excitatory agents, and GABA, glycine and possibly taurine as inhibitory agents. (For reviews see Krnjevik, 1974; Curtis & Johnston, 1974.) It is the wider involvement of amino acids in mainstream biochemistry rather than their involvement in synaptic transmission which is the factor determining their tissue concentration, and consequently it is not a surprising finding that they are not specially concentrated on the presynaptic side of the synapse since relatively small amounts are required for transmission purposes. A potent, high-affinity, uptake system operates to keep amino acid levels in the extracellular space at a minimum. In the cerebral cortex as in many other brain regions most neurons will respond with firing to glutamate iontophoresed from micropipettes, and such iontophoresis is commonly employed by electrophysiologists to detect the presence of active cells. Concentrations as low as 10–100 μM glutamate are probably sufficient to initiate action potential discharge, and the glutamate ion probably exerts its effect through interaction with glutamate receptors in glutamate-ergic synapses. Tissue levels of glutamate and GABA may be 3–10 mM, and therefore a very large concentration gradient exists across the cell membrane. Since these amino acids are being continually released during activity at the synapse, and possibly across other membranes, both during activity and at rest (Bradford, 1975; Weinreich & Hammerschlag, 1975), great importance must be attached to the maintenance of low levels of these compounds in the extracellular space, particularly in the region of the synapse. Glutamine on the other hand is physiologically inactive, and in addition, readily moves between cells and their extracellular space being the most highly concentrated amino acid in CSF (300–500 μM; see Table 7). It may in fact be providing a means of recycling glutamic acid which is released from the presynapse, or other sites, and subsequently absorbed by neurones and glial cells. After conversion to glutamine the glutamate moiety could be released to CSF and then taken up preferentially by compartments containing high glutaminase activity such as presynaptic nerve terminals (500 μmol glutamine/100 mg protein/hr.), where glutamate would be regenerated (Bradford et al., 1975) (Fig. 3.4). A compartment actively absorbing glutamate and converting it to glutamine has long been known. It forms the basis of the so-called 'Waelsch effect' and is likely to comprise the glial cells (see Balazs & Cremer, 1973). The higher level of glutamine commonly found in CSF is usually attributed to glutamine synthetase activity directed towards removing ammonia, but the removal of the potent excitatory glutamate ion must be regarded as an equally important activity for this enzyme.

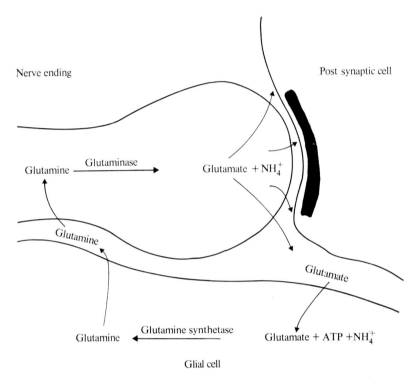

Fig. 3.4. Proposed recycling of glutamate between cells and extracellular space in brain tissue. (From Bradford & Ward, 1975.)

B. THE AMINO ACID CONTENT OF EPILEPTIC FOCI

Since the early observation by Tower and his co-workers that glutamate and GABA are diminished in concentration in slices of excised human temporal lobe foci incubated in nutrient salines, the picture has enlarged to include specific changes in the levels of several other amino acids in epileptogenic cortex both in humans and in a wide range of experimental epilepsy (Fig. 3.3; Tables 3.1 and 3.5). These include losses of glutamate, glutamine, aspartate and taurine and increases in glycine and sometimes in serine in tissue samples isolated from the spiking focus and compared with normal tissues from the same preparation. Not all these changes have been consistently observed by all investigations but where they have been detected, most have been in the direction indicated.

C. AMINO ACID CHANGES IN WHOLE BRAIN

Many systemic convulsant agents are able to change whole brain levels of certain of these amino acids in the same direction as occurs in foci (Table 3.3). A lowering of GABA and sometimes of glutamate is most commonly observed. It is not surprising that agents such as allylglycine, 3-mercaptopropionic acid, semi-carbazides, hydrazides and other antipyridoxal agents which inhibit glutamic acid decarboxylase (GAD) cause lowered GABA levels (Table 3.4) but this is sometimes accompanied by lowered glutamate levels (De Lorez Arnaiz et al., 1971, 1972, 1973 (Table 3.4); Alberici et al., 1969). A similar linkage between GABA and glutamate depletion is seen in whole brain following seizures induced by methionine sulphoximine, a compound which prevents glutamine formation and interferes with methionine metabolism, including protein synthesis (Hrebicek et al., 1971). Tower (1960) has argued that since transamination of GABA leads to glutamate production via α-oxoglutarate, diminished GABA levels and lowered flux through this 'GABA shunt' pathway would necessarily lead to a fall in glutamate concentration (Fig. 3.4). However, a loss of glutamate under conditions where GABA was substantially reduced was not reported by Roa et al. (1964) and may not be a consistent correlation. Other convulsant agents administered systemically which have been reported to change brain GABA, glutamate, glutamine and aspartate levels are metrazole, picrotoxin, strychnine and oxygen at high pressure (Table 3.4) though in many cases the changes are detectable only after onset of seizure and cannot therefore be initiating factors. Many of these amino acids are important intermediates in mainstream carbohydrate and energy metabolism in addition to their role in protein synthesis and other biosynthetic pathways. For this reason it is important when attempting to connect changed amino acid levels with seizure initiation or propagation that aberrations in these processes are given equal consideration alongside any direct influence amino acids may have on excitation and inhibition.

D. AMINO ACID CHANGES: METABOLIC IMPLICATIONS

Abnormal pool sizes of glutamate, aspartate, GABA and glutamine usually indicate changed patterns of carbohydrate metabolism. In the epileptic brain this appears to be occurring both locally in the spiking focus and in a more general way throughout the brain during and after seizures. In the spiking focus these changes might bear a causal relationship with the hyperactivity, for instance, by diminishing the availability of GABA for inhibition, or of ATP for ion transport. Deficiencies could arise through a decreased incorporation of glucose carbon into amino acids, and although available evidence indicates that glucose uptake and incorporation into alanine and lactate is stimulated during seizures, incorporation into other amino acids is reduced (Dunn & Guidditta, 1971;

Table 3.3. Some common systemic chemical convulsants and their likely mode of action

Convulsant	Chemical structure	Proposed mode of action	References
3-mercaptopropionic acid	$SH-CH_2-CH_2-COOH$	Inhibition of glutamate decarboxylase and GABA amino transferase	Lamar (1970); De Lorez Arnaiz (1971, 1973).
Allylglycine	$CH_2=CH.CH_2.CH.NH.COOH$	Inhibits glutamate decarboxylase	Alberici et al. (1969); De Lorez Arnaiz et al. (1971).
Pentylenetetrazole (Metrazole)	(tetrazole ring structure)		Stone (1957); Hawkins & Sarett (1957).
Methionine sulphoximine	$CH_3.S.CH_2.CH_2.CH.NH_2-COOH$, $=NH$	Inhibits glutamine synthetase and protein synthesis	Hrebicek (1971); Sellinger & Weiler (1963).
Thiosemicarbazide	$NH_2.CS.NH.NH_2$	Pyridoxal phosphate antagonists causing glutamate decarboxylase inhibition	Wood & Peesker (1974); Killam & Bain (1957); Roa et al. (1964); Collins (1973).
Semicarbazide, and other pyridoxal antagonists	$NH_2.CO.NH.NH_2$		
Strychnine	Alkaloid	Blocks post-synaptic inhibition at glycine receptors?	Curtis et al. (1968, 1968a); Haber & Saidel (1948).
Picrotoxin	Alkaloid	Blocks presynaptic inhibition or GABA receptors?	Curtis et al. (1968, 1968a).
Bicuculine	Alkaloid		

Table 3.4. Changes in whole-brain amino acid levels during experimental seizures induced by systemic convulsants

Convulsant	Animal	Dose (mg/Kg)	GABA	Glutamate	Glutamine	Aspartate	Taurine	References
Metrazole	Rat	100	- -	13↓	13↑	- -	- -	1
Metrazole	Rat	100	17↑	- -	- -	- -	- -	2
Metrazole	Rat	100	a	a	- -	a	a	9
Metrazole	Rat	70	15↑	25↓	- -	15↓	- -	4
Thiosemicarbazide	Dog	20	34↓	a	- -	- -	- -	5
Thiosemicarbazide	Rat	7.5	20↓	a	- -	- -	- -	10
Semicarbazide	Rat	200	32↓	a	- -	a	a	9
Semicarbazide	Rat	500	38↓	- -	- -	- -	- -	1
3-mercaptopropionic acid	Rat	150	31↓	a	a	a	- -	3
Strychnine	Rat	4	15↑	- -	- -	- -	- -	1
Strychnine	Rabbit	0.1	21↑	18↑	13↓	- -	- -	6
Strychnine			- -	30↓	- -	- -	- -	8
Picrotoxin	Rat	10	a	- -	- -	- -	- -	2
Picrotoxin	Rabbit	3	a	a	13↓	- -	- -	6
Picrotoxin	Mouse*	2–3	16↑	a	27↑	- -	- -	6
Hyperbaric oxygen	Rat	6 atmos	35↓	a	- -	a	- -	7
Hyperbaric oxygen	Rat	5 atmos	38↓	- -	- -	- -	- -	2

* Intracerebral injection

Values are those from maximum convulsive stage; ↑ level raised, ↓ level lowered; - - not measured, a = value unchanged or less than 10% changed from control.

References 1, DeRopp & Snedeker, 1961; 2, De Fendis & Elliott, 1968; 3, De Lorez Arnaiz et al., 1973; 4, Nahorski, 1970; 5, Roa et al., 1964; 6, Saito & Tokanager, 1967; 7, Wood & Watson, 1963; 8, Haber & Saidel, 1948; 9, Killam & Baim, 1957; 10, Collins, 1973.

Yoshino & Elliot, 1970). Pool sizes, however, are either unchanged, rise, or show a slight fall (Sactor et al., 1966; Nahorsky et al., 1970; Leonard & Palfryman, 1972; King et al., 1974). Of course these are seizures induced by electroshock or drugs in otherwise normal animals, and measurement of the response in chronically epileptic animals would give a more relevant answer. Restricted availability of glucose and other substrates, or of oxygen in epileptogenic tissue, could lead to the observed fall in pool size of glutamate, aspartate and GABA since they would continue to supply Krebs cycle intermediates without being replaced (Tower 1960; Bradford 1968). However, slices of human temporal lobe, and slices of the spiking foci from cobalt epileptic rats show no obvious change in respiratory rates from normal (Pappius & Elliott, 1954; Colestani & Craig, 1972), though the critical test of performance would be under conditions accelerating activity to rates closer to those found *in vivo*, and this was not examined. Moreover, since any hyperactivity in the focus would not persist *in vitro*, an abnormality may not be detected. However, Tower and his co-workers have shown that tissue levels of glutamate and GABA in excised temporal lobe foci fall during an hour's incubation whilst normal tissue shows these substances increasing under similar conditions. Significantly, the fall in glutamate is apparently prevented by the addition of 10 mM GABA or asparagine to the nutrient glucose—salines, indicating that the pathways linking GABA, asparagine and glutamate via the Krebs cycle and GABA shunt are sufficiently active to replenish the glutamate losses (Table 3.5). Intravenous injections of GABA (125 mg/Kg), and less effectively of glutamate (100 mg/Kg), into cats with active freeze-foci which showed lowered pool sizes of glutamate and GABA, suppressed

Table 3.5. Glutamate metabolism in incubated slices of human cerebral cortex (from Tower, 1960) ($\mu M/g$)

Conditions	Incubation time (min)	Glutamic acid level	Glutamine level	γ-Aminobutyric acid level
Normal	0	8·1	2·7	4·75
	60	10·0	4·55	6·1
Epileptogenic	0	7·05	3·0	4·35
	60	5·15	5·1	3·25
+L-Asparagine or γ-Aminobutyrate	60	9·5	4·75	—
Patient C.G., ♂, 29 Normal	0	7·35	2·2	
	60	10·35	3·75	
Epileptogenic	0	7·35	2·85	
	60	6·0	4·2	

the epileptiform activity but also raised the amino acid levels at the focus to within the normal range. There was no correlation in time between the two events. Suppression of abnormal electrical activity occurred 15 sec after GABA administration, but recovery of amino acid levels in the focus was not detectable until a further 15 min had elapsed. Following serial infusion, levels of GABA in the focus were 3–4 fold higher than control. The authors interpret the changes in glutamate levels following GABA administration in terms of increased GABA shunt activity, and the effects on electrical activity of the focus in terms of a direct inhibitory action of GABA rather than via any readjustment in metabolism it might cause (Table 3.6, Fig. 3.8; Berl et al., 1958; Berl et al., 1961). Since GABA, in the presence of glutamate, is an extremely good substrate for ATP synthesis by brain mitochondria, but is ineffective for liver and probably other tissue except kidney, the effects of GABA in refurbishing amino acid levels and suppressing seizures could well be through a special metabolic role this substance might perform (Lee et al., 1974; Schriver & Whelan 1969).

Table 3.6. Changes in amino acid content of freeze-focus following systemic GABA and glutamate administration

	% of control level			No. of experiments
	Glutamate	Glutamine	GABA	
Untreated	54	51	95	6
After glutamate injection (100 mg/Kg)	80	81	121	5
After GABA injection (125 mg/Kg)	73	70	244	6

Values are means for the number of experiments quoted. Cats were the animals employed. Injections were via an indwelling femoral vein cannula (from Berl et al., 1961).

In exploring the possible causal link between diminished amino acid levels and seizure initiation, some authors have emphasized the inhibitory effect which such depletion might have on the synthesis of proteins such as the S100 protein which could be important in controlling neuronal activity. It is suggested that the changes in glycine and taurine imply alterations in the patterns of protein synthesis at the focus (van Gelder et al., 1972; van Gelder & Courtois, 1972). Electrical stimulation both *in vitro* (Orrego & Lipman, 1969; Jones & McIlwain, 1971) and *in vivo* (Dunn et al., 1972; Essmann, 1973) causes lowered protein synthesis rates in whole brain or brain slices though this is not associated in these experiments with lowered pool sizes of key amino acids and is probably irrelevant to present arguments. Indeed, at the synapse itself, protein synthesis rates are substantially increased by electrical stimulation (Wedege & Bradford, 1975).

In summary, current knowledge does not yet provide an adequate metabolically-based explanation of the abnormal amino acid pool sizes which would provide the basis of a general theory of seizure initiation. Agents such as allylglycine, methionine sulphoximine and antipyridoxal agents with known actions on glutamate decarboxylase or glutamine synthetase activity (Table 3.4) are special exceptions, and since reduced activity of these enzymes is not a general and well established observation in focal tissue from animals or humans, its relevance must remain restricted. However, the effects of these agents *do indicate* that lowering GABA or glutamate concentrations from their normal range precipitates seizures, though it remains unresolved whether this is due to interference with energy metabolism or is via a more direct influence on the excitation-inhibition balance by, for instance, restricting the availability of GABA for its inhibitory role at the synapse.

E. GLUTAMATE, GABA, EXCITATION AND INHIBITION

It now seems fairly certain that synapses carrying post-synaptic receptors for GABA or for glutamate and which mediate inhibition and excitation respectively, occur widely in the CNS together with a more restricted distribution of equivalent sites for glycine and aspartate. The synapses which are operated by these amino acids are likely to constitute a large proportion of the total in the CNS, which means that the high concentrations and ubiquitous presence of these compounds, particularly glutamate, aspartate and GABA leads to a high risk of anomalous contact between the amino acids and their synapses. The enzyme producing glutamate from α-oxoglutarate, glutamate dehydrogenase, displays a largely mitochondrial location and is found in all the cells of the nervous system. Glutamate decarboxylase (GAD), the enzyme producing GABA is found concentrated in the presynaptic cytoplasm (Weinstein *et al.*, 1963, Salganicoff & De Robertis, 1965; Neal & Iversen, 1969) yet GABA is present in both glial cells and neuronal cell bodies at levels equivalent to those initially present (or generated from glucose) in isolated nerve terminals (Morgan & Whittaker, 1965; Bradford & Thomas, 1969; Rose, 1968; Watkins, 1972). Significant proportions of GAD activity, therefore, occur in these other compartments, unless their strong capacity to take up GABA effectively redistributes GABA originating in the presynapse (Iversen *et al.*, 1973; Hokfelt & Ljungdahl, 1972, 1972a; Snodgrass & Iversen, 1974). These same high affinity uptake systems probably rapidly remove physiologically-active amino acids from the extracellular space when they are released during activity at the nerve-terminal or from other regions of the neuronal surface (Wheeler *et al.*, 1966; De Feudis, 1971; Logan & Snyder, 1971; Iversen, 1971; Bradford, 1975). This is reflected in the low CSF concentrations of these compounds (low μM levels,

Table 3.7), the levels in interstitial fluid in the immediate vicinity of the cells being, no doubt, considerably lower than that in CSF removed by lumbar puncture.

Table 3.7. The free amino acid content of CSF of patients with epilepsy (from Plum, 1974)

	Control n=15		Epilepsy n=15	
	μmol/l.	S.D.	μmol/l.	S.D.
Alanine	26·9	10·3	16·8*	5·9
Arginine	16·9	5·9	16·9	6·3
Asparagine	trace	trace	trace	trace
Aspartic acid	2·4	0·8	3·8*	0·5
Cystine	trace	trace	trace	trace
Glutamic acid	11·2	7·2	19·2*	2·1
Glycine	6·9	1·3	6·5	1·9
Histidine	10·8	3·2	14·6*	2·4
Isoleucine	5·2	1·2	5·7	1·3
Leucine	13·1	2·7	12·2	1·9
Lysine	19·7	5·8	25·7*	4·3
Methionine	2·8	0·6	2·0†	0·3
Phenylalanine	8·3	1·4	9·1	1·5
Proline	trace	trace	trace	trace
Serine	28·9	6·7	39·3*	5·2
Threonine	28·5	5·6	23·8†	3·1
Tyrosine	7·3	1·6	6·8	2·3
Valine	14·0	2·8	16·9	3·2

$*\ P<0.005.$
$†\ 0.01>P>0.005.$

1. GABA AND ITS ENZYMES

The three enzymes glutamate decarboxylase (GAD) GABA-transminase (GABA-T) and succinic semialdelhyde dehydrogenase (SSDH) are responsible for the production and removal of GABA (Fig. 3.5). Agents which reduce GAD activity often markedly lower brain GABA concentrations (e.g. 40%), and in parallel causes seizures e.g. hydrazides, allylglycine, 3-mercaptoptopionic acid (Killam & Baim, 1957; De Lorez Arnaiz et al., 1971, 1972, 1973). Although there is no clear correlation between brain GABA content and seizure onset, one research group have developed an equation relating excitable state to both the level of GABA and the degree of GAD inhibition where hydrazine and its derivatives are the convulsant agents (Wood & Peesker, 1974). Because of the

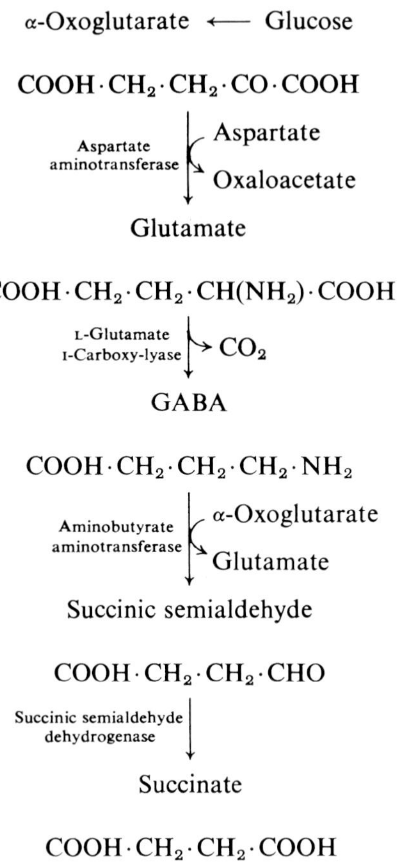

α-Oxoglutarate ⟵ Glucose

$COOH \cdot CH_2 \cdot CH_2 \cdot CO \cdot COOH$

Aspartate aminotransferase ⤹ Aspartate
⤷ Oxaloacetate

Glutamate

$COOH \cdot CH_2 \cdot CH_2 \cdot CH(NH_2) \cdot COOH$

L-Glutamate 1-Carboxy-lyase ⤷ CO_2

GABA

$COOH \cdot CH_2 \cdot CH_2 \cdot CH_2 \cdot NH_2$

Aminobutyrate aminotransferase ⤹ α-Oxoglutarate
⤷ Glutamate

Succinic semialdehyde

$COOH \cdot CH_2 \cdot CH_2 \cdot CHO$

Succinic semialdehyde dehydrogenase ↓

Succinate

$COOH \cdot CH_2 \cdot CH_2 \cdot COOH$

Fig. 3.5. The GABA shunt pathway.

likely involvement of GABA in both metabolism and inhibitory neurotransmission, changes in compartmentation of GABA must be judged as more critical than overall changes in tissue content; changed *location* could occur without detectable change in tissue *content*. Inhibition of GABA-T or SSDH levels leads to a rise in brain GABA level, and simultaneously protects against seizures, examples being the effects of α-amino oxyacetic acid (AAOA) (Snodgrass & Iversen, 1973, Collins, 1973; Perry *et al.*, 1974) and *n*-dipropyl acetate (Epilim) (Simler *et al.*, 1968, 1973; Harvey *et al.*, 1974). Whilst AOAA is unfortunately toxic when administered to humans, and has convulsant action above a certain dosage (Perry *et al.*, 1974; Tapia *et al.*, 1969), Epilum, whose passage across the blood–brain barrier is facilitated by its fatty nature, is proving a most useful anticonvulsant in the clinic (Jeavons & Clark, 1974). There is evidence that some commonly employed barbiturate anticonvulsants raise brain GABA levels (Saad *et al.*, 1972) which, if substantiated, should be taken into account in

Table 3.8. Free amino acid concentration in plasma of patients with epilepsy (from Plum, 1974)

	Control n=60		Epilepsy n=60	
	μmol/l.	S.D.	μ/mol./l	S.D.
Alanine	330	78	312	78
Arginine	80	26	65*	19
Aspartic acid	7	4	7	4
Cystine	61	21	72	25
Glutamic acid	83	42	71	37
Glycine	249	81	227	82
Histidine	85	34	78	35
Isoleucine	61	21	59	23
Leucine	109	48	95	47
Lysine	158	28	148	36
Methionine	41	15	44	15
Phenylalanine	71	24	66	21
Proline	212	64	199	55
Serine	149	55	139	56
Threonine	142	46	148	48
Tryptophane	62	26	60	27
Tyrosine	70	32	69	34
Valine	222	71	212	88
Taurine	78	25	80	28

* $P < 0.005$.

assessing mechanism of actions of these substances. Conflicting with this generalization is the report that AOAA at doses which more than doubled brain GABA gave no protection against audiogenic seizures during barbiturate withdrawal (Crossland & Turnbull, 1972).

Restricted availability of GABA could reduce transmission, reduce the rate of flow through the 'GABA-shunt', or both. However, since the 'shunt' is unlikely to deal with more than 10% of the Krebs cycle flux (Machiyama et al., 1965) a reduction in the inhibition produced by GABA seems the more forceful argument. The anticonvulsant Epilim blocks the shunt pathway and elevates brain GABA levels and the implication appears to be that raising brain intracellular GABA above normal, reduces the incidence of fits perhaps by increasing GABA output at its synapse. This anticonvulsant could equally well produce its effects by blocking GABA uptake and increasing *extracellular* GABA levels in regions containing GABA-ergic synapses. Since there is evidence that AAOA and Epilim are both inhibitors of GABA uptake, this feature of their action might at least assist any effect due to raise cellular levels of GABA (Snodgrass &

Iversen, 1973, Johnston & Baker, 1974; Harvey *et al.*, 1975). The potent inhibition of neuronal discharge produced by extracellular GABA can be seen following its iontophoresis locally in cerebral cortex and elsewhere *in vivo* (Krnjevik & Phillis, 1963) or when it is present in salines bathing slices of cerebellar cortex *in vitro* (Okamoto & Quastel, 1973). There has been a report of the successful clinical use of GABA as an anticonvulsant against *petit mal* seizure in particular (Tower, 1960), but this has not been pursued further probably because GABA does not appear to penetrate normal brain very well. However, where the blood–brain barrier is ineffective, local entry of GABA could then be expected to occur (see also Hayashi, 1966).

2. GLUTAMATE AND ITS ENZYMES

The considerations brought to bear on GABA apply equally to glutamate, though its more direct relationship with the Krebs cycle makes distinction between its metabolic and transmission involvement much more difficult. In contradistinction to GABA, glutamate is a potent excitatory agent and readily causes firing of most neurones when iontophoresed into cerebral cortex and other CNS regions (see Johnston, 1974, for review). It increases the firing rate of neurones present in cerebellar slices *in vitro* (Okamoto & Quastel, 1973) and can be shown to depolarize and increase the sodium permeability of cells in cerebral cortex slices (Gibson & McIlwain, 1965; Bradford & McIlwain, 1966). Glutamate (together with smaller amounts of aspartate, lysine and taurine) is released from the cortical surface, following stimulation of the reticular formation, suggesting that it could be the transmitter which mediates reticulo-cortical activation (Jasper & Koyama, 1969).

The glutamate ion is, therefore, well qualified to be considered as an excitatory agent in the seizure-initiating mechanisms. Small amounts of the substrate introduced intracerebrally readily produce convulsions (Hayashi, 1954, 1956; Wiechert & Gollnitz, 1968; Hennecke & Weichert, 1970). Systemically administered glutamate produces seizures in 10-day-old rats (Johnstone, 1972) and also, at higher doses, in adult rats (Bhagaran *et al.*, 1971; Bradford & Dodd, 1975). Raised whole brain levels of glutamate and of glutamate dehydrogenase have been reported in the preconvulsive state before onset of seizures initiated by a range of agents (Weichert & Gollnitz, 1968, 1969, 1969a), and a release of glutamate has been observed to occur from cobalt-foci during the onset of seizures as observed either by EEG or as motor response (Figs. 3.6 & 3.7; Koyama 1972; Dodd & Bradford 1975). The complementary change was observed to occur in the freeze-focus, namely a reduction in glutamate content correlating with the time of onset of epileptiform EEG activity (Fig. 3.8; Berl *et al.*, 1959). In view of their high concentration and their soluble state, intermittent leakage of glutamate or other amino acids from the focus could account for part or all of their lowered tissue levels, and leakage from incubated human

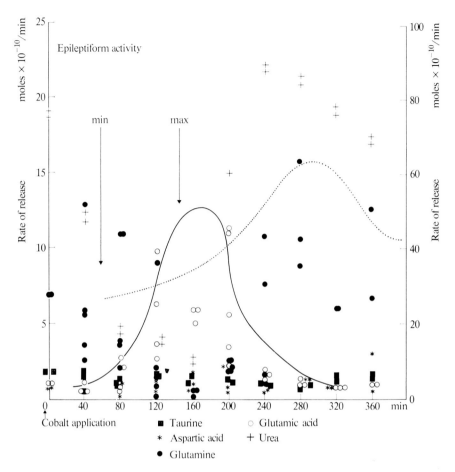

Fig. 3.6. Glutamate release from cobalt-focus *in vivo* ——— glutamate patterns; ········ glutamine pattern (adapted from Koyama, 1972).

focal tissue to the glucose-salines might partly explain the inability of the tissue to maintain or increase these levels (Table 3.5). The authors did, however, consider this possibility and reported amino acids to be absent from the incubation medium, so the 'leakage' theory would require that the relatively small amount of amino acid involved having been released to the extracellular space, and in contact with receptors, was reabsorbed by the tissue into compartments where it is rapidly metabolized. GABA and asparagine (added to 10 mM), which readjusted tissue levels, could exert their effect through acceleration of the activities of the Krebs cycle and the shunt pathway, to rates of glutamate and GABA formation in excess of the rate of loss from the tissue. The theory must also account for the effects of extracellular glutamate would predominate in the experiments of Berl *et al.*, 1961, glutamate as well as GABA preventing much of the epileptiform activity of the freeze-focus in cats when administered

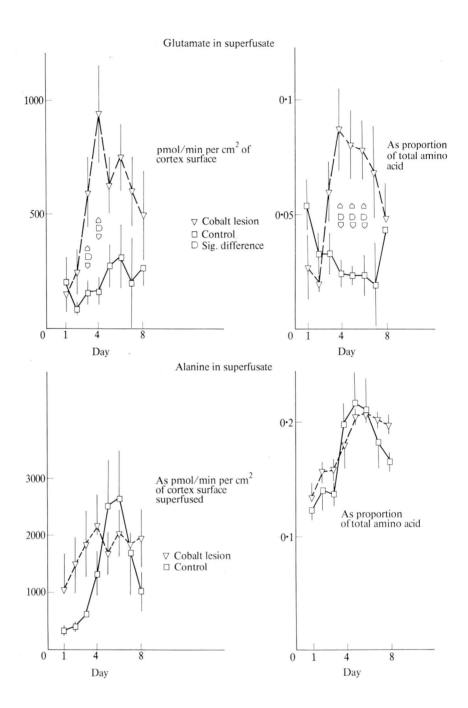

Fig. 3.7. Patterns of release of glutamate and alanine *in vivo* from cobalt-focus and cerebral cortex of rat. The animals were awake and unrestrained and collection was via a superfusion system employing artificial CSF (flow rate 5 ml per hour; see Dodd *et al.*, 1974; 1974a).

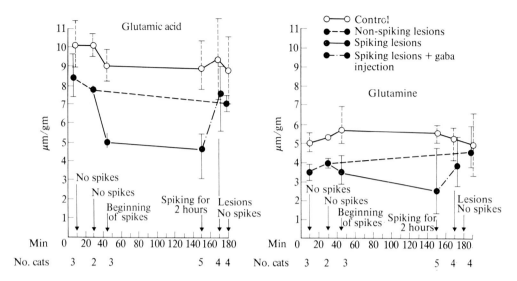

Fig. 3.8. Changes in glutamate and glutamine levels in freeze-focus of cats. Values are mean ±S.D. for number of cats indicated. Note that onset of spiking correlates with fall in tissue content of the amino acids (from Berl et al., 1960).

systematically (100 mg/Kg). However, the effect often required several injections, and involved a delay period not seen with GABA. Moreover, it was often preceded by activation of the focus. Similarly, cobalt foci in rats may be activated by subseizure-threshold doses of glutamate-administered systemically (Bradford & Dodd, 1975). Arguing against a leakage of glutamate as a basic excitatory mechanism is the work of Brown and Stone (1973) who could not detect an increase of glutamate or aspartate in fluid samples withdrawn from the subarachnoid space of dogs experiencing convulsions induced by thiosemicarbazide, metrazole, picrotoxin or methionine sulphoximine. However, a rather wide range of control glutamate levels (measured before injections of convulsant) were reported for the group of animals studied, making firm conclusions difficult. In addition, superfusion systems may have to be employed to overcome the great efficiency of the glutamate re-uptake system. The theory also requires that excitation by released glutamate would predominate over inhibition by any released GABA though lowered GABA may simply be due to the absence of glutamate. This could be encompassed by the fact that more glutamate than GABA is released, and by the possibility that GABA functions primarily at picrotoxin-sensitive presynaptic (axo-axonic) sites modulating the release of excitatory transmitters including glutamate. In the latter arrangement, the action of *extracellular* glutamate could not be damped by the presence of extracellular GABA in regions where glutamate synapses predominated (e.g. cerebral cortex). Abnormally high glutamate accumulation in the extracellular space might well be reflected as raised levels of this amino acid in the CSF at epileptic patients, and several reports which substantiate this expectation now exist in the literature (see Plum, 1974, for refs.; Table 3.7). Caution must, however, be exercised in interpreting these data because of the ease with which glutamine (present in CSF in 100-fold greater abundance than other amino acids) is converted to glutamate during extraction and analysis (Blanshard *et al.*, 1975). Potassium elevation in CSF has also been reported to occur during convulsions in monkeys (Meyer *et al.*, 1970) and may be linked with the rise in CSF glutamate.

If glutamate release were to occur as readily from central nerve fibre tracts as is reported for peripheral nerves and spinal nerve roots (Wheeler *et al.*, 1966; De Feudis, 1971; Weinreich & Hammerschlag, 1975) and its accumulation were to ensue to the appropriate extent, then some participation of glutamate in the slower mechanisms of seizure propagation could be envisaged. In addition, extracellular glutamate may mediate the considerable brain oedema which accompanies convulsion (e.g. see De Lorez Arnaiz 1972, 1973) since incubating neural tissue with this amino acid has long been known to promote tissue swelling (Terner *et al.*, 1960; Bradford & McIlwain, 1966, van Harreveld & Fifkova, 1971). Such an effect may, however, be exerted through the excitatory property of glutamate, with convulsive hypoxia being the final *common mechanism* in seizure-induced oedema.

F. GLUTAMATE AND SPREADING DEPRESSION

Strong evidence is accumulating which suggests that glutamate is also involved in the propagation of spreading depression in cerebral cortex and retina. This condition results from strong stimulation at one point and manifests itself as a general neuronal inhibition together with the disappearance of spontaneous electrical activity, and it spreads slowly (3 mm/min) in all directions from the initial point (Leao, 1964). It appears paradoxical that a convulsant such as glutamate should mediate a state of general inhibition, and seems inconsistent with a likely excitant role in epilepsy. However, the depression appears to result initially from gross depolarization of the neurones and its propagation is associated with the release of both potassium and glutamate at the advancing front; this causes further depolarization and thereby triggers further release. The neurones recover their membrane potential in about 2 min but the depression of excitability continues for 10 or 15 min. There is evidence that GABA is released during the period of depolarization and may accumulate in the extracellular space causing the more prolonged period of inhibition (van Harreveld & Fifkova, 1970, 1973; van Harreveld, 1972; Phillis & Ochs, 1971; Kaczmarek & Adey, 1974). In this way both glutamate and GABA would appear to be central to the mechanisms of propagation and maintenance of spreading depression. The brief burst of action potentials which is often recorded in the advancing front would be expected if rapid and gross depolarization were lowering the membrane potential to the critical firing level before causing a brief period of cathodal block. Thus, a role of glutamate in the spread of this essentially inhibitory condition is compatible with its established excitatory action, but suggests that spreading depression rather than convulsion should follow intracerebral injection of glutamate or its leakage in an epileptic focus (see Section V, E 2). This anomaly remains to be fully answered but is probably related to the extracellular concentrations of glutamate present in the different situations and to the secondary involvement of GABA or other inhibitory agents in spreading depression. The paroxysmal depolarization shifts (PDS) seen in many experimental foci do, after all, involve a massive depolarization, with a brief discharge of action potentials followed by cathodal block (see Section III, B 1) much as is seen during the invasion of spreading depression before the second and much longer period of inhibition commences. Also, it is notable that, the condition is much easier to evoke in small mammals such as rat and rabbit with smooth cortices and a high neuronal population density, than in cats and primates which have the opposite kind of cortical structure, and spreading depression has yet to be demonstrated in man (Phillis & Ochs, 1971). Thus in primates and man it may not be possible for glutamate to set in train the events which lead to spreading depression.

G. TAURINE AND EPILEPSY

Taurine, which remains something of a mystery substance, is currently being linked with epilepsy in the scientific literature. It is present in tissues, particularly muscle and brain, in relatively large quantities (5–10 μmol/g wet wt), but has no generally accepted function. It has been found very concentrated in the neurohypophysis and in the pineal gland, particularly in the rat (60–110 μmol/g wet wt; Crabai et al., 1974), and appears as a constituent of a group of small peptides in the brain which also contain N-acetylaspartate (Reichelt et al., 1974). It has been postulated that taurine is an important counter ion controlling tonicity of the cytoplasm and possibly the movement of calcium and other cations across membranes (Colley et al., 1974; Huxtable & Bressler, 1973). It might be well suited to this role in view of its combination of strong negative charge and metabolic stability. Many observations also point to a possible function for taurine as an inhibitory neurotransmitter. For instance, taurine iontophoresed onto neurons or applied to the retina or brainstem produces inhibition, which is blocked by strychnine but not bicuculline (Curtis et al., 1968; 1968a; Bonaventure et al., 1974; Haas & Hösli, 1973). The enzyme which produced taurine, cysteine sulphinate decarboxylase, is found concentrated in nerve terminals (Agrawal et al., 1971) and taurine itself is concentrated in synaptic vesicles (de Belleroche & Bradford, 1973) and in retinal nerve-endings (Kennedy & Voaden, 1974). Also, cerebral tissues in vitro release taurine by calcium-dependent processes when exposed to depolarizing stimuli (Kaczmarak & Davison, 1971; Davison & Kaczmarak, 1972; de Belleroche & Bradford, 1973,

Wheler et al., 1975). However, only a small proportion of the large tissue content of taurine is likely to be involved in neurotransmission, and other functions must be attributed to the greater part. Several workers have reported that taurine along with other amino acids (see Table 3.1 and Section VB above) is depleted in epileptogenic tissue from humans (van Gelder et al., 1972) and from several species of laboratory mammals with cobalt-induced epilepsy (van Gelder & Courtois, 1972; Craig & Hartman, 1973; Emson & Joseph, 1975; van Gelder, 1972), though such changes were not found in baboons with photogenic epilepsy (Hansen et al., 1973) and excised human foci (Perry et al., 1972), and were not detected by Koyama (1972) in cobalt-induced foci. Van Gelder (1972) also reported that taurine administered systemically to epileptic cats and mice not only had a strong anticonvulsant effect but restored taurine and other amino acids to normal levels in the epileptic foci (Table 3.9). The anticonvulsant effects

Table 3.9. Influence of taurine on amino acid content of epileptic mouse cortex (from van Gelder, 1972)

Amino acid	Control (17–19)	Untreated (7)	Taurine (9)
GABA	1·53 ± 0·25	1·25 ± 0·33**	1·40 ± 0·22*
Tau	11·37 ± 0·69	9·90 ± 1·68*	11·81 ± 1·93
Asp	2·94 ± 0·48	2·83 ± 0·50	3·37 ± 0·43*
Glu	12·07 ± 0·94	10·64 ± 1·47**	12·78 ± 1·30
Gln	3·00 ± 0·40	2·94 ± 0·70	2·80 ± 1·13
Ser	0·84 ± 0·13	1·35 ± 0·25***	1·31 ± 0·21***
Gly	0·93 ± 0·13	1·23 ± 0·30**	1·18 ± 0·32**
Ala	0·35 ± 0·13	0·59 ± 0·11***	0·88 ± 0·17***

Figures in parentheses refer to number of animals. All values expressed as μmoles/g fresh cortex ± S.D.
* $P = 0.05$.
** $0.02 \leq P < 0.05$.
*** $P < 0.01$.

of taurine have since been confirmed in a range of experimental epilepsy (Fig. 3.1; Mutani et al., 1974, 1974a; Wheler et al., 1974, 1975a; Derouaux et al., 1973) and at least two groups have now reported successful clinical trials with taurine (Bergamini et al., 1974; Barbeau & Donaldson, 1974). It is the extremely low toxicity of taurine which has spurred the interest in its possible clinical use as an anticonvulsant, but until more clinical trials are performed its true value in the treatment of human epilepsy remains uncertain.

At present little is understood of the mode of action of taurine as an anticonvulsant, or whether its diminished level in active focal tissue is a critical factor in generating the epileptogenic state. Systemic injections of [^{35}S]-taurine into normal rats show that taurine slowly penetrates the brain but there is little change in the overall brain content (Wheler et al., 1975a; Urquhart et al., 1974). However, since the blood–brain barrier is likely to be ineffective in the region of an epileptic focus, taurine probably enters the brain with relative ease at this point in epileptic animals (cf. folate, Obbens, 1973) and this may account for the restoration of normal tissue taurine levels in the focal region (van Gelder, 1972). There is evidence that taurine is able to regulate the excitability of the myocardium (in which it is highly concentrated) by reducing the loss of potassium which occurs under various conditions such as glycoside treatment or electrical stimulation (Chazov et al., 1973; Read & Wiley, 1963), and incubation of sarcoplasmic reticulum with taurine increased its ability to bind calcium (Huxtable & Bressler, 1973). Regulation of the movements of potassium and calcium across neuronal membranes might be part of the normal function of taurine and also underlie its therapeutic action.

VI Folate and epilepsy

Over the past 15 years observations have accumulated which indicate that some connection might exist between folic acid and the epileptic condition. Folate coenzymes are principally involved in 'one-carbon' transfers and are essential for taurine synthesis, pyrimidine nucleotide synthesis, certain amino acid interconversions and formate production and utilization.

Anticonvulsants were found to lower serum folate in proportion to their dosage and sometimes precipitated megaloblastic anaemia which is classically associated with folate deficiency and is probably linked to its role in nucleic acid synthesis (Klipstein, 1964; Reynolds et al., 1966). Diphenyl hydantoin, phenobarbital and primidone are particularly effective in this respect (Houben et al., 1971), and their anticonvulsant action may be related to their antifolate effect (Reynolds, 1974). Administration of folate to epileptic patients to counteract these effects has been reported to increase fit frequency but this conclusion remains contentious (Chanarin et al., 1960; Dennis & Taylor, 1969; Jenson & Olesen, 1970; Richens, 1971). In experiments with laboratory mammals small (0·5–1·0 mg/Kg) intracerebral or large (225–625 mg/Kg) systemic injections of folate or its derivatives certainly cause convulsions (Hommes & Obbens, 1972; Baxter et al., 1973; Bradford & Dodd, 1975) and at sub-threshold dose for normal animals (0·05–9·25 mg/Kg) can activate existing foci in experimental animals such as those produced by cobalt metal implants (Hommes & Obbens, 1972; Bradford & Dodd, 1975). Thus although there appears to be an effective blood–brain barrier to folate and its derivatives (Rall et al., 1960; Blair, 1970) epileptic foci (and also heat lesions) seem to produce regions where this barrier is breached, as was elegantly demonstrated by Obbens (1973). Actually, in spite of a restriction on entry to the brain, the CSF appears to concentrate folate and normally displays levels 4-fold higher than serum (Herbert & Zalusky, 1961;

Boykin & Hooshmand 1970). Moreover, folate has been reported to be greatly concentrated in experimental cobalt lesions (Meyersdorf et al., 1971).

All these observations prompt the question of whether folate is widely involved in seizure precipitation mechanisms, and whether any such involvement is due to a direct effect of folate itself, or is through derangement of a metabolic pathway which involves folate. At present no unequivocal answers are available. However, a striking feature of the folate molecule is the presence of a chain of

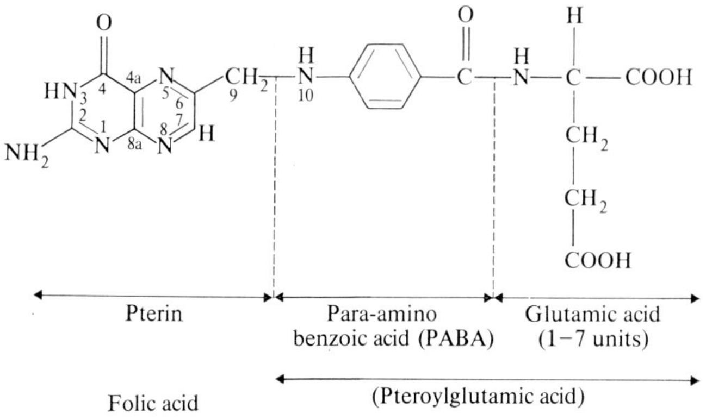

Fig. 3.9. The folate molecule.

1–7 glutamate residues (Fig. 3.9) which might allow folate to mimic the convulsant effect of glutamate itself. Another possibility is that this structural affinity with glutamate might make the folate molecule an effective glutamate-uptake blocker. Folate accumulating locally in the brain would therefore allow glutamate to accumulate in the extracellular space until it reached the threshold for convulsion. There is already one report of the inhibition of glutamate transport by folate (Roberts, 1974) and although its potency is not marked, it is in a range consistent with this hypothesis.

In brain, folate is found localized in synaptosome fractions where it is membrane-bound (Bridges & McClain, 1972). It exists here mainly as 5-methyl-tetrahydrofolate and its polyglutamate derivatives suggesting involvement of these compounds in neurotransmission. Abnormal functioning at this level would provide another mode of involvement of folate in epilepsy.

VII Summary and speculation

It is the amino acids rather than acetylcholine or the monoamines which are currently emerging as the compounds of special interest in relation to epilepsy. High-lighted among these are glutamate, GABA and taurine and speculation continues on their mode of involvement. Our own view has for some years focused on the excitatory properties of glutamic acid. Its high tissue content appears to have stimulated the evolution of high-affinity transport processes which ensure a very low concentration of glutamate in the region of synapses operated by this amino acid. Changes which reduce the efficiency of this transport activity through inhibition or oversaturation will lead first to raised levels of extracellular glutamate, and then to neuronal hyperactivity followed by cathodal block (cf. spreading depression, Section V, F) depending on the levels reached. Golgi staining reveals the presence of dendritic deafferentiation in both human and experimental epilepsy and this could occur on the hyperactive neurons detected in these foci (see Section III, B 1 and 4). Should such deafferentiation involve the widespread appearance of active glutamate receptors on the dendritic surfaces of epileptogenic cortical neurons, as occurs in denervated insect muscle (Usherwood, 1969; Cull-Candy & Usherwood, 1973), supersensitivity to glutamate would ensue and might explain the abnormally high synaptic activity which appears to generate the paroxysmal depolarization shifts commonly observed (Section III, B 1). This situation would allow even low levels of extracellular glutamate to become convulsive, and should folate accumulation in epileptic foci be a widespread phenomenon (Section VI) glutamate transport block could promote glutamate accumulation locally. The capacity of non-synaptic regions of nerves to release glutamate (Section V, E 2) suggests that this ion may be held within neurones by the membrane potential and briefly effluxed from both non-synaptic and synaptic regions during depolarizing wave of the action potential. If this were so, the apparently random and episodic occurrence of many epileptic fits might be explained by the glutamate-releasing effect of certain activity patterns in the region of the epileptic focus. Damaged neurones present in the active focus and displaying lowered membrane potential would more readily reach the threshold level at which glutamate was released. However, so simple a model does not encompass the many other factors described above, including the observed changes in taurine and aspartate. Accommodating all the strands of current knowledge of the pathophysiology and biochemistry of human and experimental epilepsy in attractive and testable hypotheses for 'seizure-initiating' mechanisms remains an impossible task. The period of accumulation, consolidation and digestion of facts must continue, and current hypotheses are likely to be short lived.

References

AGRAWAL H.C., DAVISON A.N. & KACZMAREK L.K. (1971) Subcellular distribution of taurine and cysteine sulphinate decarboxylase in developing rat brain. *Biochem J.* **122**, 759–763.

ALBERICI M., DE LOPEZ ARNAIZ R.G. & DE ROBERTIS E. (1969) Glutamic acid decarboxylase inhibition and ultrastructural changes by the convulsant drug allylglycine. *Biochem Pharmacol.* **18**, 137–143.

AJMONE-MARSAN C. (1969) Acute effects of topical epileptogenic agents. In Jasper

H.H., Ward A.A. and Pope A. (Eds.) *Basic Mechanisms of the Epilepsies*. London, Churchill.

BALAZS R. & CREMER J.E. (1973) (Eds.) *Metabolic Compartmentation in the Brain*. London, Macmillan.

BALAZS R., DAHL D. & HARWOOD J.F. (1966) Subcellular distribution of enzymes of glutamate metabolism in brain. *J. Neurochem.* **13**, 897–905.

BARBEAU A. & DONALDSON J. (1974) Zinc, taurine and epilepsy. *Arch. Neurol.* **30**, 52–58.

BATES J.A.U. (1962) The surgery of epilepsy. In Williams, D. (Ed.) *Modern Trends in Neurology* 3, London, Butterworths.

BAXTER M.G., MILLER A.A. & WEBSTER R.A. (1973) Some studies on the convulsant action of folic acid. *Brit. J. Pharmacol.* **48**, 350–351.

BERGAMINI L., MUTANI R., DELSEDIME M. & DURELLI L. (1974) First clinical experience on the antiepileptic action of taurine. *Europ. Neurol.* **11**, 261–269.

BERL S., PURPURA D.P., GIRADO M. & WAELSCH H. (1959) Amino acid metabolism in epileptogenic and non-epileptogenic lesions of the neocortex (cat). *J. Neurochem.* **4**, 311–317.

BERL S., PURPURA D.P., GONZALEZ-MONTEAGUDO O. & WAELSCH H. (1960) In Roberts E (Ed.) *Inhibition in the CNS and GABA*, pp. 445–453. Oxford, Pergamon.

BERL S., TAKAGAKI G. & PURPURA D.P. (1961) Metabolic and pharmacological effects of injected amino acids and ammonia on cortical epileptogenic lesions. *J. Neurochem.* **7**, 198–209.

BHAGARAN H.N., COURSIN D.B. & STEWART C.N. (1971) Monosodium glutamate induces convulsive disorders in the rat. *Nature, London,* **232**, 275–276.

BLAIR J.A. (1970). Toxicity of folic acid. *Lancet* i, 360.

BLANSHARD K.C., BRADFORD H.F., DODD P.R. & THOMAS A.J. (1975) A simple semi-automatic apparatus for the purification of amino acid extracts from tissues and physiological fluids. *Anal. Biochem.*, **67**, 233–244.

BONAVENTURE N., WIOLAND N. & MANDEL P. (1974) Antagonists of the putative inhibitory transmitter of taurine and GABA in the retina. *Brain Res.* **80**, 281–289.

BOWEN F.P., KARPIAK S.E., DEMIRJIAN C. & KATZMAN R. (1975) Sprouting of noradrenergic nerve-terminals subsequent to freeze-lesions of rabbit cerebral cortex. *Brain Res.* **83**, 1–14.

BOYKIN M.E. & HOOSHMAND H. (1970) CSF and serum folic acid and protein changes with diphenyl hydantoin treatment: laboratory and clinical correlations. *Neurology* **20**, 403.

BRADFORD H.F. (1968) Carbohydrate and energy metabolism. In Davison A.N. & Dobbing J. (Eds.) *Applied Neurochemistry*, pp. 222–250. Oxford, Blackwell.

BRADFORD H.F. (1975) Isolated nerve-terminal as an *in vitro* preparation for the study of dynamic aspects of transmitter metabolism and release. In Iversen & Snyder (Eds.) *Handbook of Psychopharmacology*, Vol. 1. New York, Plenum.

BRADFORD H.F. & DODD P.R. (1975) Convulsive effects of amino acids and folic acid. In preparation.

BRADFORD H.F. & McILWAIN H. (1966) Ionic basis for depolarization of cerebral tissues by excitatory acidic amino acids. *J. Neurochem.* **13**, 41–51.

BRADFORD H.F. & THOMAS A.J. (1969) Metabolism of glucose and glutamate by synaptosomes from mammalian cerebral cortex. *J. Neurochem.* **16**, 1495–1504.

BRADFORD H.F. & WARD H.K. (1975). On glutaminase activity in synaptosomes. *Brain Res.* In press.

BRIDGERS, W.F. & MCCLAIN L.D. (1972) Role of Vitamin B_6 in neurobiology. In

Costa E. & Greengard P. (Eds.) *Adv. Biochem. Psychopharmacol.* **4**, 81–92. New York, Raven Press.

BROWN D.J. & STONE W.E. (1973) Glutamate and aspartate in cortical subarachnoid fluid in relation to convulsive activity. *Brain Res.* **54**, 143–148.

BROWN W.J. (1973) Structural substrates of seizure foci in the human temporal lobe. In Brazier M.A.B. (Ed.) *Epilepsy. Its Phenomena in Man*. London, Academic Press.

CHANARIN I., LAIDLAW J., LONGHRIDGE L.W. & MOLLIN D.L. (1960) The convulsant action of therapeutic doses of folic acid. *Brit. Med. J.* **1**, 1099–1105.

CHEN G., ENSOR C.R. & BOHNER B. (1954) A facilitative action of reserpine on the CNS. *Proc. Soc. exp. Biol. Med.* **86**, 507–510.

CHUSID J. & KOPELOFF L. (1962) Epileptogenic effects of pure metals implanted in motor cortex of monkeys. *J. Appl. Physiol.* **17**, 697–700.

CLAYTON P.R. & EMSON P.C. (1975) Changes in monoamine related enzymes in cobalt-induced epilepsy. *Biochem. Soc. Trans.*, in press.

COLASANTI B.K. & CRAIG C.R. (1973) Brain concentrations and synthesis rates of biogenic amines during chronic cobalt experimental epilepsy in the rat. *Neuropharmacology* **12**, 221–231.

COLLEY L., FOX F.R. & HUGGINS A.K. (1974) The effect of changes in the external salinity on the non-protein nitrogenous constituents of parietal muscle from *Agonus cataphractus*. *Comp. Biochem. Physiol.* **48A**, 757–763.

COLLINS A., POSNER J. & PLUM F. (1970). Cerebral energy metabolism during electroshock seizures in mice. *Am. J. Physiol.* **218**, 943–952.

COLLINS G.G.S. (1973) Effect of amino oxyacetic acid, thiosemicarbazide and haloperidol on the metabolism and half-lives of glutamate and GABA in rat brain. *J. Neurochem.* **22**, 101–111.

COLLINS R.L. (1972) Audiogenic Seizures. In Purpura D.P., Perry J.K., Tower D., Woodbury D.M. & Walter R. (Eds.) *Experimental Models of Epilepsy*, pp. 347–372. New York, Raven Press.

CORRIOL J., PAPY J., ROHNER J. & JOANNY P. (1969) Electroclinical correlations established during tetani manifestations induced by parathyroid removal in the dog. In Gastaut H., Jasper H., Baucaud J. & Waltreggy A. (Eds.) *Physiopathogenesis of the Epilepsies*. Springfield, Illinois, C. C. Thomas.

CORSELLIS J.A.N. (1970) The Neuropathology of Temporal Lobe Epilepsy. In Williams D. (Ed.) *Modern Trends in Neurology* **5**, p. 254. London, Butterworths.

CRABAI F., SITZIA A. & PEPEU G. (1974) Taurine concentration in the neurohypophysis of different animal species. *J. Neurochem.* **23**, 1091–1092.

CRAIG C.R. & HARTMANN E.R. (1973) Concentration of amino acids in the brain of cobalt-epileptic rat. *Epilepsia* **14**, 409–414.

CROSSLAND J. & TURNBULL M.J. (1972) Gamma-aminobutyric acid and the barbiturate abstinence syndrome in rats. *Neuropharmacol.* **11**, 733–738.

CULL-CANDY S.G. & USHERWOOD P.N.R. (1973) Two populations of L-glutamate receptors on locust muscle fibres. *Nature New Biol.* **246**, 62–64.

CURRIE S., HEATHFIELD K.W.G., HENSON R.A. & SCOTT D.F. (1971) Clinical course and prognosis of temporal lobe epilepsy—a survey of 666 patients. *Brain* **94**, 173–190.

CURTIS D.R. & JOHNSTON G.A.R. (1974) Amino acid transmitters in the mammalian CNS. *Ergebnisse der Physiologie* **69**, 97–188.

CURTIS D.R., HOSLI L., JOHNSTON G.A.R. & JOHNSTON I.H. (1968) The hyperpolarization of spinal motoneurones by glycine and related amino acids. *Exp. Brain Res.* **5**, 235–258.

CURTIS D.R., HOSLI L. & JOHNSTON G.A.R. (1968) A pharmacological study of the depression of spinal neurones by glycine and related amino acids. *Exp. Brain Res.* **6**, 1–18.

DAVISON A.N. & KACZMERAK L.K. (1971) Taurine—a possible neurotransmitter? *Nature, London*, **234**, 107–108.

DAWSON G.D. & HOLMES O. (1966) Cobalt applied to the sensorimotor area of the cortex cerebri of the rat. *J. Physiol., London*, **185**, 455–470.

DE BELLEROCHE J.S. & BRADFORD H.F. (1973) Amino acids in synaptic vesicles from mammalian cerebral cortex: a reappraisal. *J. Neurochem.* **21**, 441–451.

DE FEUDIS F.V. & ELLIOT K.A.C. (1968) Convulsions and the GABA content of rat brain. *Canad. J. Physiol. Pharmacol.* **46**, 803–804.

DE FEUDIS V. (1971) Effects of electrical stimulation on the efflux of L-glutamate from peripheral nerve *in vitro*. *Exp. Neurol.* **30**, 291–296.

DE LA TORRE J.C. & MULLAN S. (1970) A possible role for 5-HT in drug-induced seizures. *J. Pharm. Pharmac.* **22**, 858–859.

DE LOREZ ARNAIZ G.R., ALBERICI DE CANAL M. & DE ROBERTIS E. (1971) 2 amino-4-pentonoic acid (allyglycine): a proposed tool for the study of GABA mediated systems. *Int. J. Neurosci.* **2**, 137–144.

DE LOREZ ARNAIZ G.R., ALBERICI DE CANAL M. & DE ROBERTIS E. (1972) Alteration of the GABA system and Purknje cells in rat cerebellum by the convulsant 3-mercaptopropionic acid. *J. Neurochem.* **19**, 1379–1386.

DE LOREZ ARNAIZ G.R., ALBERICI DE CANAL M., ROBIOLO B. & MISTRORGIO DE PACHEO M. (1973) The effects of the convulsant 3-mercaptopropionic acid on enzymes of the γ-amino butyrate system in rat cerebral cortex. *J. Neurochem.* **21**, 615–724.

DE ROPP R.S. & SNEDEKER E.H. (1961) Effects of drugs on amino acid levels in brain: Excitants and Depressants. *Proc. Soc. Exp. Biol. Med.* **106**, 696–700.

DENNIS J. & TAYLOR D.C. (1969) Epilepsy and folate deficiency. *Brit. Med. J.* **4**, 807–808.

DERONAUX M., PUIL E. & NAQUET R. (1973) Antiepileptic effect of taurine in photosensitive epilepsy. *Electroencephalogr. Clin. Neurophysiol.* **34**, 770.

DODD P.R. & BRADFORD H.F. (1974a) Release of amino acids from the chronically superfused mammalian cerebral cortex. *J. Neurochem.* **23**, 289–292.

DODD P.R. & BRADFORD H.F. (1975) The release of amino acids from a maturing epileptogenic focus. In preparation.

DODD P.R., PRITCHARD M.J., ADAMS R.C.F., BRADFORD H.F., HICKS G. & BLANSHARD K.C. (1974) A method for the continuous, long-term superfusion of the cerebral cortex of unanaesthetized, unrestrained rats. *Journal of Physics E: Scientific Instruments* **7**, 897–901.

DOW R., GUARDIOLA A. & MANNI E. (1962) The production of cobalt experimental epilepsy in the rat. *Electroencephelogr. Clin. Neurophysiol.* **14**, 399–407.

DOW R.C., MCQUEEN J.K. & TOWNSEND H.R.A. (1972) The production and detection of epileptogenic lesions in rat cerebral cortex. *Epilepsia* **13**, 459–465.

DUNN A. & GIUDITTA A. (1971) A long-term effect of electroconvulsive shock on the metabolism of glucose in the mouse brain. *Brain Res.* **27**, 418–421.

DUNN A., GUIDDITA A. & PAGLIUCA N. (1972) The effects of electroconvulsive shock on protein synthesis in mouse brain. *J. Neurochem.* **18**, 2093–2099.

EARLE K.M., BALDWIN M. & PENFIELD W. (1953) Incisural sclerosis and temporal lobe seizures produced by hippocampal herniation at birth. *Archs. Neurol. Psychiat.* **69**, 27–42.

EMSON P.C. & JOSEPH M.H. (1975) Neurochemical and morphological changes during the development of cobalt-induced epilepsy in rat. *Brain Res.* In press.
ESCUETA A.V. & APPEL S.H. (1969) Biochemical studies of Synapses II. K transport. *Biochemistry* 8, 725–733.
ESCUETA A.V. & APPEL S.H. (1970) The effects of electrically induced seizures on K transport within isolated nerve-terminals. *Neurology* 20, 392.
ESCUETA A.V. & APPEL S.H. (1970a) Diphenylhydantoin and K transport in isolated nerve-endings. *J. Clin. Invest.* 49, 27a–28a.
ESCUETA A.V. & REILLY E.C. (1971) The effects of diphenylhydantoin on Potassium transport within synaptic terminals of the epileptogenic foci. *Neurology* 21, 418.
ESSMAN W.V. (1973) *Neurochemistry of Electroshock.* Flushing, New York, Spectrum Publications Inc.
FALCONER M.A. (1971) Genetic and related aetiological factors in temporal lobe epilepsy. *Epilepsia* 12, 13–31.
FALCONER M.A., SERAFETINIDES E.A. & CORSELLIS J.A.N. (1964) Aetiology and pathogenesis of temporal lobe epilepsy. *Archs. Neurol.* 10, 232–248.
FALCONER M.A. & TAYLOR D.C. (1968) Surgical treatment of drug-resistant temporal lobe epilepsy due to medial temporal sclerosis: Etiology and significance. *Archs. Neurol.* 18, 353–361.
FERGUSON J.H. & JASPER H.H. (1971) Laminar DC studies of acetylcholine-activated epileptiform discharge in cerebral cortex. *Electroencephalogr. Clin. Neurophysiol.* 30, 377–390.
FERGUSON J.H. & CORNBLATH D.R. (1974) Acetylcholine epilepsy: modification of DC shift in chronically undercut cat cortex. *Electroencephalogr. Clin. Neurophysiol.* In press.
FERRIER D. (1876) The Functions of the Brain. London: Smith Elder & Co. Facsimile reprint by Dawsons of Pall Mall, London 1966.
FISCHER J., HOLUBAR J. & MALIK V. (1967) A new method of producing chronic epileptogenic cortical foci in rats. *Physiol. Bohem* 16, 272–277.
GASTAUT H., NAQUET R., MEYER A., CAVANAGH J.B. & BECK E. (1959) Experimental psychomotor epilepsy in the cat. Electroclinical and anatomopathological correlations. *J. Neuropathol. Exp. Neurol.* 18, 270–293.
GIBSON I.M. & MCILWAIN H. (1965) Continuous recording of changes in membrane potential in mammalian cerebral tissues *in vitro*; recovery after depolarization by added substances. *J. Physiol.* 176, 261–283.
GODDARD G.V., MCINTYRE D.C. & LEECH C.K. (1969) A permanent change in brain function resulting from daily electrical stimulation. *Exp. Neurol.* 25, 295–330.
GOTTESFELD Z. & ELAZAR Z. (1972) GABA and glutamate in different EEG stages of the penicillium focus. *Nature, London*, 240, 478–479.
GREEN J.D., CLEMENTS C.D. & DE GROOT J. (1957) Experimentally Induced Epilepsy in the Cat with Injury of Cornu Ammonis. *Archs. Neurol. Psychiat.* 78, 259–263.
GUERRERO-FIGUEROA R., DE BALBIAN-VERSTER F., BARROS A. & HEATH R.G. (1964) Cholinergic mechanisms in subcortical mirror focus and effects of topical application of GABA and acetylcholine. *Epilepsia* 5, 140–155.
HAAS H.L. & HÖSLI L. (1973) The depression of brain stem neurons by taurine and its interaction with strychnine and bicuculline. *Brain Res.* 52, 399–402.
HABER C. & SIDEL L. (1948) Glutamic acid in neural activity. *Fed. Proc.* 7, 47.
HAGGENDAL H. (1968) Effect of hyperbaric oxygen on monoamine metabolism in central and peripheral tissues of rat. *Eur. J. Pharmacol.* 2, 323–325.

HANSEN S., PERRY T.C., WADA J.A. & SOKOL M. (1973) Brain amino acids in baboons with light-induced epilepsy. *Brain Res.* **50**, 480–483.

HARMONY T., URBA-HOLMGREN R., URBAY C.M. & SZAVA S. (1968) (Na-K)—ATPase activity in experimental epileptogenic foci. *Brain Res.* **11**, 672–680.

HARVEY P.K.P., BRADFORD H.F. & DAVISON A.W. (1975) Further studies on the mode of action of n-dipropyl acetate. In preparation.

HARVEY P.K.P., BRADFORD H.F. & DAVISON A.N. (1975) The inhibitory effects of Sodium dipropyl acetate on the degradative enzymes of the GABA shunt. *FEBS letters*, **52**, 251–254.

HAWKINS & SARETT (1957) On the efficacy of asparagine, glutamine, CABA and 2-Pyrrolidinone in preventing chemically induced seizures in mice. *Clin. Chim. Acta* **2**, 481.

HAYASHI T. (1954) The effects of Sodium glutamate on the nervous system. *Keio J. Med.* **3**, 183–192.

HAYASHI T. (1966) GABA and its derivatives in mental health. In Martin G.J. & Kisch B. (Eds.) *Enzymes in Mental Health*, pp. 160–170. Philadelphia, J.B. Lippincott and Co.

HENNECKE A. & WEICHERT G. (1970) Seizures and the dose of L-glutamic acid in rats. *Epilepsia* **11**, 327–331.

HERBERT V. & ZALUSKY R. (1961) Selective concentration of folic acid activity in CSF. *Fed. Proc.* **20**, 453.

HOKFELT T. & LJUNGDAHL A. (1972) Autoradiographic identification of cerebral and cerebellar cortical neurons accumulating labelled GABA. *Exp. Brain Res.* **14**, 354–362.

HOKFELT T. & LJUNGDAHL A. (1972a) Application of cytochemical techniques to the study of suspected transmitter substances in the Nervous System. *Advances in Biochemical Psychopharmacology* **6**, 1–36.

HOMMES O.R. & OBBENS E.A.M.T. (1972) The epileptogenic action of Na-folate in the rat. *J. neurol. Sci.* **16**, 271–281.

HOUBEN P.F.M., HOMMES O.R. & KNAVEN P.J.H. (1971) Anticonvulsant drugs and folic acid in young mentally retarded epileptic patients. *Epilepsia* **12**, 235–247.

HREBICEK J., KOLOUSEK J., WEDERMAN M., & CHARAMZA O. (1971) Changes in the incorporation of (^{75}Se)-methionine and of the electrical activity in various brain structures of the cat after administration of methionine sulphoximine. *Brain Res.* **28**, 109–117.

HUNT W.A. & CRAIG C.R. (1973) Alterations in Cation levels and Na-K ATPase activity in rat cerebral cortex during the development of cobalt-induced epilepsy. *J. Neurochem.* **20**, 559–567.

IVERSEN L.L. (1971) Role of tissue uptake mechanisms in synaptic neurotransmission. *Brit. J. Pharmacol.* **41**, 571–591.

IVERSEN L.L., KELLY J.S., MINCHIN M., SCHON F. & SNODGRASS S.R. (1973) Role of amino acids and peptides in synaptic transmission. *Brain Res.* **62**, 567–576.

JASPER H.H. (1969) Mechanisms of propagation: Extracellular studies. In Jasper H.H., Ward A.A. & Pope A. (Eds.) *Basic Mechanisms of the Epilepsies*. London, Churchill.

JASPER H.H. & KOYAMA I. (1969) Rate of release of amino acids from the cerebral cortex in the cat as affected by brain stem and thalamic stimulation. *Canad. J. Physiol. Pharmacol.* **47**, 889–905.

JASPER H.H., WARD A.A. & POPE A. (Eds.) (1969) *Basic Mechanisms of the Epilepsies*. London, Churchill.

JEAVONS P.M. & CLARK J.E. (1974) Sodium valproate in treatment of epilepsy. *Brit. Med. J.* **ii**, 584–586.

JENSEN O.N. & OLESEN O.V. (1970) Subnormal serum folate due to anticonvulsive therapy. *Arch. Neurol.* **22**, 181–182.

JOHNSTON G.A.R. (1972) Convulsions induced in 10-day-old rats by intraperitoneal injection of monosodium glutamate and related excitant amino acids. *Biochem. Pharmacol.* **22**, 137–139.

JOHNSTON G.A.R. & BALCAR V.J. (1974) Amino oxyacetic acid: a relatively non-specific inhibitor of uptake of amino acids and amines by brain and spinal cord. *J. Neurochem.* **22**, 609–610.

JONES D.A. & McILWAIN H. (1971) Amino acid distribution and incorporation into proteins in isolated electrically stimulated cerebral tissues. *J. Neurochem.* **18**, 41–51.

KACZMERAK L.K. & DAVISON A.N. (1972) Uptake and release of taurine from rat brain slices. *J. Neurochem.* **19**, 2355–2362.

KACZMERAK L.K. & ADEY W.R. (1974) Some chemical and electrophysiological effects of glutamate in the cerebral cortex. *J. Neurobiol.* **5**, 231–241,

KETY S.S., JAVOY F., THIERRY A.M., JULON L. & GLOWINSKI J. (1967) A sustained effect of electro convulsive shock on the turnover of norepinephrine in the central nervous system of the rat. *Proc. Nat. Acad. Sci.* **58**, 1249–1254.

KILLAM K.F. & BAIN J.A. (1957) Convulsant hydrazides I: *in vitro* and *in vivo* inhibition of vitamin B_6 enzymes by convulsant hydrazides. *J. Pharmacol. exp. Therap.* **119**, 255–262.

KING L.J., CARL J.L. & LAO L. (1974) Brain amino acids during convulsions. *J. Neurochem.* **22**, 307–310.

KLIPSTEIN F.A. (1964) Subnormal serum folate and macrocytosis associated with anticonvulsants during therapy. *Blood* **23**, 68–78.

KOPELLOFF L.M. & CHUSID J.G. (1967) Indoklon and metrazol convulsions in epileptic and control monkeys. *Int. J. Neuropsychiat.* **3**, 174–178.

KOPELLOFF L.M., CHUSID J.G. & KOPELLOFF N. (1954) Chronic experimental epilepsy in Macaca mulatta. *Neurology* **4**, 218–227.

KOYAMA I. (1972) Amino acids in the cobalt-induced epileptogenic and non-epileptogenic cat's cortex. *Canad. J. Physiol. Pharmacol.* **50**, 740–752.

KRNJEVIC K. & PHILLIS J.W. (1963) Iontophoretic studies of neurones in the mammalian cerebral cortex. *J. Physiol., London,* **165**, 274–304.

KRNJEVIC K. (1974) The chemical nature of synaptic transmission in vertebrates. *Physiol. Rev.* **54**, 418–540.

KUFFLER S.W. (1967) Neuroglial cells: physiological properties and a potassium mediated effect of neuronal activity on the glial membrane potential. *Proc. Roy. Soc. B,* **168**, 1–91.

LAMAR C. (1970) Mercaptopropionic acid: a convulsant that inhibits glutamate decarboxylase. *J. Neurochem.* **17**, 165–170.

LEAO A.A.P. (1944) Spreading a depression of activity in the cerebral cortex. *J. Neurophysiol.* **7**, 359–390.

LEE L.W., LIAO G.L. & YATSU F..M (1974) The effects of GABA on brain mitochondrial ATP synthesis. *J. Neurochem.* **23**, 721–724.

LEHMAN A. (1967) Audiogenic seizure data in mice supporting new theories of biogenic amine mechanisms in the CNS. *Lief Sci.* **6**, 1423–1431.

LEONARD B.E. & PALFREYMAN M.G. (1972) Effect of experimentally induced seizures on some amino acids and ammonia in rat brain. *Biochem. Pharmacol* **21**, 1206–1209.

LEWIN E. (1972) The production of epileptogenic cortical foci in experimental animals by freezing. In Purpura D.P., Perry J.K., Tower D.B., Woodbury D.M. & Walter R. (Eds.) *Experimental Models of Epilepsy*. New York, Raven Press.

LEWIN E. & McCRIMMON A. (1967) ATPase activity in discharging cortical lesions induced by freezing. *Arch. Neurol.* **16**, 321–325.

LEWIN E. & McCRIMMON A. (1968) The intralaminar distribution of sodium-potassium-ATPase activity in discharging cortical lesions induced by freezing. *Brain Res.* **8**, 291–297.

LINDSAY J.M.N. (1971) Genetics and epilepsy: a model from critical path analysis. *Epilepsia* **12**, 47–54.

LOGAN W.M. & SNYDER S.H. (1971) Unique high affinity uptake systems for glycine, glutamic acid and aspartate in CNS tissue of the rat. *Nature (London)* **234**, 297–299.

MACHIYAMA Y., BALAZS R. & JULIAN T. (1965) Oxidation of glucose through the GABA pathway in brain. *Biochem. J.* **96**, 688.

MANGAN J.L. & WHITTAKER V.P. (1965) The distribution of free amino acids in subcellular fractions in guinea-pig brain. *Biochem. J.* **98**, 128–137.

MARCUS E.M. (1972) Experimental models of Petit Mal epilepsy. In Purpura D.P., Perry J.K., Tower D.B., Woodbury D.M. & Walter R. (Eds.) *Experimental Models of Epilepsy*. New York, Raven Press.

MAYERSDORF A., STRIEFF R.R., WILDER B.J. & HAMMER R.H. (1971) Folic acid and vitamin B_{12} alterations in primary and secondary epileptic foci induced by metallic cobalt powder. *Neurology* **21**, 417.

MELDRUM B.S. (1975) Present views on hippocampal sclerosis and epilepsy. In Williams D. (Ed.) *Modern Trends in Neurology*, **10**, pp. 223–237. London, Butterworths.

MELDRUM B.S. & BALZAMO E. (1972) Epilepsy in the photosensitive baboon. *Papio papio*, and drugs acting on cerebral GABA and serotonin mechanisms. In Goldsmith E-I & Moor-Jankowski J. (Eds.) *Medical Primatology*. Basel, Karger.

MELDRUM B.S., BALZAMO E., HORTON R.W., LEE G. & TRIMBLE M. (1975) Photically-induced epilepsy in *Papio papio* as a model for drug studies. In Meldrum B.S. & Marsen C.D. *Primate Models of Neurological Disorders*. New York, Raven Press.

MELDRUM B.S., HORTON R.W. & BRIERLY J.B. (1974) Epileptic brain damage in adolescent baboons following seizures induced by allylglycine. *Brain* **97**, 407–418.

MELDRUM B.S., HORTON R.W. & TOSELAND P.A. (1975a) A primate model for the acute testing of anticonvulsant drugs. *Arch. Neurol.* in press.

MELDRUM B.S., NAQUET R. & BALZAMO E. (1970) Effects of atropine and eserine on the EEG, behaviour and photically-induced epilepsy in the adolescent baboon (*Papio papio*). *Electroencephalogr. clin. Neurophysiol.* **31**, 563–572.

MEYER J.S., KANDA T., SHINOHARA Y. & FUKUNCHI Y. (1970) Changes in cerebrospinal fluid sodium and potassium concentrations during seizure activity. *Neurology* **20**, 1179–1184.

MORRELL F. (1960) Secondary epileptogenic lesions. *Epilepsia* **1**, 538–560.

MUTANI R. (1967) Cobalt experimental amygdaloid epilepsy in the cat. *Epilepsia* **8**, 73–92.

MUTANI R. (1967a) Cobalt experimental hippocampal epilepsy. *Epilepsia* **8**, 223–240.

MUTANI R., BERGAMINI L., DELSEDIME M. & DURELLI L. (1974) Effects of taurine in chronic experimental epilepsy. *Brain Res.* **79**, 330–332.

MUTANI R., BERGAMINI L., FARIELLO R. & DELSEDIME M. (1974a) Effects of taurine on cortical acute epileptic foci. *Brain Res.* **70**, 170–173.

NAHORSKY S.R., ROBERTS D.J. & STEWART G.G. (1970) Some neurochemical aspects of pentamethylenetetrazole convulsive activity in rat brain. *J. Neurochem.* **17**, 621–631.

NAQUET R. (1969) Photogenic seizures in the baboon. In Jasper H.H., Ward A.A. & Pope A. (Eds.) *Basic Mechanisms of the Epilepsies.* London, Churchill.

NAQUET R. & MELDRUM B.S. (1972) Photogenic seizures in baboons. In Purpura D., Perry J.K., Tower D., Woodbury D.M. & Walter R. (Eds.) *Experimental Models of Epilepsy.* New York, Raven Press.

NEAL M.J. & IVERSEN L.L. (1969) Subcellular distribution of endogenous and (^{3}H)-GABA in rat cerebral cortex. *J. Neurochem.* **16**, 1245–1252.

NETTHAUS G. (1970) Relationship of brain serotonin to convulsions. *Neurology* **18**, 298–299.

NEWCOMBE F. (1969) *Missile Wounds of the Brain*, p. 31. Oxford, Oxford University Press.

NIE V. & ETTLINGER G. (1974) Ablation of the primary infero-temporal epileptogenic focus in rhesus monkeys with independent secondary spike discharges. *Brain Res.* **69**, 149–152.

NIE V., MACCABE J.J., ETTLINGER G. & DRIVER M.V. (1974) The development of secondary epileptic discharges in rhesus monkey after commissure section. *Electroencephalogr. Clin. Neurophysiol.* **37**, 473–481.

OBBENS E.A.M.T. (1973) *Experimental Epilepsy induced by Folate Derivatives.* Nijmegen, Drukkerij Gebr. Janssen.

OKAMOTO K. & QUASTEL J.H. (1973) Spontaneous action potentials in isolated guinea-pig slices: effects of amino acids and conditions affecting sodium and water uptake. *Proc. Roy. Soc.* **B, 184**, 83–90.

OUNSTED C., LINDSAY J. & NORMAN R.M. (1966) Biological factors in temporal lobe epilepsy. In *Clinics in Developmental Medicine*, Vol. 22. London, Heinemann.

ORREGO F. & LIPMANN F. (1967) Protein synthesis in brain slices. *J. Biol. Chem.* **242**, 665–671.

OSBORNE R.H. & BRADFORD H.F. (1973) Tetanus toxin inhibits amino acid release from nerve-endings *in vitro. Nature New Biol.* **244**, 157–158.

PAPPIUS H. & ELLIOT K.A.C. (1954) Adenosine triphosphatase, electrolytes and oxygen uptake rates of human normal and epileptogenic cerebral cortex. *Canad. J. Biochem. Physiol.* **32**, 484–490.

PAYAN H.M. (1971) Morphology of cobalt experimental epilepsy in rats. *Exp. Mol. Pathol.* **15**, 312–319.

PERRY T.L., HANSEN S., SOKOI M. & WADA J.A. (1972) Amino acids in brain biopsies of epileptic foci. *Clin. Res.*, **20**, 949.

PERRY T.L., URQUHART N., HANSEN S. & KENNEDY J. (1974) GABA: drug-induced elevation in monkey brain. *J. Neurochem.* **23**, 443–446.

PHILLIS J.W. (1970) *The Pharmacology of Synapses.* London, Pergamon Press.

PHILLIS J.W. & OCHS S. (1971) Excitation and depression of cortical neurones during spreading depression. *Exp. Brain. Res.* **12**, 132–149.

PLUM C.M. (1974) Free amino acid levels in the CSF of normal humans and their variation in cases of epilepsy and Spielmeyer–Vogt–Batten disease. *J. Neurochem.* **23**, 595–600.

POSNER J. & PLUM F. (1968) Cerebral metabolism during electrically-induced seizures in Man. *Trans. Amer. Neurol. Ass.* **93**, 84.

PRINCE D.A. (1969) Microelectrode studies of penicillin foci. In Jasper H.H., Ward A.A. & Pope A. (Eds.) *Basic Mechanisms of the Epilepsies.* London, Churchill.

PRINCE D.A. & FUTAMACHI K.J. (1970) Intracellular recordings from chronic epileptic foci in monkey. *Electroencephalogr. Clin. Neurophysiol.* **29**, 496–510.

PRINCE D.A., LUX H.D. & NEHER E. (1973) Measurement of extracellular potassium activity in cat cortex. *Brain Res.* **50**, 489–495.

PROCKOP D.J., SHORE P.A. & BRODIE B.B. (1959) An anticonvulsant effect of monoamine oxidase inhibitors. *Experientia* **15**, 145–150.

PURPURA D.P., PERRY J.K., TOWER D.B., WOODBURY D.M. & WALTER R.D. *Experimental Models of Epilepsy.* New York, Raven Press.

RALL D.P., RIESELBACH R.E., OLIVERIO V.T. & MORSE E. (1962) Pharmacology of folic acid antagonists related to brain and CSF. *Cancer Chemother. Reports* **16**, 187–190.

RASMUSSEN T. (1969) The role of surgery in the treatment of focal epilepsy. *Clin. Neurosurg.* **16**, 288–314.

RASMUSSEN T. (1974) Cortical excision for medically refractory focal epilepsy. In HARRIS P. & MAWDSLEY C. (Eds.) *Epilepsy.* London, Churchill-Livingstone.

REYNOLDS E.H., MILNER G., MATTHEWS D.M. & CHANARIN I. (1966) Anticonvulsant therapy, megaloblastic haemopoesis and folic acid metabolism. *Quart. J. Med.* **140**, 521–537.

REYNOLDS E.H. (1974) The relationship between the antifolate and antiepileptic action of phenobarbitone, diphenylhydantoin and primidone. In Harris P. & Mawdsley C. (Eds.) *Epilepsy*, London, Churchill-Livingstone.

RICHENS A. (1971) Folic acid in epilepsy. *Brit. Med. J.* **1**, 109.

RICHTER D. & CROSSLAND J. (1949) Variation in acetylcholine content of brain with physiological state. *Am. J. Phisiol.* **159**, 247–255.

RIECHELT K.L. & EDMINSON P.D. (1974) Biogenic amine specificity of cortical peptide synthesis in monkey brain. *FEBS Letters* **47**, 185–192.

ROA P.D., TEWS J.K. & STONE W.E. (1964) A neurochemical study of thiosemicarbazide seizures and their inhibition by amino-oxyacetic acid. *Biochem. Pharmacol.* **13**, 477–487.

ROBERTS P.J. (1974) Inhibition of high-affinity glial uptake of ^{14}C-glutamate by folate. *Nature, London*, **250**, 429–430.

Rose S.P.R. (1968) Glucose and amino acid metabolism in isolated nuronal and glia cell fractions *in vitro. J. Neurochem.* **15**, 1415–1429.

SAAD S.F., ELMASRY A.M. & SCOTT P.M. (1972) Influence of certain anticonvulsants on the concentration of GABA in the cerebral hemispheres of mice. *Europ. J. Pharmacol.* **17**, 386–392.

SACKTOR B., WILSON J.E. & TIEKERT C.G. (1966) Regulation of glycolysis in brain, *in situ*, during convulsions. *J. Biol. Chem.* **241**, 5071–5075.

SAITO S. & TOKUNAGA Y. (1967) Some correlations between picrotoxin-induced seizures and GABA in animal brain. *J. Pharm. exp. Therap.* **157**, 546–554.

SALGANICOFF L. & DE ROBERTIS E. (1965) Subcellular distribution of enzymes of the glutamic acid, glutamine and GABA cycles in rat brain. *J. Neurochem.* **12**, 287–309.

SCHEIBEL M.E. & SCHEIBEL A.B. (1973) Hippocampal pathology in temporal lobe epilepsy: A Golgi survey. In Brazier M.A.B. (Ed.) *Epilepsy: Its Phenomena in Man.* New York, Academy Press.

SCHLESINGER K., BOGGAN W. & FREEDMAN D.X. (1965) Genetics of audiogenic seizures: I. Relation to brain serotonin and norepinephrine in mice. *Life Sci.* **4**, 2345–2351.

SCHLESINGER K., BOGGAN W. & GRIEK B. (1968) Pharmacogenetic correlates of penty-

lene tetrazole and electroconvulsive seizure thresholds in mice. *Psychopharmacologia* **13**, 181–188.
SCHMIDT R.P., THOMAS L.B. & WARD A.A. (1959) Microelectrode studies of chronic epileptic foci in monkey. *J. Neurophysiol.* **22**, 285–296.
SCRIVER C.R. & WHELAN D.T. (1969) Glutamic acid decarboxylase (GAD) in mammalian tissue outside the central nervous system and its possible relevance to hereditary vitamin B_6 dependency with seizures. *Ann N.Y. Acad. Sci.* **166**, 83–96.
SELLINGER O.Z. & WEILER P. (1963) The nature of the inhibition *in vitro* of cerebral glutamine synthetase by the convulsant methionine sulphoximine. *Biochem. Pharmacol.* **12**, 989–997.
SHANES A.M. (1959) Electrochemical aspects of physiological and pharmacological action in excitable cells. *Pharmacol. Rev.* **10**, 59–164.
SHARPLESS S.K. (1969) Isolated and deafferented neurons: disuse supersensitivity. In Jasper H.H., Ward A.A. & Pope A. (Eds.) *Basic Mechanisms of the Epilepsies.* London, Churchill.
SIMLER S., KANDRIANANOSA H., LEHMANN A. & MANDEL P. (1968) Effects of *n*-dipropyl acetate on audiogenic seizures in the mouse. *J. Physiol. Paris.* Suppl. **60**, 547.
SIMLER S., CIESIEISLI L., MAITRE M., RANDRIAMARIOSA H. & MANDEL P. (1973) Effect of sodium *n*-dipropylacetate on audiogenic seizures and brain γ-aminobutyric acid level. *Biochem. Pharmacol.* **22**, 1701–1708.
SNODGRASS S.R. & IVERSEN L.L. (1973) Effects of amino-oxyacetic acid on [^{3}H] GABA uptake by rat brain slices. *J. Neurochem.* **20**, 431–439.
SNODGRASS S.R. & IVERSEN L.L. (1974) Amino acid uptake into human brain tumours. *Brain Res.* **76**, 95–107.
SOKOLOFF L. (1969) Cerebral blood flow and energy metabolism. In Jasper H.H., Ward A.A. & Pope A. (Eds.) *Basic Mechanisms of the Epilepsies.* London, Churchill.
SPINKS A. & WARING W.S. (1963) Anticonvulsant drugs. *Prog. Med. Chem.* **3**, 261–331.
STONE W.E. (1957) The role of acetylcholine in brain metabolism and function. *Am. J. Phys. Med.* **36**, 222–225.
SUTHERLAND J.M. & TAIT H. (1971) *The Epilepsies.* London, Churchill-Livingstone.
TAPIA R., PEREZ DE LA MORA M. & MASSIEU G.H. (1969) Correlative changes of pyridoxal kinase, pyridoxal-5-phosphate and glutamate decarboxylase in brain, during drug-induced convulsions. *Ann. N.Y. Acad. Sci.* **166**, 257–266.
TERNER C., EGGLESTON L.V. & KREBS H.A. (1950) The role of glutamic acid in the transport of potassium in brain and retina. *Biochem. J.* **47**, 139–149.
TOWER D.B. (1957) Glutamic acid and GABA in seizures. *Clin. Chim. Acta* **2**, 397–402.
TOWER D.B. (1960) *Neurochemistry of Epilepsy.* Springfield, Illinois, Charles C. Thomas.
TOWER D.B. (1960a) Administration of GABA to man: systemic effects and anticonvulsant action. In Roberts E. (Ed.) *Inhibition in the Nervous System and GABA.* New York, Pergamon Press.
TOWER D.B. (1969) Neurochemical mechanisms. In Jasper H.H., Ward A.A. & Pope A. (Eds.) *Basic Mechanisms of the Epilepsies.* London, Churchill.
TOWER D.B. & ELLIOT K.A.C. (1952) Activity of acetyl choline system in human epileptogenic focus. *J. Appl. Physiol.* **4**, 669–676.
URQUHART N., PERRY T.L., HANSER S. & KENNEDY J. (1974) Passage of taurine into adult mammalian brain. *J. Neurochem.* **22**, 871–872.

USHERWOOD P.N.R. (1969) Glutamate sensitivity of insect muscle fibres. *Nature, London* **223**, 411–413.

VAN GELDER, N.M. & COURTOIS A. (1972) Close correlation between changing content of specific amino acids in epileptogenic cortex of cats and severity of epilepsy. *Brain Res.* **43**, 477–484.

VAN GELDER N.M., SHERWIN A.L. & RASMUSSEN T. (1972) Amino acid content of epileptogenic human brain: focal versus surrounding regions. *Brain Res.* **40**, 385–393.

VAN HARREVELD A. (1972) The extracellular space in the vertebrate CNS. In Bourne G.H. (Ed.) *Structure and Function of Nervous Tissue*. New York, Academic Press.

VAN HARREVELD A. & FIFKOVA E. (1970) Glutamate release from the retina during spreading depression. *J. Neurobiol.* **2**, 13–29.

VAN HARREVELD A. & FIFKOVA E. (1971) Effects of glutamate and other amino acids on the retina. *J. Neurochem.* **18**, 2145–2154.

VAN HARREVELD A. & FIFKOVA E. (1973) Mechanisms involved in spreading depression. *J. Neurobiol.* **4**, 375–387.

WEINSTEIN H., ROBERTS E. & KAKEFUDA T. (1963) Studies of sub-cellular distribution of γ-amino butyric acid and glutamic decarboxylase in mouse brain. *Biochem. Pharmacol.* **12**, 503–509.

WARD A.A. (1969) The epileptic neuron: chronic foci in animals and man. In Jasper H.H., Ward, A.A. & Pope A. (Eds.) London, Churchill.

WARD A.A. (1972) Topical convulsant metals. In Purpura D.P., Perry J.K., Tower D.B., Woodbury D.M. & Walter R. (Eds.) *Experimental Models of Epilepsy*. New York, Raven Press.

WATKINS J. (1972) Metabolic regulation in the release and action of excitatory and inhibitory amino acids in the CNS. In Smellie R.M.S. (Ed.) *Transmitters and Metabolic Regulation. Biochem. Soc. Symp.* **36**, 33–47.

WEDEGE E. & BRADFORD H.F. (1975) Acceleration of protein synthesis at the nerve terminal induced by depolarizing stimulation. *FEBS Letters*. In press.

WEICHERT P. & GOLLNITZ G. (1969) Metabolic investigations of epileptic seizures: the role of amino-transferases and glutamate dehydrogenase in convulsions. *J. Neurochem.* **16**, 689–693.

WEICHERT P. & GOLLNITZ G. (1970) Metabolic investigations of epileptic seizures: investigations of glutamate metabolism in regions of the dog brain in preconvulsive states. *J. Neurochem.* **17**, 137–147.

WEICHERT P. & GOLLNITZ G. (1968) Metabolic investigations of epileptic seizures: The activity of glutamate decarboxylase prior to and during experimentally produced convulsions. *J. Neurochem.* **15**, 1265–1270.

WEICHERT P. & GOLLNITZ G. (1969) Metabolic investigations of epileptic seizures: the concentration of free amino acids in cerebral tissue prior to and during cerebral seizures. *J. Neurochem.* **16**, 1007–1016.

WEICHERT P. & HERBST A. (1966) Provocations of cerebral seizures by derangement of the natural balance between glutamic acid and GABA. *J. Neurochem.* **13**, 59–64.

WEINREICH D. & HAMMERSCHLAG R. (1975) Nerve-impulse enhanced release of amino acids from non-synaptic regions of peripheral and central nerve trunks of bullfrog. *Brain Res.* **84**, 137–142.

WESTRUM L.E., WHITE L.E. & WARD A.A. (1965) Morphology of the experimental epileptic focus. *J. Neurosurg.* **21**, 1033–1044.

WHEELER D.D., BOYARSKY L.L. & BROOKS W.H. (1966) The release of amino acids from nerve during stimulation. *J. cell Physiol.* **67**, 141–147,

WHELAN D.T., SCRIVER L.R. & MOHGADDIN F. (1969) Glutamic acid decarboxylase and GABA in mammalian kidney. *Nature, London* **224**, 916–217.

WHELER G.H.T., OSBORNE R.H., BRADFORD H.F. & DAVISON A.N. (1974) Treatment of experimental epilepsy with taurine. *Biochem. Soc. Trans.* **2**, 285–286.

WHELER T., BRADFORD H.F. & DAVISON A.N. (1975) A study of the uptake and release of taurine by brain slices. In preparation.

WHELER T., BRADFORD H.F. & DAVISON A.N. (1975a) Further studies on the treatment of experimental epilepsy with taurine. In preparation.

WILDER B.J. (1972) Projection phenomena and secondary epileptogenesis: Mirror foci. In Purpura D.P., Perry J.K., Tower D., Woodbury D.M. & Walter R. (Eds.) *Experimental Models of Epilepsy*. New York, Raven Press.

WILLIAMS D. (1970) The propagation of epileptic events. In Williams D. (Ed.) *Modern Trends in Neurology*. **5**, 289–295. London, Butterworths.

WOOD J.D. & PEESKER S.J. (1974) Development of an expression which relates the excitable state of the brain to the level of GAD activity and GABA content, with particular reference to the action of hydrazine and its derivatives. *J. Neurochem.* **23**, 703–712.

WOOD J.D. & WATSON W.J. (1963) GABA levels in the brain of rats exposed to oxygen at high pressures. *Canad. J. Biochem. Physiol.* **41**, 1907–1013.

WOODBURY D.M. & KEMP J.W. (1970) Some possible mechanisms of action of anti-epileptic drugs. *Pharmakopsychiat. Neuro-psychopharmakol.* **3**, 201–228.

YOSHINO Y. & ELLIOT K.A.C. (1970) Incorporation of carbon atoms from glucose into free amino acids in the brain under normal and altered conditions. *Canad. J. Biochem.* **48**, 228–235.

ZUCKERMAN E.C. & GLASER G.H. (1970) Activation of experimental epileptogenic foci. *Arch. Neurol.* **23**, 258–364.

4 Transmitter amines in brain disease

G. Curzon

I General, 170

    A. Introduction, 170
    B. Amine metabolism, 170
    C. Brain amine neuronal systems, 174
        1. Noradrenaline, 175
        2. Dopamine, 175
        3. 5-Hydroxytryptamine, 176
        4. Adrenaline, 177
    D. Study of brain amine abnormalities in man, 179
        1. General, 179
        2. Cerebrospinal fluid, 179
        3. Brain, 182

II Parkinsonian states, 183

    A. Introduction, 183
    B. Biochemical and neuropathological findings, 185
        1. Dopamine, 185
        2. Other amines, 187
        3. Brain enzymes, 188
        4. Other biochemical changes, 189
    C. Relevant aspects of treatment—L-DOPA and other drugs, 189

III Other dyskinesias, 194

    A. Huntington's chorea, 194
    B. Drug-provoked dyskinesias, 197
    C. Manganese dyskinesia, 199

IV Psychological disorders, 199

 A. Depression, 199
 B. Schizophrenia, 204
 C. Other disorders involving mental dysfunction, 206
   1. Down's syndrome, 206
   2. Infantile autism, 206
   3. Defects of amino acid metabolism, 207

V Migraine, 208

 A. Introduction, 208
 B. Amine changes in attacks, 209
 C. Amine containing foods as precipitants, 210

VI Comments, 212

I General

A. INTRODUCTION

This chapter will deal largely with the catecholamines dopamine and noradrenaline and the indolealkylamine 5-hydroxytryptamine (serotonin). It is similar in subject and form to a previous review published three years ago (Curzon, 1972). The rapid development of the field is indicated by the reference list as more than 60% of it was not included in the previous article.

The principal amines to be discussed have the following characteristics consistent with transmitter action in the brain.

(a) They are present and synthesized within specific neuronal tracts.
(b) They are released at nerve terminals following stimulation of cell bodies.
(c) They are retaken up into terminals from the synaptic cleft.
(d) They activate or inhibit specific neuronal populations.
(e) Alteration of their concentration or disposition affects brain function.

An enormous literature exists on the metabolism of these substances and their neurophysiological, behavioural and pathological relationships. Only those topics which are more relevant to the understanding of evidence of amine involvement in brain disease will be discussed or indicated here. Broader accounts are available, e.g. Bacq, 1971; Ciba, 1974; Costa et al., 1974a, b; Usdin and Snyder, 1973. Mechanisms of control of brain amine metabolism have been reviewed by Mandell (1973) and Costa and Meek (1974).

B. AMINE METABOLISM

The amines under consideration are synthesized *in vivo* from the aromatic L-amino acids L-tyrosine and L-tryptophan. Tryptophan is an essential amino acid but not tyrosine which can be formed by *p*-hydroxylation of phenylalanine. These amino acids have many routes of metabolism. Pathways leading to amine synthesis only account for a small fraction of their total turnover.

Routes of synthesis of the catecholamines and of 5-hydroxy-tryptamine have many similarities (Fig. 4.1). In both cases the initial step is ring hydroxylation of an amino acid—of tyrosine by tyrosine hydroxylase and of tryptophan by tryptophan hydroxylase (review, Kaufman, 1974) to give L-3,4-dihydroxyphenylalanine (L-DOPA) and L-5-hydroxytryptophan. These ring hydroxylated amino acids are then decarboxylated to give dopamine and 5-hydroxytryptamine. While the two hydroxylases are different enzymes and substrate specific, the possibility of a single L-aromatic amino acid decarboxylase or separate DOPA and 5-hydroxytryptophan decarboxylases remains controversial (Sims & Bloom, 1973; Dairman *et al.*, 1975).

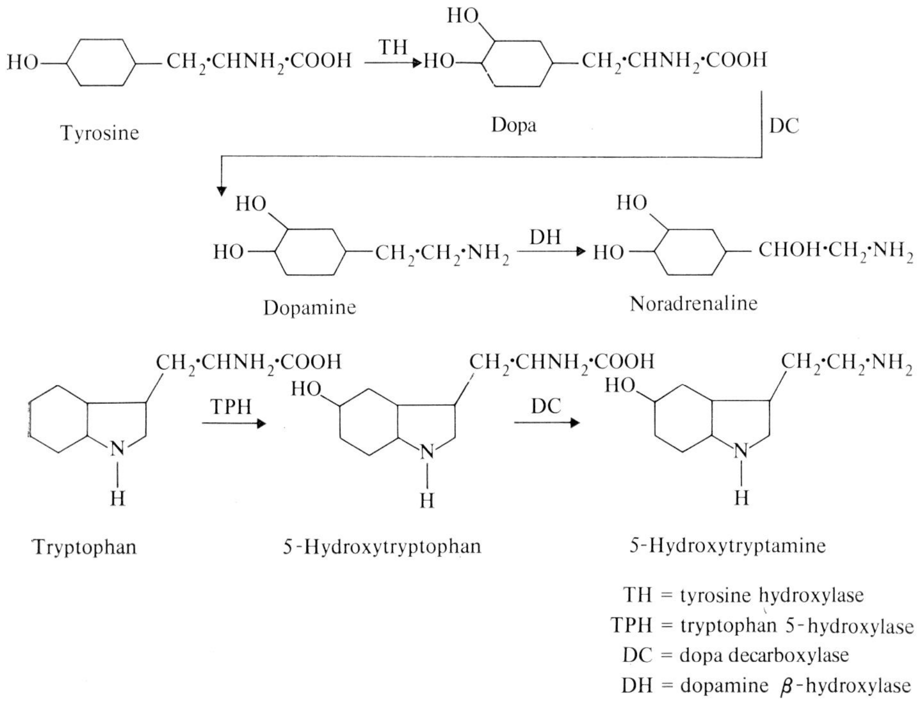

Fig. 4.1. Amine synthesis.

TH = tyrosine hydroxylase
TPH = tryptophan 5-hydroxylase
DC = dopa decarboxylase
DH = dopamine β-hydroxylase

The two hydroxylases are rate limiting so that the amount of amine synthesized depends on their activity rather than that of the decarboxylating enzymes. Therefore the intermediates, L-DOPA and L-5-hydroxytryptophan are not normally detectable. However, tryptophan hydroxylase, though rate limiting, is normally not saturated with its substrate (Kaufman, 1974) and therefore brain 5-hydroxytryptamine turnover is under the influence not only of hydroxylase activity but also of brain tryptophan (Eccleston et al., 1955; Tagliamonte et al., 1971) and hence of plasma tryptophan concentration. Tryptophan hydroxylase activity provides a 'marker' for 5-hydroxytryptamine neurones.

Tryptophan, uniquely among the amino acids, is largely bound to plasma protein. Therefore, only the small unbound fraction is directly available to the brain (Knott & Curzon, 1972) and this is influenced by the concentration of fatty acids (Curzon & Knott, 1974) and other agents (Gessa & Tagliamonte, 1974) which alter plasma tryptophan binding. Unlike tryptophan hydroxylase, tyrosine hydroxylase has been considered to be normally saturated with substrate although recent evidence suggests that rat brain catecholamine synthesis is increased somewhat after injecting considerable amounts of tyrosine (Wurtman et al., 1974).

In noradrenergic neurones, dopamine is side chain hydroxylated by dopamine β-hydroxylase to give noradrenaline. Thus activity of this enzyme provides a 'marker' for these neurones. Adrenaline is formed through N methylation of noradrenaline by phenylethanolamine N-methyltransferase which is present in high activity in the adrenal medulla but has only slight activity in the brain. This, however, is neuronally localized (Hokfelt et al., 1974a) which indicates that the small amounts of adrenaline previously reported in the brain probably occur in specific adrenaline neurones. Indolethylamine N-methyl transferase activity is present in the brain as both a non-specific N-methyltransferase (Saavedra et al., 1973) and an indoleamine specific enzyme (Hsu & Mandell, 1974) but it is not clear to what extent the products N-methyl 5-hydroxytryptamine and N-dimethyl 5-hydroxytryptamine (bufotenine) occur endogenously.

Inside the neurones the amines are largely contained in vesicles within which they are inaccessible to destructive enzymes. This stored material is in equilibrium with free amine in the neuronal cytoplasm where it can be destroyed by mitochondrial monoamine-oxidase. A number of different mitochondrial monoamineoxidase forms can be detected in brain by electrophoresis. These have broad specificities but different relative activities towards different substrates and different sensitivities towards inhibitors (Sandler & Youdim, 1974).

The initial products of monoamineoxidase action are various aldehydes (Fig. 4.2). Normally these are not detectable being largely oxidized further to acids. They may also be reduced to alcohols (Eccleston et al., 1966; Breese et al., 1969) but these pathways are quantitatively much less important (except in the case of noradrenaline—see below). By the action of monoamineoxidase and aldehyde dehydrogenase, dopamine is converted to 3,4-dihydroxyphenylacetic acid (dopacetic acid), noradrenaline to 3,4-dihydroxymandelic acid and 5-hydroxytryptamine to 5-hydroxyindolylacetic acid (5HIAA). Only in the latter case is the acid the major metabolite as a second enzyme catechol-O-methyltransferase also acts on the catecholamines (Axelrod, 1971) to give the O-methylated derivatives 3-methoxytyramine and normetanephrine. The two enzymes have different principal sites of action. Thus monoamine-oxidase occurs in mitochondria within the neuronal cytoplasm and acts on amines released intraneuronally while catechol-O-methyltransferase acts on catecholamines released into the synaptic cleft (Kopin, 1966). Monoamineoxidase also occurs extraneuronally, e.g. in glia.

By the sequential action of these enzymes 3-methoxy 4-hydroxyphenylacetic acid (homovanillic acid, HVA) is formed as the terminal metabolite of dopamine. There are two possible routes to homovanillic acid—dopamine can be converted to 3,4-dihydroxyphenylacetic acid by intraneuronal monoamineoxidase and then O-methylated extraneuronally or alternatively it may be released into the synaptic cleft, O-methylated to 3-methoxytyramine which on re-entering the neurone is oxidized by monoamineoxidase. Drug experiments in the mouse indicate that the second route predominates there being but little O-methylation of dopacetic acid (Roffler-Tarlov et al., 1971). Therefore the *ratio* of the methylated to the non-methylated acid should provide some index

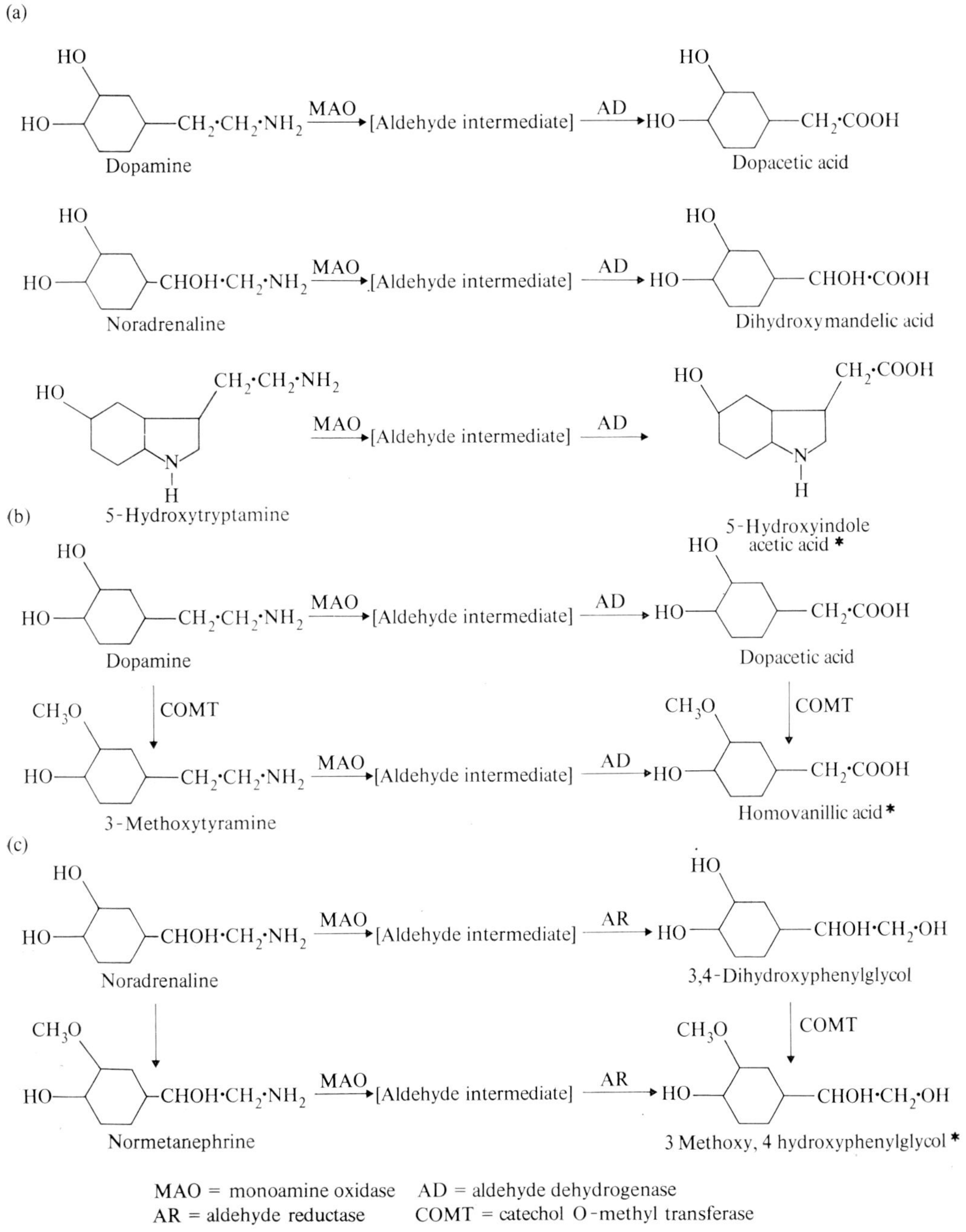

(a)

Dopamine —MAO→ [Aldehyde intermediate] —AD→ Dopacetic acid

Noradrenaline —MAO→ [Aldehyde intermediate] —AD→ Dihydroxymandelic acid

5-Hydroxytryptamine —MAO→ [Aldehyde intermediate] —AD→ 5-Hydroxyindole acetic acid *

(b)

Dopamine —MAO→ [Aldehyde intermediate] —AD→ Dopacetic acid

↓ COMT   ↓ COMT

3-Methoxytyramine —MAO→ [Aldehyde intermediate] —AD→ Homovanillic acid *

(c)

Noradrenaline —MAO→ [Aldehyde intermediate] —AR→ 3,4-Dihydroxyphenylglycol

↓   ↓ COMT

Normetanephrine —MAO→ [Aldehyde intermediate] —AR→ 3 Methoxy, 4 hydroxyphenylglycol *

MAO = monoamine oxidase   AD = aldehyde dehydrogenase
AR = aldehyde reductase   COMT = catechol O-methyl transferase

Fig. 4.2. Amine destruction. * = major terminal metabolite

of the relative amounts of extraneuronal (functional) to intraneuronal (non-functional) dopamine metabolism. The molar *concentration* of homovanillic acid does not give a direct measure of dopamine molecules released in neuronal firing as most of these are retaken up by the neurone unmethylated.

Outside the brain the main route of noradrenaline and adrenaline metabolism parallels that of dopamine, the two enzymes acting on either substrate to give the same product, 3-methoxy 4-hydroxy-mandelic acid (vanillomandelic acid, VMA). However, this substance is not formed to any extent in the brain as the aldehydic immediate products of monoamineoxidase action are not oxidized to acids but reduced to alcohols (Breese *et al.*, 1969; Maas & Landis, 1966) so that the characteristic metabolite of noradrenaline in the brain is 3-methoxy 4-hydroxyphenylglycol (MHPG).

5-Hydroxytryptamine, catecholamines and other known more or less well-authenticated transmitters probably account for quite a small fraction of the brain neurones and therefore other amines detectable in the brain are worth some attention as possible neurotransmitters. A considerable number of these substances are detectable. Thus tryptophan can be directly decarboxylated to tryptamine in the brain (Saavedra & Axelrod, 1972), though it is normally present at much lower concentration than 5-hydroxytryptamine. While evidence is against tryptamine having the specific neuronal and vesicular localization to be expected for a transmitter (Marsden & Curzon, 1974) its concentration increases greatly after giving tryptophan with a monoamineoxidase inhibitor and it might have important effects at 5-hydroxytryptamine receptors in these circumstances.

Other amines occur in still smaller amounts in brain, e.g. phenylethylamine formed by the action of decarboxylase on phenylalanine, phenylethanolamine formed by the action of dopamine β hydroxylase on phenylethylamine and octopamine formed from tyrosine by the action of both enzymes (Axelrod & Saavedra, 1974). These substances have various properties consistent with transmitter function but they have not yet been detected in neurones containing no other amine and they may well merely be the products of side reactions in catecholamine or other transmitter containing neurones.

As well as tryptamine, 5-methoxytryptamine is present in the brain (Green *et al.*, 1973) and also melatonin (Koslow, 1974) though the latter occurs mainly in the pineal gland. Finally, ethanolamine, piperidine, putrescine, spermidine, spermine, histamine and methylhistamine also occur in human brain (Perry *et al.*, 1967) and histamine in particular has many transmitter-like properties (Snyder & Taylor, 1972).

C. BRAIN AMINE NEURONAL SYSTEMS

Biochemical studies on the regional distribution of amines in human brain (Bertler, 1961; Hornykiewicz, 1964a) are consistent with more detailed animal findings which show two prominent features—high concentrations of both noradrenaline and 5-hydroxytryptamine in the hypothalamus and extremely high dopamine concentration in the striatum. These results reflect high densities of corresponding amine containing nerve terminals. Recent methodological advances enabling determinations to be made on minute amounts of brain, reveal marked differences of amine concentration in different small nuclei of the same gross region (Brownstein *et al.*, 1974; Saavedra *et al.*, 1974). In agreement with previous findings on gross brain regions these results indicate many differences between the distributions of noradrenaline and dopamine and support the concept of different and manifold functional roles of the brain amines. On the whole they confirm and extend the detailed neuroanatomical brain amine investigations in which the Falck and Hillarp histochemical fluorescence method was used. These earlier findings demonstrated clearly the specific neuronal localizations of the amines (Fuxe *et al.*, 1970b). By the application of the fluorescence method to the brains of animals with various experimental lesions the courses of aminergic tracts of the rat brain have been mapped out as indicated in Figs. 4.3 and 4.4.

1. NORADRENALINE

Cell bodies in the medulla oblongata (A1, 2) give rise to noradrenaline fibres which descend to the spinal cord. There are also two principal ascending systems. Firstly, a ventral system with cell bodies in the lower brain stem in the medulla and pons (A1, 2, 5, 7) and terminals in the lower brain stem, midbrain and diencephalon. The second system with cell bodies in the locus caeruleus (A6) contains tracts to the lower brain stem and the cerebellum and ascending or dorsal tracts to the cortex and hippocampus. The locus caeruleus system is particularly important giving rise to terminals in all brain areas although most midbrain and diencephalon terminals originate from the ventral system.

2. DOPAMINE

Cell bodies in the zona compacta of the substantia nigra (A9) give rise to a major group of fibres ascending through the hypothalamus and terminating in the two striatal regions—the caudate nucleus and putamen which contain a very large fraction of the dopamine content of the brain. Other axons from cell bodies caudal to the substantia nigra (A8) probably also contribute to this group. Another group from cell bodies dorsal to the nucleus interpeduncularis make up the mesolimbic dopamine system and terminate in the nucleus accumbens and olfactory tubercule. The tubero-infundibular dopamine system with cell bodies in the hypothalamus (A12) innervates the external layer of the median eminence. Dopamine containing terminals are also present in the cortex (Thierry *et al.*, 1973; Hokfelt *et al.*, 1974b).

3. 5-HYDROXYTRYPTAMINE

Axons from cell bodies in the midbrain raphe (B7, 8) and the pontine raphe (B5, 6) form a medial ascending pathway with terminals in the hypothalamus and preoptic area. Also lateral ascending pathways terminating in the cortex and caudate nucleus originate from cell body groups B7, 8, 9. The cerebellum is innervated from cell body groups B5, 6, 7, 8 while cell bodies in the medulla (B1, 2, 3) give rise to fibres descending to the spinal cord.

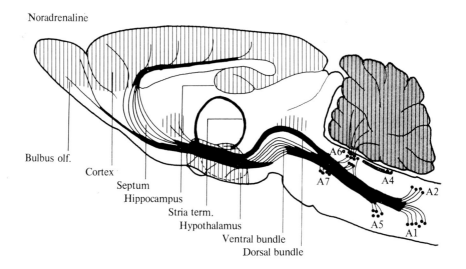

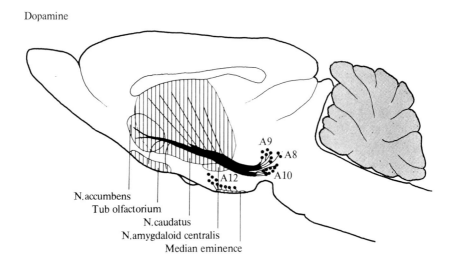

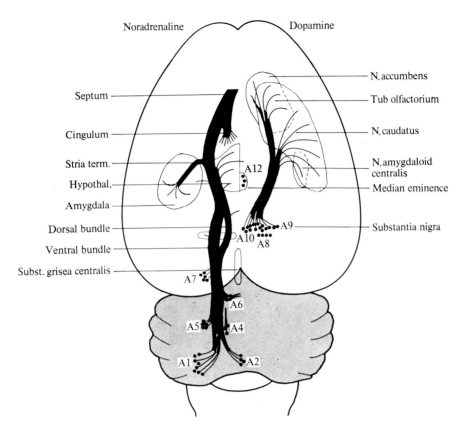

Fig. 4.3. Dopamine and noradrenaline pathways shown in horizontal and sagittal projection. The stippled area indicates nerve terminal areas. From Ungerstedt (1971a) with acknowledgements. See text for details.

4. ADRENALINE

Rat brain phenylethanolamine-N-methyltransferase is localized in specific tracts (Hokfelt *et al.*, 1974a). As this enzyme is rate limiting for adrenaline synthesis the small amounts of adrenaline in the brain are probably similarly localized. The tracts derive from two groups of cell bodies in the medulla and consist of long ascending and descending fibres to the brain stem and spinal cord. Terminals are found in the lower brain stem, the locus caeruleus, parts of the hypothalamus and in the periventricular grey matter.

Technical problems now being surmounted (see later) have impeded the application of the histochemical fluorescence methods to human post-mortem brain. However, a detailed study on human fetal brains (Nobin & Bjorklund, 1973) demonstrated that although there are many differences of detail neuronal localization of amines is largely similar to that found in the rat. Thus major catecholamine systems with cell bodies in the medulla, locus caeruleus and

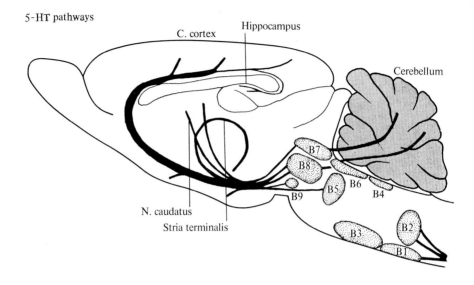
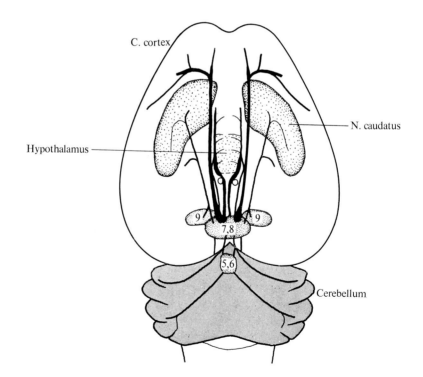

Fig. 4.4. 5-Hydroxytryptamine pathways shown in horizontal and sagittal projection. From Fuxe and Jonsson (1974) with acknowledgements. See text for details.

substantia nigra were detected. The latter included a nigrostriatal system—presumably containing dopamine. Evidence for a tuberoinfundibular dopamine system was also found. Indoleamine cell bodies were detected in the raphe, the lateral reticular formation of the pons and midbrain and in the roof of the fourth ventricle.

Functions of the various brain amines are commented on in later sections of this chapter specifically in relation to brain disease. More general reviews on functional aspects are given by Fuxe et al., (1970b) and Barchas et al. (1972).

D. STUDY OF BRAIN AMINE ABNORMALITIES IN MAN

1. GENERAL

Important clues about the role of brain amines in human disease have derived as often as not from accidental observations rather than from systematic biochemical investigation. Thus mood elevation when iproniazid was used to treat tuberculosis led to its recognition as a monoamineoxidase inhibitor and to attention being focused on amines in relation to brain disease. Similarly, the accidental ingestion of LSD led to the whole field of indoleamine studies in mood and behaviour disorders. Such accidental findings have on the whole proved more fruitful than the many studies of amine metabolites in urine and blood samples from patients with brain disorders. The disadvantage of the latter approach is that extracerebral metabolism accounts for almost all the amine metabolite content of blood and urine. Therefore detection of urinary abnormalities does not prove abnormal brain metabolism and conversely normality of urinary amine metabolites can occur in the presence of a gross abnormality occurring specifically in the brain. Urinary 3-methoxy 4-hydroxyphenylglycol was thought to be an exception to this generalization as this substance is of major importance only in brain noradrenaline metabolism. However, even here experiments in which rat brain noradrenaline neurones were destroyed by intraventricular injection of 6-hydroxydopamine showed that little if any of the urinary 3-methoxy-4-hydroxyphenylglycol came from the brain (Karoum et al., 1974; Bareggi et al., 1974).

Another indirect way of studying brain amine biochemistry has been the use of blood platelets as a partial model of the neurone (Pletscher, 1968). Both cells take up amine (in particular, 5-hydroxytryptamine) by an energy-dependent process, store it in subcellular organelles and metabolize it by monoamineoxidase. Thus abnormal platelet amine disposition and metabolism suggests (though it by no means proves) a similar brain abnormality.

2. CEREBROSPINAL FLUID

Cerebrospinal fluid (CSF) amine metabolite concentrations are more relevant to brain metabolism than are urine values. Therefore in recent years there have been many investigations of these substances in lumbar CSF from patients with brain disease. This work rests on the assumption that concentrations of homovanillic acid and 5-hydroxyindolylacetic acid in the lumbar CSF give information on brain metabolism of their parent amines dopamine and 5-hydroxytryptamine. The assumption is supported by the following findings.

(a) Amine metabolite concentrations in the lateral ventricular CSF of the dog show some correlation with concentrations in adjacent brain areas (Guldberg, 1969).

(b) Acidic amine metabolites given peripherally to dogs hardly penetrate to lateral ventricle CSF (Guldberg & Yates, 1968; Ashcroft et al., 1968) and the noradrenaline metabolite 3-methoxy-4-hydroxyphenylglycol when given to man did not penetrate to CSF (Chase et al., 1973).

(c) Drugs altering brain amine turnover in animals cause appropriate alteration of the acidic metabolites in CSF from animals (Guldberg and Yates, 1968) and humans (Chase et al., 1970; Fyro et al., 1974; Post & Goodwin, 1974).

(d) In Parkinsonism and in senile and pre-senile dementia defective turnover of brain amines is indicated by determinations at autopsy and low CSF concentrations of the acidic amine metabolites are found during life (Gottfries et al., 1969).

However, a number of limitations restrict the deduction of brain amine turnover and function from amine metabolite concentration in the lumbar sac (the usual and most readily available source of human CSF). These limitations have been recently reviewed (Curzon, 1973, 1975; Garelis et al., 1974) and may be discussed in terms of the diagram in Fig. 4.5.

(a) The first problem (fundamental not only to CSF studies but to the whole brain amine field) is the uncertain relationship between net brain amine metabolism and aminergic receptor response.

(b) Another uncertainty is the fraction of brain amine metabolites which enter the CSF. About 90% of rat brain 5-hydroxyindolylacetic acid is transported directly to the blood (Meek & Neff, 1973) and it has been calculated that the homovanillic acid entering the CSF represents at the most a third of the dopamine turnover of the caudate (Sourkes, 1973).

(c) Amine metabolite concentrations in CSF vary with dilution and transport within the CSF space. For example ventricular volume can increase considerably on brain shrinkage and lead to dilution. Also, lumbar metabolite concentrations are influenced by the rate at which they reach the sac from the ventricules and by their transport from CSF to blood. The latter process can be inhibited by probenecid but this can introduce a new variable as inhibition depends on CSF probenecid concentration which can vary widely (Korf & Van Praag, 1971; Sjostrom, 1972). It is also possible that the movement of the metabolites to the sac is altered by locomotion. Thus Post et al. (1973) find that increased locomotion leads to increased concentrations while Claveria et al., (1974) find low concentrations in patients with severely impaired movement. However, altered neuronal activity in these situations could also be involved.

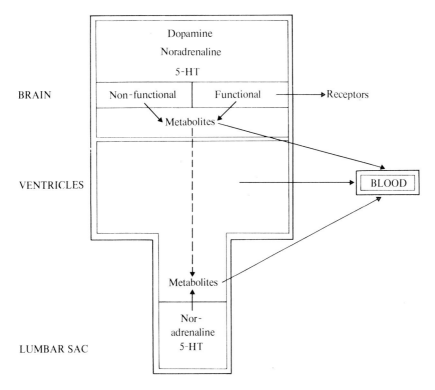

Fig. 4.5. Transport of amine metabolites from the brain.

(d) The relative contributions of different parts of the nervous system to CSF amine metabolite concentrations is unclear. A disproportionate amount presumably comes from periventricular regions. Also while homovanillic acid originates essentially completely from the brain, both animal experiments (Bulat & Zivkovic, 1971) and determinations on lumbar CSF from patients in whom CSF flow from the ventricles to the sac is impeded (Curzon et al., 1971; Garelis et al., 1974) indicate that much if not all lumbar 5HIAA may be of local origin.

These factors and also methodological limitations (Bowers, 1972) limit the interpretation of lumbar CSF findings. It is rash to assume that differences between concentrations in different individuals or in the same subject at different times necessarily imply differences of brain amine metabolism. On the other hand CSF concentrations may be more reliably used to investigate whether drugs altering brain amine metabolism in animals appreciably alter it when given therapeutically to humans (Chase et al., 1970; Fyro et al., 1974; Post & Gordon, 1974) though in some circumstances non-neural structures may contribute to CSF values, e.g. following L-DOPA treatment (Bartholini et al., 1971). However, even taking all these uncertainties into account, CSF amine metabolite determination remains the only way by which we can at present readily obtain information on amine metabolism in the human brain during life and it is probable that recent advances in methodology, for example the mass fragmentographic methods used by Sedvall's group (e.g. Fyro et al., 1974), will increase the value of this approach.

The above discussion has largely concerned homovanillic acid and 5-hydroxyindolylacetic acid. The major brain noradrenaline metabolite, 3-methoxy-4-hydroxyphenylglycol differs from these substances in not being a carboxylic acid and it is not transported from the CSF by the probenecid inhibitable transport process. Unlike them it does not show a downward concentration gradient from the ventricles to the lumbar sac (Chase et al., 1973). Spinal noradrenaline neurones may make a considerable contribution to its concentration.

3. BRAIN

Direct investigations on human brain biopsy or autopsy material obviously provide the most concrete evidence of brain amine abnormalities. It is more than likely that these are implicated in many brain disorders. However, it is symptomatic of the difficulties in this field that more than a decade elapsed between the demonstration of grossly disturbed dopamine metabolism in a small number of brains of subjects with Parkinson's disease and the accumulation of data required for the comprehensive report by Bernheimer et al. (1973) on their findings in this common disorder. When brain amine abnormalities are less striking, e.g. in depressive illness, then difficulties of interpretation are considerable because differences between control and abnormal groups might reflect chronic drug effects and agonal mechanisms as much as underlying pathological differences. Problems due to post-mortem changes are probably less severe. Thus human brain 5HT does not vary greatly with time between death and autopsy (Maclean et al., 1965) and dopamine and noradrenaline concentrations were similar in brains taken at autopsies performed either at 3 or 20 h after death ((Ehringer & Hornykiewicz, 1960). Animal work suggests that the activity of many brain enzymes involved in amine metabolism is remarkably stable post-mortem (Vogel et al., 1969).

In principle the histochemical fluorescence method of Falck and Hillarp is ideal for work on human brain amine pathology and has been used to study noradrenaline terminals in biopsy material (Nystrom et al., 1972) but diffusion during the first hour after death results in disappearance of most evidence of localization from post-mortem material (De La Torre, 1972). However, neuronal uptake mechanisms remain effective for longer periods so that tracts can be visualized up to 7 h after death by incubating brain slices in α-methylnoradrenaline, which is taken up by catecholamine terminals and gives a fluorescence reaction. 5-Hydroxytryptamine containing tracts could be seen by incubating slices with 5-hydroxytryptamine. Visualization was facilitated by previous destruction of catecholamine fibres by incubation with 6-hydroxydopamine

II Parkinsonian states

A. INTRODUCTION

Patients with idiopathic Parkinson's disease typically exhibit rigidity, akinesia and tremor of gradually increasing severity. An additional group with post-encephalitic Parkinsonism have symptoms which may develop many years after encephalitic infection. Most of these patients were victims of the epidemic of encephalitis after the First World War. Aged patients with rigidity as the main symptom are classified by some neurologists as having arteriosclerotic Parkinsonism. A still smaller and heterogenous group have Parkinsonian symptoms associated with exposure to poisonous substances, in particular to manganese.

Twenty years ago Greenfield (1955) concluded in a review article on Parkinson's disease that classical anatomical and histological studies were unlikely to reveal much more and that a biochemical approach was more likely to lead to important advances. At that time, no biochemical abnormalities of any significance had been reported in Parkinson's disease. In the same year a development in biochemical instrumentation, the spectrophotofluorometer was to make possible the findings which led to our present knowledge of the biochemistry of Parkinson's disease.

This understanding stems from the recognition of dopamine as a likely neurotransmitter. The importance of dopamine as a precursor of noradrenaline had been long recognized (Blaschko, 1939) but because its activity in pharmacological assay systems is slight its intrinsic significance was only appreciated when it could be determined fluorimetrically and its distribution in the brain was found not simply to parallel that of noradrenaline (Bertler, 1961). As particularly high concentrations were found in the striatum which is involved in the control of movement it seemed likely that dopamine might be a transmitter of importance for central mechanisms subserving movement.

Almost concurrently with the above findings Ehringer and Hornykiewicz (1960) reported low dopamine concentrations in the basal ganglia of a small group of patients with Parkinson's disease or post-encephalitic Parkinsonism. Low dopamine concentration was also found in a related structure, the substantia nigra (Hornykiewicz, 1963). This linked the new biochemical findings with classical neuropathology as loss of melanin containing pigmented cells in the substantia nigra had long been recognized as characteristic of Parkinson's disease (e.g. Greenfield and Bosanquet, 1953). It therefore was of great interest when Anden et al. (1964) showed that the dopamine concentration of the rat caudate nucleus decreased if lesions were made in the substantia nigra.

These findings and developments stemming from them are discussed below.

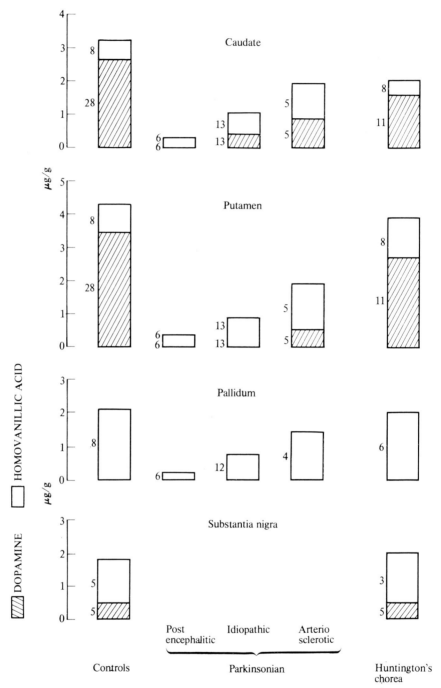

Fig. 4.6. Dopamine and homovanillic acid concentrations in basal ganglia of patients with various dyskinesias. Homovanillic acid concentrations are represented by the full height of the columns. Numbers of subjects are shown. Dopamine concentrations in the putamen in postencephalitic and idiopathic patients are too low to be shown on this scale (<0·05 μg/g). Dopamine was not determined in the pallidum. (Data from Bernheimer et al., 1973.)

B. BIOCHEMICAL AND NEUROPATHOLOGICAL FINDINGS

1. DOPAMINE

The abnormally low basal ganglia dopamine and homovanillic acid concentrations in Parkinsonian states is shown in Fig. 4.6. These results were obtained in a detailed study in which they were related to the substantia nigra lesions and symptoms of subjects with various Parkinsonian syndromes (Bernheimer et al., 1973). Three main groups were studied with postencephalitic, idiopathic and arteriosclerotic-senile forms of the disease. In the first two groups the substantia nigra showed atrophy and glial scarring but damage was more severe and diffuse in the post encephalitic disease and less severe and focal in the idiopathic condition. The arteriosclerotic-senile patients showed generally milder symptoms with especially mild hypokinesia and substantia nigra lesions were focal and clearly related to cerebrovascular abnormalities.

Cell loss in the substantia nigra and dopamine concentration in the caudate nucleus and putamen were significantly and negatively correlated (Fig. 4.7). The concentration of the dopamine metabolite homovanillic acid also fell indicating a defect of dopamine synthesis rather than of its storage. Although substantia nigra cell loss and homovanillic acid concentration correlated negatively this biochemical change was less marked than that of dopamine, especially in the less severely impaired patients. Similarly in monkeys with experimental nigral lesions caudate nucleus dopamine concentration falls more markedly

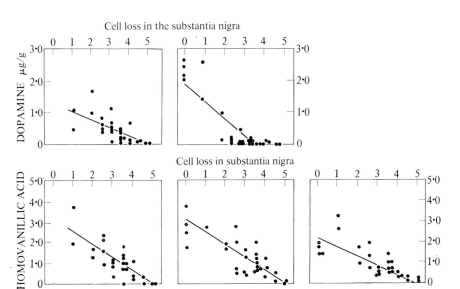

Fig. 4.7. Relationships between cell loss in the substantia nigra and basal ganglia dopamine and homovanillic acid concentrations (in µg/g) in patients with various Parkinsonian disorders. (Data from Bernheimer et al., 1973).

than that of homovanillic acid (Sharman *et al.*, 1967). These results suggest that when some dopamine containing neurones are lost there is a compensatory increase of activity of survivors so that the homovanillic acid/dopamine ratio increases.

The relationship between substantia nigra lesions and decreased striatal dopamine agrees with findings in animals with experimental lesions in this nucleus (Anden *et al.*, 1964). Similarly, lesions of nerve fibres between the substantia nigra and the striatum also caused depletion and this occurred specifically where the particular fibres damaged entered the striatum. Conversely, removal of part of the striatum increased the dopamine fluorescence in the substantia nigra and in the nerve fibres between it and the striatum (Anden *et al.*, 1965). These experiments led to the recognition of nigro-striatal dopamine containing neurones.

Results in Figs 4.6 and 4.7 and the decreasing severity of substantia nigra lesions in the order postencephalitic > idiopathic > arteriosclerotic indicate a general parallelism between overall severity of the illness and dopamine and homovanillic acid deficiencies. In the postencephalitic disease deficiencies are gross so that regional differences in their severity are not apparent, but in the idiopathic disease deficiencies are most severe in the putamen. Results by Fahn *et al.* (1971) are in agreement. This regional difference may reflect a preponderance of lesions in the parts of the substantia nigra connected to the putamen and may be relevant to differences between the symptomatology of the postencephalitic and idiopathic diseases, e.g. to the prominence of aphonia in postencephalitic Parkinsonism. This kind of study of pathological material should shed light on the normal roles of the different parts of the basal ganglia. This in turn may suggest how dopamine disturbances can be involved in non-Parkinsonian disorders.

Relationships between severity of individual Parkinsonian symptoms and dopamine changes are shown in Fig. 4.8. It can be seen that striatal dopamine is markedly decreased even when symptoms are mild. This agrees with the concept that early stages of nigro-striatal degeneration lead to compensatory changes so that surviving neurons become functionally more effective. Increased arborization could enable surviving neurones to partially take over the functions of their dead neighbours while the smaller fall of homovanillic acid than of dopamine indicates that dopamine turnover is increased in the surviving neurons. Also dopamine receptors may become more sensitive. Bernheimer *et al.* (1973) suggest that clinically manifest Parkinsonism represents a late stage in the disease process in which compensatory changes are no longer sufficient to maintain normal neuronal function.

Results in Fig. 4.8 also suggest that akinesia and tremor are more obviously related to the degree of dopamine deficiency than is rigidity. The validity of this and indeed other clinical-biochemical correlations found by Bernheimer *et al.* (1973) depend on their evidence that the symptoms of a heterogenous group of

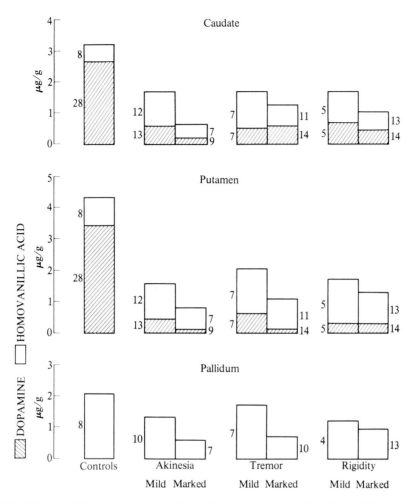

Fig. 4.8. Relationships between severity of symptoms and basal ganglia dopamine and homovanillic acid concentrations in patients with various Parkinsonian disorders. Homovanillic acid concentrations are represented by the full height of the columns. Numbers of subjects are shown. (Data from Bernheimer *et al.*, 1973.)

Parkinsonian subjects reflect different severities of essentially similar pathological processes in the substantia nigra.

2. OTHER AMINES

Not only are dopamine concentrations low in the brains of Parkinsonian patients but also concentrations of noradrenaline and 5-hydroxytryptamine. Results suggest that these deficiencies are not restricted to the basal ganglia (Table 4.1). The noradrenaline deficiency agrees with the degeneration of the locus caeruleus which has been noted (Greenfield & Bosanquet, 1953). However,

Table 4.1. Noradrenaline and 5-hydroxytryptamine in the brain in Parkinson's Disease

	Noradrenaline μg/g		5-Hydroxytryptamine μg/g	
Region	Normal	Parkinson	Normal	Parkinson
Caudate	0·07 (8)	0·03 (12)	0·33 (6)	0·12 (5)
Putamen	0·11 (10)	0·03 (12)	0·32 (6)	0·14 (5)
Globus pallidus	0·09 (6)	0·11 (7)	0·23 (6)	0·13 (5)
Thalamus	0·09 (4)	0·05 (2)	0·26 (4)	0·13 (4)
Hypothalamus	1·29 (7)	0·67 (9)	0·29 (6)	0·12 (5)
Substantia nigra	0·04 (11)	0·02 (10)	0·55 (6)	0·26 (5)

Numbers in parentheses indicate the number of brains for which the value given represents a mean. Data from Hornykiewicz (1964b).

most attention has been given to dopamine and the striatum because of the special significance of the striatum with respect to movement, the high dopamine concentration there, its proportionately greater decrease than of other substances and the relative success of DOPA treatment. The importance of dopamine rather than 5-hydroxytryptamine is also suggested by work on the brain of a senile patient with unilateral Parkinsonian tremor as while striatal dopamine concentrations was lowest on the side contralateral to the tremor that of 5-hydroxytryptamine concentration was similar on both sides (Barolin et al., 1964). Nevertheless, defects of amines other than dopamine could have pathological consequences in Parkinsonism and should not be disregarded (See II C, d, e; III B, a).

3. BRAIN ENZYMES

Any amine defect caused by tract lesions may be expected to be associated with a decrease of intraneuronal enzymes involved in synthesis of amines contained in the tracts. Thus, experimental lesions in the ventromedial tegmentum lead to depletion not only of amines but also of the enzymes necessary for their synthesis (Poirier et al., 1969; Goldstein et al., 1969). Similarly, L-aromatic amino acid decarboxylase activity is strikingly impaired in striatal regions of brains of Parkinsonian patients (Lloyd et al., 1973). Results suggest that decarboxylase activity also falls in the hypothalamus and median raphe which may be related to the deficiencies of noradrenaline and 5-hydroxytryptamine respectively in Parkinsonian brains. Activities of the enzymes involved in dopamine and noradrenaline destruction—monoamineoxidase and catechol-O-methyltransferase—were normal in both striatal regions and elsewhere.

Glutamic acid decarboxylase activity is reported to be low in the substantia nigra and globus pallidus in Parkinson's disease (McGeer et al., 1971) (see later for discussion).

4. OTHER BIOCHEMICAL CHANGES.

Low CSF homovanillic acid and 5-hydroxyindolylacetic acid concentrations in Parkinson's disease (Pullar et al., 1970; Rinne & Sonninen, 1972) are consistent with the low dopamine and 5-hydroxytryptamine concentrations in the basal ganglia. CSF values, however, show considerable scatter and some overlap with the normal range. Thus Parkinson's disease is not invariably associated with low homovanillic acid in the CSF. Conversely, low values are not specific to Parkinson's disease and may occur in other akinetic states whether there is an obvious involvement of the basal ganglia or not (Curzon, 1973; Claveria et al., 1974). Severity of rigidity or akinesia but not tremor have been found in some studies to correlate with particularly low CSF homovanillic acid in patients with Parkinson's disease (Chase & Ng, 1972; Papeschi et al., 1970; Rinne & Sonninen 1972). This agrees with findings in the caudate nucleus (Fig. 4.8). The lumbar homovanillic acid concentration has been more markedly and consistently found low than has that of 5-hydroxyindolylacetic acid which is only partially derived from striatal metabolism (see 1D, 2). Lateral ventricular CSF more closely reflects striatal metabolism and here 5-hydroxyindolylacetic acid concentration is markedly low in Parkinson's disease (Guldberg et al., 1967). Neither of the above amine metabolites is markedly low in CSF of aged patients with akinesia accompanying cerebral arteriosclerosis or senile dementia (Parkes et al., 1974). This is in reasonable agreement with brain findings (Fig. 4.6).

The many studies of urinary amine metabolites in Parkinson's disease have been discussed in earlier reviews (Curzon, 1967, 1972). It is likely that the reported abnormalities largely reflect secondary factors such as diminished exercise or altered gut flora or stasis rather than the primary disturbance of brain metabolism.

C. RELEVANT ASPECTS OF TREATMENT—L-DOPA AND OTHER DRUGS

The demonstration of dopamine deficiency led immediately to attempts to treat Parkinson's disease by replacement therapy. Dopamine itself could not be given as it does not readily pass into the brain. Therefore its immediate precursor DOPA was used. Early work in which relatively small amounts were given led to only moderate and transient improvement. The widespread use of DOPA treatment dates from Cotzias et al. (1967) who gave large and repeated doses of DL-DOPA (3–16 g/day orally for up to 350 days) to 16 patients of whom 8 were markedly improved. As dosage gradually increased so rigidity decreased. Tremor was less affected though it decreased in some patients especially at higher dosages. These results were confirmed (e.g. Calne *et al.*, 1969a; Godwin-Austen *et al.*, 1969) in numerous subsequent trials of L-DOPA.

The utility of L-DOPA in the treatment of Parkinson's disease has provoked much work on its biochemical effects. Some of this has involved giving DOPA to animals at dosages per unit body weight even larger than the dosages given therapeutically. Such studies may not be very relevant to the therapeutic action and are even less likely to be relevant to the physiological role of DOPA.

L-DOPA and related treatments may be discussed under the following five headings.

(a) *Metabolism and mechanism of action*. Replacement of deficient dopamine at receptors is the most obvious mechanism for L-DOPA action and it still seems the most likely mechanism. A very low efficiency of replacement is indicated by the need to give daily doses of the order of a thousand times the amount of striatal dopamine synthesized per day. This is hardly surprising—most orally administered L-DOPA is decarboxylated even before it reaches the circulation (Granerus *et al.*, 1973) and much of the residue is decarboxylated extracerebrally or in the cerebral vasculature (Bartholini *et al.*, 1971) or is 3-O-methylated (see below). Furthermore, of the L-DOPA which reaches the brain a considerable fraction appears to be metabolized to dopamine by L-aromatic amino acid decarboxylase of noradrenaline rich regions (Lloyd *et al.*, 1973). Even the small fraction converted to striatal dopamine is not functionally comparable to normal brain dopamine because of the defective machinery for uptake and release. Some striatal dopamine formed from exogenous DOPA may be made not in surviving dopamine neurones but in 5-hydroxytryptamine neurones by the action of L-aromatic amino acid decarboxylase (Ng *et al.*, 1970). These and other aspects of L-DOPA action are shown in Fig. 4.9.

Various metabolites other than dopamine have been suggested to be involved in the therapeutic action of DOPA. For example it has been proposed that the metabolite 3-O-methyldopa which is eliminated very slowly during L-DOPA treatment and thus gradually accumulates may act as a DOPA reservoir being slowly demethylated to L-DOPA. However, L-3-O-methyldopa is an inadequate treatment for Parkinson's disease (Calne *et al.*, 1973) and a much less effective catecholamine precursor than L-DOPA (Bartholini *et al.*, 1974). It has also been suggested that the tetrahydroisoquinoline alkaloids salsolinol and tetrahydropapaveroline which are detectable in the urine of patients during L-DOPA treatment (Sandler *et al.*, 1973) may have therapeutic effect. While these substances are structurally similar to the dopamine receptor agonist apomorphine evidence is against appreciable stimulatory action as unlike apomorphine they do not stimulate rat striatal cyclic AMP synthesis (Iversen *et al.*, 1975).

A possible influence on the effectiveness of L-DOPA is the brain S-adenosylmethionine deficiency which is demonstrable in DOPA-treated rats. As S-adenosylmethionine is a co-factor for the extraneuronal O-methylation of

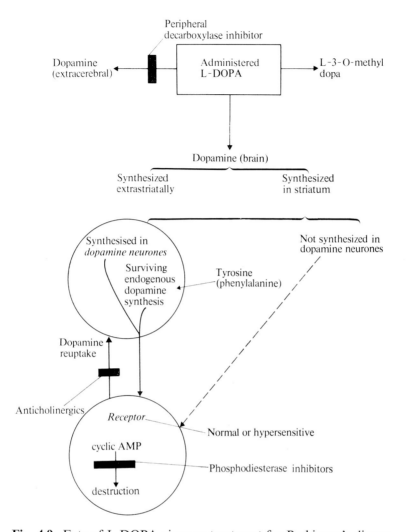

Fig. 4.9. Fate of L-DOPA given as treatment for Parkinson's disease.

catecholamines its deficiency might enhance their synaptic action (Wurtman and Romero, 1972), especially in the Parkinsonian brain where neuronal reuptake is presumably defective.

(b) *Slowness of action.* While L-DOPA may be dramatically effective, improvement more commonly develops over some weeks. However animal experiments indicate maximal conversion of exogenous L-DOPA to brain catecholamines within an hour or so. This has encouraged speculations that either increased catecholamine formation was not responsible for the therapeutic action or that some chronic change in their metabolism was involved. As discussed above it seems unlikely that the gradual accumulation of 3-O-methyldopa explains this phenomenon. Chronic L-DOPA treatment of animals does lead to altered activity of some enzymes required for catecholamine metabolism but findings in different species are not in complete agreement (see Roberge & Poirier, 1973, for discussion).

(c) *Inadequacy of response.* While L-DOPA often leads to marked and prolonged benefit in idiopathic Parkinson's disease it is of little value in the postencephalitic disease as though improvement is often striking, unpleasant and bizarre responses are frequent on more prolonged treatment. A fascinating description of the effect of L-DOPA on post-encephalitics who had been in a state of profoundly diminished motor activity and psychomotor responsiveness for many years is given by Sacks (1973). Presumably the harmful effects of L-DOPA in these patients is related to their brain damage being more extensive and long standing than in most patients with the idiopathic disease. At the biochemically opposite extreme, L-DOPA has little beneficial effect in elderly akinetic subjects where autopsy (Bernheimer *et al.*, 1973) and CSF findings (Godwin-Austen *et al.*, 1971; Parkes *et al.*, 1973) indicate only moderate or negligible deficiencies of brain dopamine metabolism. Failures of response by patients with the idiopathic disease may simply be due to defective gastrointestinal absorption of L-DOPA, but this is rare (Bergmann *et al.*, 1974) while limited benefit is common. Not only in the arteriosclerotics but also in idiopathic patients limited response is significantly associated with relatively high pretreatment CSF homovanillic acid concentration (Curzon, 1973; Gumpert *et al.*, 1973; Jequier & Dufresne, 1972) though its determination is of little utility to the clinician in deciding whether to give L-DOPA to any individual patient. Results suggested an association between coexistence of relatively high pretreatment homovanillic acid and low 5-hydroxyindolylacetic acid in the CSF and limited beneficial effect of L-DOPA. This may indicate that a brain 5-hydroxytryptamine defect has a role in the symptomatology of some Parkinsonians.

The most serious inadequacy of the L-DOPA treatment is that frequently its beneficial effect is maintained only for about two years after which time disabilities and harmful side effects continue to increase though the drug may still have a limited alleviatory effect (Hunter *et al.*, 1973).

(d) *Harmful effects.* Harmful effects of L-DOPA treatment are hardly surprising considering the large dosage given, the role of dopamine in functions other than locomotion (e.g. Fuxe *et al.*, 1973) and the access of L-DOPA to non-dopaminergic neurones. Nausea, postural hypotension and mood changes may occur in the early stages of treatment. These often decline but as treatment continues abnormal movements (dyskinesia), extremely sudden akinetic episodes (so-called 'on–off' effects) and other unpleasant consequences become more common (Barbeau, 1972; Damasio *et al.*, 1973). 'On–off' effects at present defy explanation though there is some suggestion of an association with high plasma DOPA levels (Claveria *et al.*, 1973; Bergmann *et al.*, 1974). The abnormal movements may be considered in terms of compensatory changes during the course of the disease in which some dopamine receptors become hypersensitive so that they respond in an exaggerated manner to dopamine formed from exogenous DOPA. Although decreasing L-DOPA dosage may cause the dyskinesia to disappear the therapeutic effect often also disappears. This suggests that the same dosage provides sufficient dopamine for normal response at some receptors but abnormal response at others. This interpretation agrees with results of drug experiments on rats with unilateral lesions in the substantia nigra resulting from unilateral stereotactic injection of 6-hydroxydopamine (Ungerstedt, 1971b). Thus drugs such as apomorphine which acts on dopamine receptors have more powerful effects on receptors on the lesioned side resulting in circling movements contralateral to the lesion. (Conversely amphetamine releases dopamine from intact nerve endings and activates receptors on the intact side so that circling is ipsilateral to the lesion.) These findings and the stereotyped movements of rats given L-DOPA after bilateral destruction of dopamine neurones are suggested to be analogous to the dyskinesia of some L-DOPA treated patients (Ungerstedt et al., 1973).

Mental symptoms in Parkinsonian patients may be either alleviated or precipitated by L-DOPA treatment (Marsh & Markham, 1973). Intellectual performance is frequently impaired in Parkinson's disease and may be improved by L-DOPA (Marsh et al., 1971; Loranger et al., 1972) analogously to the reversal by L-DOPA of the impairment of learning shown by rats after destruction of dopamine neurones (Zis et al., 1974). Psychotic behaviour of some patients on L-DOPA has been suggested to be due to a resultant decrease of brain 5-hydroxytryptamine. While there is no direct evidence for this relationship affected patients had low plasma tryptophan either when fasting or following an oral tryptophan load (Lehmann, 1973) and tryptophan is claimed to alleviate the psychotic symptoms (Birkmayer et al., 1972; Lehmann, 1973).

(e) *Modified L-DOPA treatments and other drug therapies.* The anticholinergic drugs long used in the treatment of Parkinsonism were proposed initially because patients exhibit signs of excessive cholinergic activity, e.g. hypersalivation. These drugs also inhibit neuronal reuptake of released dopamine and thus may enhance its action at receptors (Coyle & Snyder, 1969; Fuxe et al., 1970a). The deterioration of patients on L-DOPA + anticholinergics when the latter are withdrawn (Hughes et al., 1971) possibly involves the above effect.

L-DOPA is often given together with a peripheral decarboxylase inhibitor. These drugs prevent extracerebral decarboxylation to dopamine so that extracerebral effects of the latter are prevented. Also more of the DOPA given becomes available to the brain thus enabling reduction of dosage (Yahr, 1973). If L-DOPA is given with pyridoxine so that pyridoxal phosphate (the cofactor for decarboxylases) is increased then the beneficial effect is antagonized. Antagonism does not occur when a peripheral decarboxylase inhibitor is also given (Klawans et al., 1971) which suggests that the pyridoxal caused increased extracerebral decarboxylation and hence less L-DOPA was available to the brain. The effect of these drugs on rat brain dopamine is consistent with this mechanism though formation of complexes between DOPA and pyridoxal phosphate has also been invoked (Pfeiffer & Ebadi, 1972).

Because of the various inadequacies of L-DOPA treatment much effort is being made to develop dopamine agonists as possible alternatives. This work is leading to increased understanding of the nature of the dopamine receptor but not (as yet) to drugs which are clearly superior to DOPA (Calne et al., 1975). Another hopeful approach is to increase the effectiveness of L-DOPA or of surviving dopaminergic neurons by means of drugs which inhibit destruction of the striatal cyclic AMP which is formed by action of dopamine agonists or of dopamine itself and which probably mediates receptor responses (Kebabian et al., 1972). Thus caffeine which inhibits the destruction of cyclic AMP by phosphodiesterase potentiates in animals both the hyperkinetic effect of L-DOPA (Stromberg & Waldeck, 1973) and the stereotyped behaviour caused by amphetamine or apomorphine (Klawans et al., 1973b).

If the brain 5-hydroxytryptamine deficiency in Parkinson's disease (Bernheimer et al., 1961) has any pathological role then its replacement should have beneficial effects. However, its immediate precursor 5-hydroxytryptophan, when given with a peripheral decarboxylase inhibitor, was found by Chase et al. (1972) to exacerbate symptoms. Also three patients showed little change when given L-tryptophan and a rapid deterioration when this was supplemented with pyridoxine (Hall et al., 1972). On the other hand, Mena et al. (1970) found that 5-hydroxytryptophan alleviated Parkinsonian symptoms due to chronic manganese poisoning in a patient who did not respond to DOPA. Also, Rodriguez (1972) reported that the activity of DOPA against tremor in mice caused by tremorine was enhanced by 5-hydroxytryptophan and suggested a trial in patients whose symptoms were not controlled by DOPA. Sano and Taniguchi (1972) claim an antitremor effect of 5-hydroxytryptophan when given to subjects with Parkinson's disease, while Coppen et al. (1972) found that ability to perform functional tasks was improved significantly by L-DOPA treatment only when it was supplemented by L-tryptophan. However, they felt this was a consequence of an antidepressant effect of tryptophan. The use of tryptophan in the control of psychotic behaviour occurring on L-DOPA treatment has been discussed on page 193. Combined treatment with DOPA + tryptophan may involve problems because of mutual competition for transport mechanisms.

III Other dyskinesias

A. HUNTINGTON'S CHOREA

Huntington's chorea is a disease showing dominant inheritance with an incidence of about 1 in 10,000. To a large degree the neuropathological and motor abnormalities represent an opposite pole to Parkinson's disease with striatal nerve cell loss and striato-nigral degeneration (Vogt & Vogt, 1937) instead of the nigrostriatal degeneration of Parkinson's disease. Hyperkinesis and involuntary movement are major symptoms instead of the typical Parkinsonian hypokinesis (though occasional patients have an akinetic-rigid form of Huntington's chorea especially in the earlier stages). As well as the movement disturbances mental disorder is frequent—at least 1% of long stay mental hospital patients have Huntington's chorea.

Involuntary movement is a common result of L-DOPA treatment of Parkinson's disease (Cotzias *et al.*, 1969; Calne *et al.*, 1969b) and symptoms of Huntingtonian subjects are revealed or exacerbated by L-DOPA. Indeed it has been used as a predictive test for Huntington's chorea (Klawans *et al.*, 1972). However, excessive brain dopamine does not occur in this disease (Bird & Iversen, 1974). On the contrary, in one study dopamine and homovanillic acid concentrations were found to be moderately decreased with more marked changes in the caudate than in the putamen (Bernheimer *et al.*, 1973). (Compare these findings with those in idiopathic Parkinson's disease where the defect is more marked in the putamen (Fig. 4.6)). Atrophy of the caudate nucleus and putamen with little or no nigral change is typical of Huntington's chorea. Thus the normal or moderately low striatal concentrations together with the low weights of striatal regions lead to low net *content* of dopamine and homovanillic acid at autopsy. This is consistent with the decrease of lumbar CSF homovanillic acid with increasing severity of symptoms observed during life (Curzon *et al.*, 1972). CSF 5-hydroxindolylacetic acid did not decrease. Very low concentrations of both amine metabolites are reported in lumbar CSF of two patients with the akinetic rigid form of the disease (Curzon *et al.*, 1972; Johansson & Roos, 1974). This is superficially in agreement with the Parkinson-like symptoms but unlike in Parkinson's disease, brain dopamine content is not strikingly low and is somewhat higher than in ordinary Huntingtonians (Bird & Iversen, 1974).

Striatal shrinkage in Huntington's chorea probably involves a major decrease of non-dopamine containing neurones. Therefore although net dopamine synthesis is not abnormally great there is probably a dopaminergic predominance. This is consistent with the exacerbation of symptoms by L-DOPA and the therapeutic effect of amine depleters such as tetrabenazine (McLennan *et al.*, 1974).

Perry *et al.* (1973) in a study of brain amino acid concentrations obtained clear evidence of significantly low concentration of the transmitter γ-aminobutyric acid and of its peptide derivative homocarnosine. Furthermore, although results suggested moderate γ-aminobutyric acid deficiency throughout the brain it was most striking in the substantia nigra and other basal ganglia regions (Fig. 4.10). The low values did not simply reflect cerebral oedema or non-specific changes as 34 other amino acids or related substances were present at normal concentration.

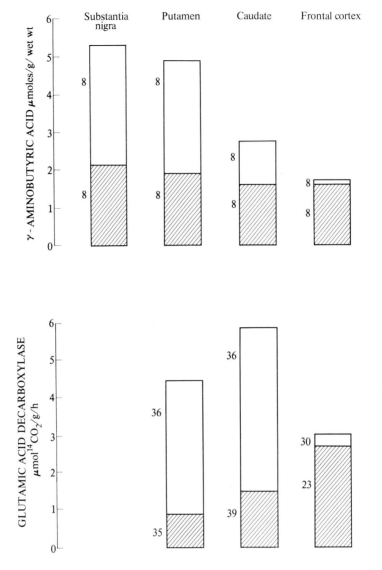

Fig. 4.10. Huntington's chorea: γ-aminobutyric acid and glutamic acid decarboxylase in brain regions. Data from Perry *et al.* (1973) and Bird *et al.* (1974) respectively. The globus pallidus was included with the putamen in the γ-aminobutyric acid studies. Results on control subjects are represented by the full height of the columns and results on those with Huntington's chorea by the shaded portions. Numbers of subjects are shown.

The γ-aminobutyric acid deficiency occurs together with even more strikingly low values in various parts of the basal ganglia of glutamic acid decarboxylase (Bird & Iversen, 1974). This enzyme is responsible for γ-aminobutyric acid synthesis and occurs mainly in inhibitory neurones which release γ-aminobutyric acid as their transmitter (Iversen, 1972). The distribution of γ-aminobutyric acid in the human substantia nigra (Kanazawa *et al.*, 1973) suggests its presence in striato-nigral neurones with terminals mainly in the pars reticulata of the substantia nigra bordering the zone containing the cell bodies of the nigro-striatal dopamine neurones, i.e. the pars compacta. The findings in Huntington's chorea therefore indicate degeneration of such inhibitory neurones in the basal ganglia.

Choline acetyltransferase activity/g tissue—an index of cholinergic neurones—was strikingly low in striata of patients with particularly marked deficits of glutamic acid decarboxylase although activity was normal in autopsy material from some subjects who had exhibited typically choreiform movements. Acetylcholine imbalance could have a role in the disease as Klawans and Rubovits (1972) found that the anticholinesterase physostigmine and the anticholinergic drug benztropine alleviated and exacerbated symptoms respectively.

Reduced γ-aminobutyric acid synthesis occurs also in the basal ganglia in Parkinson's disease (McGeer *et al.*, 1973; Lloyd *et al.*, 1973). This suggests that chorea results not from an absolute γ-aminobutyric acid deficiency but from an imbalance between γ-aminobutyric acid and dopamine neurones so that the reduced inhibitory activity of the former neurones leads to an exaggerated response to dopamine. Both the precipitation of chorea by DOPA in Huntingtonians (Klawans *et al.*, 1972) and the choreiform movements which sometimes occur when Parkinsonian patients are given DOPA are consistent with this relationship. In untreated Parkinsonians the functional deficit in dopamine predominates so that rigidity and akinesia occur. Such a generalization is in keeping with the relative deficits of the two transmitters demonstrable in autopsy material from subjects with the two disorders. Increased responsiveness of dopamine receptors may also lead to the chorea of occasional hyperthyroid patients. Thus, haloperidol a dopamine receptor antagonist alleviated hyperthyroid chorea in one patient (Klawans & Shenker, 1972) while guinea pigs made hyperthyroid by thyroxine showed increased responsiveness to the dopamine receptor agonist apomorphine (Klawans *et al.*, 1973a).

As with the degeneration of dopamine neurones in Parkinson's disease we can only speculate about the origin of the neuronal degeneration in Huntington's chorea. It suggests, however, that the development of drugs increasing brain γ-aminobutyric acid will prove valuable though present useful drug treatments decrease functional concentrations of catecholamines and 5-hydroxy-tryptamine, e.g. phenothiazines, the dopamine receptor antagonist Pimozide (Fog & Pakkenberg, 1970) and the amine depleter tetrabenazine (McLennan *et al.*, 1974).

B. DRUG-PROVOKED DYSKINESIAS

Movement disorders provoked by drug treatments were rare before the advent of modern psychotropics such as reserpine, phenothiazines and butyrophenones.

It is more than likely that these disorders which often limit the clinical use of the above drugs are due to interactions with brain amines.

(a) *Amine depleters.* Reserpine treatment of human subjects or animals may lead to various Parkinsonian symptoms. The reversal of symptoms in animals by L-DOPA and 5-hydroxytryptophan was an indication of the importance of brain amine disturbance in Parkinson's disease (e.g. Roos & Steg, 1964) and the effect of L-DOPA in particular was taken to indicate that the symptoms resulted simply from dopamine depletion. This argument was strengthened by the negligible effect of L-DOPA on mouse brain noradrenaline concentration (Everett & Borcherding, 1970) although its turnover does increase (Keller et al., 1974). It now appears that noradrenaline has a role in reserpine rigidity and therefore possibly in human Parkinsonism. Thus Marsden et al. (1974) found that reversal of reserpine rigidity in mice by L-DOPA was prevented by FLA-63, which inhibits dopamine β-hydroxylase or by phenoxybenzamine (a noradrenaline receptor antagonist) but not by pimozide (a dopamine receptor antagonist). Also Anden et al. (1973) found that the dopamine receptor agonist apomorphine alone only partly reversed the decrease of motor activity after reserpine but was potentiated by the noradrenaline receptor agonist clonidine. The latter drug alone had little effect.

(b) *Phenothiazines and butyrophenones.* These substances can cause various movement disturbance in man, e.g. dystonia in children and young adults and either Parkinsonism or restlessness in older subjects which disappear on dose reduction. The tardive dyskinesias which usually appear after at least one year of drug treatment are more serious as they are not reversed by drug withdrawal and indeed often only appear after withdrawal. A great variety of symptoms are described (review, Crane, 1973) with facial dyskinesias being common especially in older patients. The provoking drugs have little effect on brain amine levels in animals. They behave as if they block receptors so that a compensatory increase of dopamine turnover occurs (Carlsson & Lindqvist, 1963; Nyback & Sedvall, 1969) with increased but functionally ineffective release into the synaptic cleft and elevated dopamine metabolite concentrations. Animal experiments indicate some correlation between the potency of these drugs in producing movement disturbance and the increase of brain homovanillic acid they cause (O'Keefe et al., 1970). The receptor blockade can lead to a compensatory hypersensitivity of dopamine receptors which can be demonstrated in the rat after withdrawing chronic haloperidol treatment by giving the agonist apomorphine (Gianutsos et al., 1974). The role of increased activity at dopamine receptors is consistent with the similarity between the above dyskinesias and those which often limit the therapeutic utility of L-DOPA in Parkinsonians. Tardive dyskinesias may be treated with the amine depleter tetrabenazine (Kazamatsuri et al., 1972a). They also respond to partial reblockade of the hypersensitive dopamine receptors by haloperidol (Kazamatsuri et al., 1972b), though it is obvious that this treatment would eventually be harmful if it led to a further increase in sensitivity.

C. MANGANESE DYSKINESIA

Manganese miners who inhale large amounts of ore dust may develop psychiatric followed by neurological disturbances. Most affected subjects have typical Parkinsonian symptoms but about 10% exhibit muscular dystonia (Mena et al., 1967). Brain amine determinations have only been reported on a single case (Bernheimer et al., 1973). Dopamine was markedly low in striatal regions, the substantia nigra and in the hypothalamus. Noradrenaline was also low in these regions with the exception of the substantia nigra and 5-hydroxytryptamine concentrations were normal. Catecholamine abnormalities and the substantia nigra lesions in this patient were similar to those in brains of similarly impaired patients with idiopathic Parkinson's disease.

The above findings may be compared with those in squirrel monkeys which developed rigidity, tremor and dystonia after treatment with manganese dioxide. The affected animals had much less dopamine and 5-hydroxytryptamine in the caudate than did control or unaffected animals. This abnormality appeared to be specific to the basal ganglia as concentrations in the rest of the brain were normal (Neff et al., 1969). Results in rabbits after chronic manganese dioxide treatment were closer to those found in human autopsy material with significant decreases of the brain catecholamines but not of 5-hydroxytryptamine and neuronal loss and other degeneration in the basal ganglia and other regions (Mustafa & Chandra, 1971).

In consistency with the general similarity of the biochemical and morphological defect to that in Parkinson's disease, patients with manganese dyskinesia are helped by L-DOPA treatment. It is of interest that one patient whose symptoms were exacerbated by L-DOPA improved on 5-hydroxytryptophan (Mena et al., 1970).

IV Psychological disorders

A. DEPRESSION

Biochemical research on depression has largely concerned the endogenous illness in which depressed mood is not obviously related to circumstance. The importance of biochemical changes in such depressions is consistent with the frequent association with somatic symptoms and their responsiveness to physical treatments such as electroconvulsive therapy (Carney et al., 1965) or drugs. Depressions related to personality or circumstance—neurotic and exogenous depressions—respond less well to these treatments. Biochemical precipitation of endogenous depression is also suggested by the occasional patient whose depressed mood rhythmically alternates with mania as there is usually an associated rhythm of various physical parameters (e.g. Jenner et al., 1967).

Biochemical studies are not made easier by the problems of classification of depression. This has itself been the subject of much investigation (Kendell, 1968). Also depressives exhibit anxiety to widely different degrees (Roth et al., 1972) and relationships between this component of the illness and biochemical abnormalities has received little attention. Rigid mechanistic interpretations of the terms 'endogenous' and 'exogenous' have themselves led to confusions (Lewis, 1971). It is common experience that exogenous depression often has somatic concomitants and conversely abnormal biochemical adaptation to normal changes of milieu may well have a causal role in the so-called endogenous illness. How interactions between genetic, biochemical, developmental and social factors might lead to depressive illness has been discussed by Akiskal & McKinney (1973).

A causal role of brain amine deficiency in depressive illness is strongly suggested by the effects on mood of drugs which alter brain amine metabolism. Thus, amine depleters such as reserpine (Bunney & Davis, 1965) not infrequently cause depression while monoamine oxidase inhibitors alleviate depression. The latter substances increase human brain amine concentrations on a time course approximately paralleling the improvement (Bevan-Jones et al., 1972). As these drugs influence both catecholamine and 5-hydroxytryptophan metabolism it is not immediately obvious whether their effects on mood are due to effects on a specific amine, on more than one amine, on relative concentrations of amine or even on different amines in different subjects. Mendels and Frazer (1974) point out that while reserpine causes a non-specific decrease of brain amines and leads to depression this results but rarely from treatment with drugs which specifically decrease either catecholamines or 5-hydroxytryptamine. Direct evidence of amine disturbance in depression is summarized below. It is on the whole convenient to discuss evidence for disturbed indoleamine and catecholamine metabolism separately.

(a) *Evidence for indoleamine disturbance.* Early work suggested that some depressives may have an overall defect of 5-hydroxytryptamine synthesis. Thus depressives who responded to the monoamine oxidase inhibitor iproniazid had less initial urinary 5-hydroxyindolylacetic acid than did non-responders (Pare & Sandler, 1959; Van Praag & Leijnse, 1963). This was not confirmed in a third study (Burgermeister et al., 1963). However, many studies do confirm the low concentrations of 5-hydroxyindolylacetic acid in the lumbar CSF of depressives reported by Ashcroft et al. (1966). This work is reviewed by Mendels et al. (1972) and by Post et al. (1973). CSF findings in mania are less clear-cut and evidence is on the whole against a depression-mania polarity associated with low and high 5-hydroxytryptamine turnover respectively (Coppen, 1973; Post et al., 1973). Taken together with the persistence of low 5-hydroxyindolylacetic acid concentration in the CSF of depressives on recovery (Ashcroft et al., 1973a;

Coppen, 1973) these findings suggest that this metabolic abnormality is not merely due to such factors as diminished motor activity during depression though this may have some influence as simulated manic activity increased CSF 5-hydroxyindolylacetic acid values in depressives (Post et al., 1973). Low values (after probenecid treatment) were found in many endogenous depressives but not in patients classified as neurotic depressives (Van Praag, 1974).

Initially the low 5-hydroxyindolylacetic acid concentration in the CSF tended to be taken to indicate that low brain 5HT was solely responsible for the depression. This is not necessarily the case. Ashcroft's group (1972) point out that the persistence of low CSF values during recovery suggests that this might involve increased 5HT receptor sensitivity. Therefore simple parallelism between brain receptor response and CSF 5-hydroxyindolylacetic acid concentration cannot necessarily be assumed. Parallelism also breaks down on treatment with tricyclic antidepressants (Post & Goodwin, 1974), which increase synaptic cleft concentrations of 5-hydroxytryptamine and decrease its catabolism to 5-hydroxyindolylacetic acid.

Low 5-hydroxytryptamine concentration in the hind brains of depressive suicides is reported in one study (Shaw et al., 1967) while other workers found that 5-hydroxytryptamine was not significantly lower in the brain stem of 'endogenously depressed' suicides than in controls though low values were found in 'reactive personality' suicides (Pare et al., 1969). A third report gives normal hindbrain 5-hydroxytryptamine but significantly low 5-hydroxyindolylacetic acid (Bourne et al., 1968) while a detailed regional study suggests lower 5-hydroxytryptamine in the raphe nuclei (Lloyd et al., 1974). The above discrepancies presumably reflect the many variables involved in this kind of work—mode of suicide, previous medication, diagnostic problems, agonal changes, etc.

Defective brain 5-hydroxytryptamine synthesis in depression could result from decreased availability to the brain of the precursor amino acid tryptophan. Lehmann (1972, 1973) has described a wide range of psychiatric disturbance in disorders in which this may occur. Availability depends on the amount of plasma free tryptophan (see I B), and Coppen et al. (1973) report that plasma free (but not total) tryptophan is significantly low in female depressives and increases on recovery. While this would be readily explicable if plasma unesterified fatty acid concentration was low (Curzon & Knott, 1974) this is apparently not the case in depression (Cardon & Mueller, 1966; Van Praag & Leijnse, 1966). However, evidence of low sympathetic activity in depression (Perez-Reyes, 1969) suggests that further study of these relationships may be worthwhile as increased sympathetic activity leads to increased lipolysis.

Another factor which could decrease tryptophan availability in depressives is its destruction by liver pyrrolase. High plasma cortisol is commonly found in depression and may lead to increased pyrrolase activity. That this occurs is indicated by increased excretion of tryptophan metabolites on the pyrrolase pathway after loading with tryptophan, although the scatter of values is large and differences between control and test groups are not always statistically significant (Curzon & Bridges, 1970; Frazer et al., 1973). As rats injected with cortisol had increased liver pyrrolase activity and decreased brain 5-hydroxytryptamine and 5-hydroxyindoleacetic acid (e.g. Green & Curzon, 1975) it was suggested that increased catabolism of tryptophan by pyrrolase might lead to decreased brain 5HT synthesis in some depressives (Curzon, 1969). Enhanced extra-cerebral catabolism of exogenous tryptophan in depressives is suggested by the failure of CSF 5-hydroxyindolylacetic acid to increase on chronic tryptophan administration even though similarly treated schizophrenics showed considerable increases (Bowers, 1970). However, after a single dose of tryptophan Ashcroft et al. (1973b) found no difference between depressed and control groups.

A beneficial effect of tryptophan in depression would be consistent with involvement of brain 5-hydroxytryptamine deficiency in the illness. While the alleviation of depression by monoamineoxidase inhibitors is enhanced by tryptophan (Coppen et al., 1963; Glassmann & Platman, 1969) early claims that tryptophan alone is an antidepressant have been criticized (e.g. Herrington et al., 1975) though it may be of benefit in mania (Bunney et al., 1971; Prange et al., 1974). The potentiation of monoamineoxidase inhibitors by tryptophan is not necessarily mediated by increased 5-hydroxytryptamine. For example, comparably treated laboratory animals show large increases of brain tryptamine (Marsden & Curzon, 1974) and N-methylation of 5-hydroxytryptamine might also result (Morgan & Mandell, 1969). Both of these possibilities could have behavioural consequences.

There have been a number of claims that 5-hydroxytryptophan alleviated depression (e.g. Sano, 1972). As this is the immediate precursor of 5-hydroxytryptamine it might be thought to be a more effective precursor than tryptophan. However, animal experiments indicate that it only penetrates to 5-hydroxytryptamine neurones to a limited extent and that it is partly decarboxylated to 5-hydroxytryptamine in other types of cell (Fuxe et al., 1971).

The report that p-chloro-N-methylamphetamine, which *decreases* brain 5-hydroxytryptamine synthesis had an antidepressant effect on both endogenous and reactive depressions (Kits & Van Praag, 1970) seems paradoxical. However, paradoxes are readily generated by attempts to embrace within simple hypotheses the brain 5-hydroxytryptamine defect in depression and the effects of drugs altering its metabolism on depressive mood and on the behaviour of laboratory animals. It should not be taken for granted that low 5-hydroxytryptamine function = depression and increased function = alleviation.

(b) *Evidence for catecholamine disturbance.* The precipitation of depression by amine depleters and determinations of urinary catecholamines or their metabolites (reviewed Schildkraut, 1971) together encourage the hypothesis that depression and mania are associated respectively with low and high catecholamine turnover. Decreased excretion of 3-methoxy-4-hydroxyphenylglycol (Schildkraut, 1974) has received particular attention because of the possibility (which now seems unlikely) that a considerable part of it originated in the brain (see I B). Therefore the urinary abnormality may reflect consequences rather than causes of depressive illness. There are indications, however, that determination of urinary catecholamine metabolites may be of clinical value as a means of classifying patients and thereby predicting responses to antidepressant drugs (Schildkraut, 1974).

Brain determinations suggest neither a central dopamine nor a noradrenaline deficiency in depressive suicides (Bourne et al., 1968; Pare et al., 1969) and CSF homovanillic findings in depression are equivocal though there is some suggestion of rather high values in mania (reviewed Post et al., 1973). As for CSF 3-methoxy-4-hydroxyphenylglycol concentration there is a striking lack of agreement with low (Post et al., 1973) and normal values (Shaw et al., 1973; Shopshin et al., 1974) and decreased and increased values respectively on recovery (Shaw et al., 1973) or on transition from depression to mania (Shopshin et al., 1974). As the latter authors remark there is a 'consistent disparity . . . from one study to another in this field' and there is no convincing evidence of defective central catecholamine metabolism in depression. In agreement with this negative conclusion, L-DOPA, although it reverses reserpine sedation in animals (Carlsson et al., 1957) is not on the whole a useful treatment for depression (Bunney et al., 1971).

While evidence is against a persistent brain catecholamine abnormality in depression a sudden increase of functional concentrations may be involved in the transition from depression to mania. Thus L-DOPA or tricylic drugs which increase synaptic catecholamine concentration may precipitate this change while α-methyl-p-tyrosine which inhibits catecholamine synthesis may cause transition from mania to depression. The specific decrease in rapid eye movement sleep preceding mania is also consistent with increase of catecholamine (or decrease of indoleamine) function (Bunney et al., 1972).

Discussion of depressive illness in this chapter has dealt separately with the indoleamines and catecholamines largely for reasons of clarity of presentation. However, animal experiments show that these amines influence each other's behavioural effects. Thus Mabry and Campbell (1973) showed that the catecholamine induced arousal which occurred on giving rats amphetamine was increased synergistically by inhibition of 5-hydroxytryptamine synthesis while Green and Graham-Smith (1974) found that when rats were given tryptophan + a monoamineoxidase inhibitor the resultant hyperactivity depended on dopamine function. Thus relatively small functional abnormalities of single transmitter amines might together lead to considerable abnormality of mood and behaviour. Also abnormalities of different transmitters could lead to qualitatively similar mood change in different subjects. These effects could be further enhanced by altered cholinergic function as Janowsky et al. (1972) have compiled much evidence that high and low cholinergic activity respectively may be involved in depression and mania respectively.

B. SCHIZOPHRENIA

Twenty years ago when it was first hypothesized that transmitter amines had a causal role in psychoses far more attention was paid to schizophrenia than to depression. This was because of the analogies drawn between schizophrenia and hallucinations caused by LSD and mescaline and because of structural similarities between these substances and 5-hydroxytryptamine and catecholamines respectively. The inhibition by LSD of the action of 5-hydroxytryptamine on smooth muscle received particular emphasis as it suggested that a similar antiserotonin effect in the brain was involved in both LSD hallucinations and schizophrenia (Woolley & Shaw, 1954). Many urine and blood studies provoked by the above considerations led to transient flurries of excitement. The pitfalls and problems of this kind of work have been usefully described (Kety, 1959; Tanimukai & Himwich, 1970).

Many years later LSD was shown to interact with brain 5-hydroxytryptamine function behaving as if it acted on receptors so that neuronal firing and associated amine release and catabolism was decreased. Thus while stimulation of 5-hydroxytryptamine containing neurones causes 5-hydroxyindolylacetic acid to rise (Aghajanian et al., 1967) and increases 5-hydroxytryptamine synthesis (Shields & Eccleston, 1972), LSD decreases the firing rate of the neurones (Aghajanian et al., 1970) and has opposite biochemical effects (Rosecrans et al., 1967; Shields & Eccleston, 1973). NN-Dimethyltryptamine, another hallucinogen also decreases firing while 2- brom LSD has little effect. The latter substance is a powerful antagonist of the action of 5-hydroxytryptamine on muscle but has little hallucinogenic potency. Hallucinogens related to the catecholamines, e.g. mescaline, depressed firing but of only one group of the neurones affected by the indolic hallucinogens (Aghajanian et al., 1970). The more recent observations that both tryptophan a 5-hydroxytryptamine precursor and p-chlorophenylanine, an inhibitor of its synthesis also inhibit firing (Aghajanian, 1972), however, do not exhibit the satisfying symmetry of the above results.

Tryptophan is not a hallucinogen but can exacerbate schizophrenia when given with a monoamineoxidase inhibitor (Pollin et al. 1961). As the methyl donors methionine and betaine have similar effects (Brune & Himwich, 1962, 1963) it has been proposed that methylation and in particular N-methylation of an indoleamine might be involved in schizophrenia (reviewed Tanimukai & Himwich, 1970). The N-methylated tryptamine residues in many hallucinogens and the demonstration of brain N methyltransferases (see I B) add to the attractiveness of this hypothesis. However, there is direct evidence neither for

N-methylated indoleamines nor any other indoleamine abnormality in the brains of schizophrenics. The unconfirmed claim that in schizophrenia plasma tryptophan is particularly accessible to the brain (Frohman et al., 1969) is consistent with the above hypothesis as is also the claim (albeit a disputed one (Ban, 1973)) that nicotinamide has beneficial effect (Hoffer, 1972) because nicotinic acid can decrease tryptophan availability (Curzon & Knott, 1974). On the other hand reducing tryptophan and methionine intake was without beneficial effect (Berlet et al., 1965) and methionine-activating enzyme activity is normal in schizophrenic red cells (Dunner et al., 1973).

Another concept of transmitter amine disturbance in schizophrenia provides a hypothetical biochemical basis for a rather Skinnerian mechanism of the schizophrenic process (Stein, 1971; Stein & Wise, 1971). This hypothesis originates from results of self-stimulation experiments in animals which suggest that noradrenergic mechanisms are involved in pleasure or reward and thus with the development of normal purposeful thought and behaviour. If noradrenaline neurones are destroyed by 6-hydroxydopamine then self-stimulation is prevented. Stein and Wise propose that schizophrenia develops through the gradual destruction of these neurones by a 6-hydroxydopamine-like substance.

Although the validity of the animal experiments with 6-hydroxydopamine is debatable (Antelman et al., 1972) and there is no evidence for substances with similar properties in brain, degeneration of noradrenaline containing neurones in schizophrenia is consistent with the low brain dopamine-β-hydroxylase activity found at autopsy (Wise & Stein, 1973). It will be interesting to learn how specific this deficit is to schizophrenia.

Evidence also points to a possible causal role for dopaminergic hyperactivity in schizophrenia (reviews, Klawans, 1972; Snyder, 1973). This is consistent with the abnormal and stereotyped movements of many schizophrenics and also with the frequency of schizophrenia-like behaviour in Huntingtonian patients (see III A). Amphetamine, which inhibits reuptake of dopamine by striatal nerve endings, causes both stereotyped hyperactivity in animals (Munkvad et al., 1968) and psychosis in human addicts, which is very comparable with paranoid schizophrenia (Connell, 1958). Both D- and L-amphetamine are equally active (Snyder, 1973) and therefore a dopaminergic rather than a noradrenergic mechanism is suggested (Taylor & Snyder, 1970). The therapeutic potency of phenothiazines used to treat schizophrenia largely (Snyder, 1973) though not invariably (Crow & Gillbe, 1973) parallel their antagonistic effect on dopamine receptors in animals. These thereapeutic effects often occur together with Parkinsonian or other extrapyramidal side effects. Therefore dopaminergic involvement in schizophrenia is strongly suggested though it is perhaps surprising that L-DOPA precipitates schizophreniform states relatively rarely. It should not be assumed that striatal dopamine activity is necessarily involved. Hokfelt et al. (1974b) have demonstrated dopamine containing terminals in the rat cortex (especially in the limbic parts) and point out that abnormal activity here may be more relevant to schizophrenia than a striatal abnormality.

The syndrome of minimal brain dysfunction in children may be appropriately mentioned at this point as it may involve dopaminergic hypoactivity rather than the hyperactivity suggested in schizophrenia. The symptoms of restlessness, clumsiness and inattention have been compared with changes following encephalitic infection and are equally alleviated by D and L amphetamine which suggests a dopaminergic hypoactivity (Taylor & Snyder, 1970) being responsible for the disorder (Wender, 1972: 1974).

C. OTHER DISORDERS INVOLVING MENTAL DYSFUNCTION

1. DOWN'S SYNDROME

Considerable interest in the significance of transmitter amines in Down's syndrome was provoked by the demonstration that 5-hydroxytryptamine concentration in blood (and in particular in the platelets) was low (Rosner et al., 1965; Tu & Zellwerger, 1965). As well as this defect, which is by no means specific to Down's syndrome (Coleman, 1973), a defective maturation of the sympathetic nervous system is perhaps indicated by very low dopamine—β-hydroxylase activity in the plasma of affected children but not adults (Wetterberg et al., 1972). However, the normal 5-hydroxyindolylacetic acid and homovanillic acid concentrations in the CSF are evidence against abnormal central amine metabolism (Lott et al., 1972). Because of the plasma 5-hydroxytryptamine defect a very detailed study has been made of the effect of treatment with its precursor 5-hydroxytryptophan. Although this increased platelet 5-hydroxytryptamine, benefit was not noted in a double-blind trial apart from a transient improvement of muscle tone and seizures were more frequent than in untreated subjects (Coleman, 1973). Treatment with tryptophan had similar effects (Airaksinen, 1973). The low plasma 5-hydroxytryptamine content probably originates in a lack of ATP which leads to defective 5-hydroxytryptamine binding to platelets (Bouillin & O'Brien, 1971). Decreased plasma protein binding of tryptophan Airaksinen & Airaksinen, 1972) may also possibly lead to abnormal indoleamine metabolism.

2. INFANTILE AUTISM

This syndrome may be differentiated from general retardation by a characteristic failure to respond to others combined with ritualistic behaviour and abnormal language patterns. It has been debated whether infantile autism is a specific disease or merely a common response to various pathological processes. However, a common biochemical abnormality has been found by Bouillin et al. (1971), who showed that when blood platelets of autistic children were loaded with ^{14}C labelled 5-hydroxytryptamine its efflux into an incubation medium was grossly high. Prediction from efflux data agreed with psychological evaluation in nine subjects out of ten. The abnormal efflux is not obviously related to deficient platelets ATP as (unlike in Down's syndrome) this is normal. Whether any central abnormality of 5-hydroxytryptamine binding also occurs is unknown.

3. DEFECTS OF AMINO ACID METABOLISM

Inherited defects of amino acid metabolism especially those involving the aromatic acids may lead to brain amine abnormalities and mental disturbance. If the amino acid is itself an amine precursor then defective absorption or metabolism can lead to defective amine synthesis. Alternatively, non-precursor amino acids or their derivatives may inhibit the transport of an amine precursor to sites of amine synthesis in the brain or inhibit enzymes involved in amine synthesis. Thus the defective 5-hydroxytryptamine metabolism in phenylketonuria is possibly secondary to the high levels of phenylalanine. Serum 5-hydroxytryptamine, urinary 5-hydroxyindolylacetic acid (Pare et al., 1959) and CSF 5-hydroxyindolylacetic acid (Tu & Partington, 1972) are all low. When rats are fed a high phenylalanine diet then both learning and brain 5-hydroxytryptamine synthesis are both impaired (McKean et al., 1967). Similar results are obtained on feeding leucine, while conversely a high tryptophan diet improved learning and increased brain 5-hydroxytryptamine.

As animal experiments suggest that defective brain 5-hydroxytryptamine synthesis in early infancy causes irreversible impairment of brain growth (Hole, 1972) it may be that tryptophan supplementation during pregnancy of women at risk of producing phenylektonuric offspring and of the children themselves in early infancy may be useful. This speculation is also applicable to other defects of amino acid metabolism where brain 5-hydroxytryptamine deficiency is suggested.

The aromatic amino acids which are neurotransmitter precursors belong to a group of large neutral amino acids transported to the brain by a common mechanism—leucine, isoleucine, phenylalanine, tryptophan, tyrosine, valine (Blasberg & Lajtha, 1966; Kiely & Sourkes, 1972). Thus, decreased brain 5-hydroxytryptamine in rats on high leucine intake is probably due to inhibition of tryptophan transport to the brain (Ramanamurthy & Srikantia, 1970; Krishnaswamy & Raghuram, 1972). Similar biochemical changes possibly occur in maple syrup disease in which mental retardation is associated with very high plasma concentrations of branched chain amino acids and of related keto acids (Mackenzie & Woolf, 1969). Decreased brain 5-hydroxytryptamine could also be involved in mental changes in the variant of pellagra associated with a high leucine diet (Gopalan & Srikantia, 1960). Abnormal amine metabolism may play a part in the mental symptoms occurring in ordinary pellagra as the nicotinamine deficiency would tend to lead the already low tryptophan intake onto the pyrrolase pathway (Yamaguchi *et al.*, 1967) and thus away from 5-hydroxytryptamine synthesis. Again, decreased brain 5-hydroxytryptamine synthesis could have a role in causing the psychotic symptoms of some patients with Hartnup disease, as intestinal absorption of tryptophan is defective (Milne *et al.*, 1960; Lehmann, 1972). Various other rare disturbance of aromatic amino acid metabolism associated with mental defect and usually with evidence of inheritance have been reported (see Eastham & Jancar, 1968, for review) and might well also lead to disturbed brain amine metabolism, e.g. tyrosinaemia, tryptophanuria, hydroxykynureninuria and hypercalcaemia with indicanuria.

This section has stressed interference by other amino acids with the access of tryptophan to the brain. As previously discussed, this is of special importance because tryptophan hydroxylase (unlike tyrosine hydroxylase) is normally unsaturated with substrate (see I B). Various possible influences on tryptophan availability and 5-hydroxytryptamine function are summarized in Fig. 4.11.

Brain amine metabolism in senile and presenile dementia is discussed in Chapter 1.

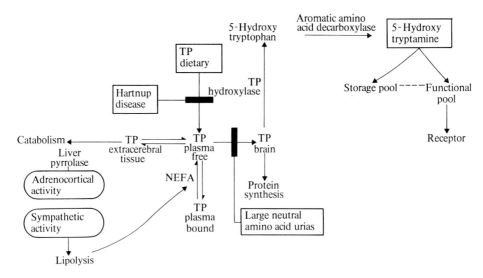

Fig. 4.11. Possible mechanisms involved in abnormal 5-hydroxytryptamine metabolism and function

TP = tryptophan; NEFA = non-esterified fatty acid.

V Migraine

A. INTRODUCTION

Migraine is defined as a familial disorder characterized by recurrent attacks of headache widely variable in intensity, frequency and duration. The name given to the disease derives from the frequently unilateral nature of attacks (migraine

= hemicrania). These are usually associated with anorexia, nausea and vomiting and may be preceded by or associated with neurological and mood disturbance. In classical migraine the headache is preceded or accompanied by transient focal neurological phenomena, e.g. visual, sensory or speech disturbance. Non-classical migraine is more common and is not associated with sharply defined focal neurological disturbance. Cluster headache (Horton's syndrome) is much rarer than migraine and unlike it has a predominantly male presentation (Lance & Anthony, 1971). Individual cluster attacks are brief but occur a few times each day.

Amines have been investigated in migraine not only because of their central transmitter actions but also because of their peripheral vasoactive properties as vascular changes are prominent. Amines could also be involved in the precipitation of attacks and in associated disturbance, e.g. gastrointestinal symptoms. A major difficulty in migraine research is the lack of closely relevant animal models.

B. AMINE CHANGES IN ATTACKS

Roles for the amines in migraine attacks are suggested not only by their vasoactive properties but also by other considerations. Altered milieu can precipitate attacks and sufferers often show 'overachieving' or 'overconscientious' characteristics. Therefore catecholamine secretion in response to stress could be important. This is consistent with the prophylactic effect of the β-adrenergic blocker propranolol (Weber & Reinmuth, 1972). Release of catecholamines might be responsible for the initial contraction of the temporal artery which occurs in the typical migraine attack. A role for 5-hydroxytryptamine is suggested by the antagonism of its pharmacological effects by the anti-migraine drug methysergide (Curran & Lance, 1964) and by the pain producing action of 5-hydroxytryptamine (Keele, 1967). Involvement of amines is also pointed to by the fairly typical attacks on giving migraine sufferers the amine depleter reserpine (Curzon et al., 1969). Subjects in whom reserpine induced attacks showed greater increases of the urinary amine metabolites 5-hydroxyindolylacetic acid and 3-methoxy-4-hydroxymandelic acid than did the (significantly older) sufferers in whom headache was not precipitated. Headaches induced by glyceryl nitrate are also claimed to be associated with an increased urinary amine metabolite excretion which does not occur in subjects in whom attacks are not precipitated (Campus et al., 1967).

Early claims of consistently large increases in excretion of the above metabolites on days when spontaneous attacks occurred were followed by other reports of less striking changes of both metabolites (Curran et al., 1965) and of one subject with striking increases of 5-hydroxyindolylacetic acid during and preceding attacks though most other subjects did not show increases (Curzon et al., 1966). Blood 5-hydroxytryptamine changes are more generally found.

Thus, plasma (Anthony et al., 1967) and whole blood concentrations (Hilton & Cumings, 1972) fall—in particular at the onset of attacks—apparently due to release of platelet 5-hydroxytryptamine (Anthony & Lance, 1975). This does not occur in cluster headache (Lance & Anthony, 1971).

The factor releasing platelet 5-hydroxytryptamine in typical migraine has been suggested to be an unesterified fatty acid as their plasma concentration increases during attacks (Anthony, 1975) and they release platelet 5-hydroxytryptamine from rabbit platelets *in vitro* (Inouye et al., 1970). It is therefore relevant to comment on the lipolytic action of catecholamines. As this is mediated by cyclic AMP, plasma unesterified fatty acid concentration can be increased both by increased catecholamine secretion and by raised cyclic AMP in fat stores (Robison et al., 1971). It may be significant that a number of dietary migraine precipitants (*vide infra*) could interact with this system to increase plasma fatty acid concentration. Thus, chocolate, tea and coffee all contain purines which inhibit cyclic AMP destruction and therefore tend to increase lipolysis. Plasma fatty acid would also tend to increase after intake of other dietary precipitants, e.g. fatty foods or alcohol. Particularly high plasma fatty acid concentrations were noted in subjects with migraine provoked by starvation (Hockaday et al. 1971). Similarly attacks provoked by glucose load occur during the reactive hypoglycaemia phase (Roberts, 1969) when plasma fatty acid would be high.

Whether the above fatty acid changes do cause release of platelet 5-hydroxytryptamine is not yet known. Presumably they also release plasma tryptophan from albumin binding and thus make it directly available for metabolism (Curzon & Knott, 1974; Curzon et al., 1974). The precipitation of migraine attacks via this effect is, however, seemingly contradicted by the claim (Sicuteri, 1973) that tryptophan alleviates migraine.

C. AMINE CONTAINING FOODS AS PRECIPITANTS

Many migraine sufferers find that particular foods or beverages precipitate attacks, i.e. in decreasing order of frequency chocolate, cheese, citrus fruits, alcohol, fried foods, tea and coffee (Hanington et al., 1969). These authors made a major advance when they reported that patients suffering from dietary migraine had attacks after taking a capsule containing tyramine much more frequently than did controls (Table 4.2). Cheese, citrus fruits and certain red wines are rich in tyramine and it had previously been found responsible for the headaches and hypertensive crises noted when patients on monamineoxidase inhibitors ate cheese (Blackwell & Mabbitt, 1965). The commonest migraine precipitant chocolate does not contain tyramine, but like some cheeses (Asatoor et al., 1963) is rich in phenylethylamine and Sandler et al. (1974) have recently reported that subjects claiming to suffer from chocolate provoked headache frequently had attacks when given phenylethylamine. A curious feature of these

Table 4.2 Precipitation of headache by oral tyramine or phenylethylamine

	*Migrainous		†Control	
	No effect	Headache	No effect	Headache
Lactose	45	3	97	4
Tyramine	18	65	97	5
Lactose	30	6	—	—
Phenylethylamine	18	18	—	—

*Migrainous subjects in the tyramine experiment claimed to have dietary migraine, while those in the phenylethylamine experiment claimed that attacks were provoked by chocolate.
†Control subjects were healthy women without history of headache.
Results from Hanington et al. (1970) and Sandler et al. (1974).

headaches was that onset was delayed for about 12 h after ingestion. This is also observed after chocolate and also in some subjects given tyramine.

In another study done under careful double-blind conditions attacks were provoked by tyramine no more readily than by placebo (Moffett et al., 1972). However, numbers of subjects studied were small and even here abnormal response to tyramine was indicated by the frequency of EEG changes in subjects claiming that attacks were provoked by tyramine containing foods.

Tyramine and phenylethylamine are mildly vasoactive and Hanington & Harper (1968) suggest that tyramine-provoked headache might involve release of noradrenaline (or perhaps 5-hydroxytryptamine) by tyramine leading to cerebral and extracerebral vasoconstriction. As tyramine releases 5-hydroxytryptamine from platelets *in vitro* (Bartholini & Pletscher, 1964) this could be important in tyramine headache. Unfortunately no platelet studies have yet been reported in tyramine headache.

There is evidence that tyramine metabolism may be defective in susceptible subjects. Thus Sandler et al. (1970) found that it led to increased output of catecholamine metabolites in tyramine sensitive migrainous subjects but not in controls. This suggested a metabolic defect in the sensitive subjects so that circulating tyramine levels become high enough to release noradrenaline from stores. Youdim et al. (1971) subsequently found that after tyramine loading sensitive subjects excreted abnormally small amounts of sulphate conjugated tyramine which suggested that tyramine migraine might be associated with a general defect in sulphate conjugation at ring hydroxyls.

A deficiency of platelet monoamineoxidase activity has also been found in migraine (Sandler et al., 1974). The deficiency was comparable in subjects both with and without a history of diet-provoked attacks. Whether the above metabolic disturbances appreciably alter blood tyramine and phenylethylamine concentrations preceding or during dietary migraine attacks is at present unknown. Also it would be of interest to know whether typical migraine can be precipitated by sufficient dosage of the amines in normal subjects. The existence of tyramine and phenylethylamine sensitive migraine raises the question of how many other dietary constituents have a similar role. Not only cheese and chocolate but any food prepared by fermentation may well contain high concentrations of vasoactive amines formed by decarboxylation of amino acids. It is also likely that individual migrainous subjects may be sensitive to many substances not necessarily all of an amine character. For example, 'hot-dog headache' has been described in a tyramine sensitive subject apparently due to nitrite, added as a meat preservative (Henderson & Raskin, 1972). It was suggested that both the nitrite and tyramine headaches might involve a common mechanism.

VI Comments

In the past decade, the sophistication of approach to transmitter amine metabolism and function has increased considerably. A radical change which began some ten years ago, has resulted from the use of stereotactic and histochemical fluorescence methods. This has led to our present knowledge of the location of the cell bodies, axons and terminal systems of aminergic neurones. Relationships between amine metabolism and function can be very different in these different parts of the neuronal tracts.

It is valuable to know not only the concentrations of the amines but also those of precursors and metabolites and thus obtain information on net regional turnover. This is, however, of uncertain value if we are concerned with aminergic function. In these circumstances, receptor sensitivity is highly relevant and may be studied by measuring the behavioural effects of agonists acting at receptors of specific amines or by flooding receptors with the amine normally acting on them. Also the fraction of amine turnover derived from material released into the synaptic cleft is obviously more relevant to function than is net turnover as intraneuronal (and, therefore, functionally irrelevant) catabolism may make a large contribution to the latter. Reliable and simple methods for distinguishing between functional and non-functional pools without considerable disturbance of neuronal activity are, however, at present not available.

These considerations stem mainly from the facts that brain amines occur in neuronal tracts and that they have roles in transmission. They are compounded by the implications of the striking localization of functions which is unique to the brain—a defect in a small part of the liver will usually have no apparent pathological effect, but a comparable brain abnormality may grossly impair a specific mechanism. We are also starting to appreciate the importance of biochemical and functional relationships between different neuronal systems. It is also apposite to remark that the serotoninergic and catecholaminergic systems discussed in this chapter contain only a small fraction of the neurones of the brain.

The above comments imply that it is no longer adequate to study brain amines exclusively from the standpoint of any single conventionally defined scientific discipline. Neuroanatomical, biochemical, pharmacological, behavioural and other approaches are all necessary. This multiplicity of procedures involves both opportunities and restrictions—it is not possible to apply more than a fraction of them directly to appropriate human material. Therefore, a fruitful approach to the pathological roles of the brain amines is likely to be one in which limited studies in the human and wider ranging work on relevant animal models mutually reinforce each other.

References

AGHAJANIAN G.K. (1972) Chemical-feedback regulation of serotonin-containing neurones in brain. *Ann. N.Y. Acad. Sci.* **193**, 86–94.

AGHAJANIAN G.K., FOOTE W.E. & SHEARD M.H. (1970) Action of psychotogenic drugs on single midbrain raphe neurons. *J. Pharmac.* **171**, 178–187.

AGHAJANIAN G.K., ROSECRANS J.A. & SHEARD M.H. (1967) Serontonin: release in the forebrain by stimulation. *Science, N.Y.* **156**, 402–403.

AIRAKSINEN E.M. (1974) Tryptophan treatment of infants with Down's syndrome. *Ann. Clin. Res.* **6**, 33–39.

AIRAKSINEN E.M. & AIRAKSINEN M.M. (1972) The binding of tryptophan to plasma proteins and the rate of the inactivation of 5HT released from platelets in Down's syndrome. *Ann. Clin. Res.* **4**, 361–365.

AKISKAL H.S. & MCKINNEY W.T. (1973) Depressive disorders: towards a unified hypothesis. *Science N.Y.* **182**, 20–29.

ANDEN N.E., CARLSSON A., DAHLSTROM A., FUXE K., HILLARP N.A. & LARSSON K. (1964) Demonstration and mapping out of nigro-neostriatal dopamine neurons. *Life Sci.* **3**, 523–530.

ANDEN N.E., DAHLSTROM A., FUXE K. & LARSSON K. (1965) Further evidence for the presence of nigro-neostriatal dopamine neurons in the rat. *Am. J. Anat.* **116**, 329–333.

ANDEN N.E., STROMBOM U. & SVENSSON T.H. (1973) Dopamine and noradrenaline receptor stimulation: reversal of reserpine-induced suppression of motor activity. *Psychopharmacologia* **29**, 289–298.

ANTELMAN S.M., LIPPA A.S. & FISHER A.E. (1972) 5-Hydroxydopamine, noradrenergic reward, and schizophrenia. *Science, N.Y.* **175**, 919–920.

ANTHONY M. (1975) Patterns of plasma free fatty acid and prostaglandin E_1 changes in migraine and stress. *Background to Migraine.* In press.

ANTHONY M. & LANCE J.W. (1975) Role of serotonin in migraine. In Pearce J. (Ed.) *Contemporary Essays on Migraine.* London, Heinemann. In press.

ANTHONY M., HINTERBERGER H. & LANCE J.W. (1967) Plasma serotonin in migraine and stress. *Arch. Neurol.* **16**, 544–552.

ASATOOR A.M., LEVI A.J. & MILNE M.D. (1963) Tranylcypromine and cheese. *Lancet* ii, 733–734.

ASHCROFT G.W., BLACKBURN J.M., ECCLESTON D., GLEN A.I., HARTLEY W., KINLOCH N.E., LONERGAN M., MURRAY L.G. & PULLAR I.A. (1973a) Changes on recovery in the concentrations of tryptophan and the biogenic amine metabolites in the cerebrospinal fluid of patients with affective illness. *Psychol. Med.* **3**, 319–325.

ASHCROFT G.W., CRAWFORD T.B.B., CUNDALL R.L., DAVIDSON D.L., DOBSON J., DOW R.C., ECCLESTON D., LOOSE R.W. & PULLAR I.A. (1973b) 5-Hydroxytryptamine metabolism in affective illness: the effect of tryptophan administration. *Psychol. Med.* **3**, 326–332.

ASHCROFT G.W., CRAWFORD T.B.B., ECCLESTON D., SHARMAN D.F., MACDOUGALL E.J., STANTON J.B. & BINNS J.K. (1966) 5-Hydroxyindole compounds in the cerebrospinal fluid of patients with psychiatric or neurological diseases. *Lancet* **ii**, 1049–1052.

ASHCROFT G.W., DOW R.C. & MOIR A.T.B. (1968) The active transport of 5-hydroxyindol-3-ylacetic acid and 3-methoxy-4-hydroxyphenylacetic acid from a recirculatory perfusion system of the cerebral ventricles of the unanaesthetized dog. *J. Physiol. (London)* **199**, 397–425.

AXELROD J. (1971) Noradrenaline: fate and control of its biosynthesis. *Science N.Y.* **173**, 598–606.

AXELROD J. & SAAVEDRA J.M. (1974) Octopamine, phenylethanolamine, phenylethylamine and tryptamine in the brain (1974). In *Aromatic Amino Acids in the Brain. Ciba Foundation Symposium* **22**, pp. 51–59.

BACQ Z.M. (1971) *Fundamentals of Biochemical Pharmacology.* Oxford, Pergamon.

BAN T.A. (1973) *Recent Advances in the Biology of Schizophrenia.* Springfield, Illinois, Thomas.

BARBEAU A. (1972) Long-term appraisal of levodopa therapy. *Neurology* **22** (Suppl.), 22–24.

BARCHAS J.D., CIARANELLO R.D., STOLK J.M., BRODIE H.K.H. & HAMBURG D.A. (1972) Biogenic amines and behaviour. In Levine S. (Ed.) *Hormones and Behaviour*, pp. 235–329. New York, Academic Press.

BAROLIN G.S., BERNHEIMER H. & HORNYKIEWICZ O. (1964) Seitenverschiedenes Verhalten des Dopamins (3-Hydroxytyramin) in Gehirn eines Falles von Hemiparkinsonismus. *Schweiz. Arch. Neurol. Neurochir. Psychiat.* **94**, 241–248.

BARTHOLINI G., LLOYD K.G. & PLETSCHER A. (1974) Regional distribution of catecholamines in rat brain after L-3-0-methyldopa and L-DOPA. *Life Sci.* **14**, 323–328.

BARTHOLINI G. & PLETSCHER A. (1964) Two types of 5-hydroxytryptamine release from isolated blood platelets. *Experientia* **20**, 376.

BARTHOLINI G., TISSOT R. & PLETSCHER A. (1971) Brain capillaries as a source of homovanillic acid in cerebrospinal fluid. *Brain Res.* **27**, 163–168.

BAREGGI S.R., MARC V. & MORSELLI P.L. (1974) Urinary excretion of 3-methoxy-4-hydroxyphenylglycol sulphate in rats after intraventricular injection of 6-OHDA. *Brain Res.* **75**, 177–180.

BERGMANN S., CURZON G., FRIEDEL J., GODWIN-AUSTEN R.B., MARSDEN C.D. & PARKES J.D. (1974) The absorption and metabolism of a standard oral dose of levodopa in patients with Parkinsonism. *Brit. J. Clin. Pharmac.* **1**, 417–424.

BERLET H.H., SPAIDE J., KOHL H., BULL C.E., HIMWICH H.E. (1965) Effects of reduction of tryptophan and methionine intake on urinary indole compounds and schizophrenic behaviour. *J. nerv. ment. Dis.* **140**, 297–304.

BERNHEIMER H., BIRKMAYER W. & HORNYKIEWICZ O. (1961) Verteilung des 5-Hydroxytryptamine (Serotonin) im Gehirn des Menschen und sein Verhalten bei Patienten mit Parkinson-Syndrom. *Klin. Wschr.* **39**, 1056–1059.

BERNHEIMER H., BIRKMAYER W., HORNYKIEWICZ O., JELLINGER K. & SEITELBERGER F.

(1973) Brain dopamine and the syndromes of Parkinson and Huntington. Clinical, morphological and neuro-chemical correlations. *J. Neurol. Sci.* **20**, 415–455.

BERTLER A. (1961) Occurrence and localization of catecholamines in the brain and other tissues. *Acta. Physiol. Scand.* **51**, 97–107.

BEVAN-JONES B., PARE C.M.B., NICHOLSON W.J.N., PRICE K. & STACEY R.S. (1972) Brain amine concentrations after monoamineoxidase inhibitor administration. *Brit. med. J.* **1**, 17–19.

BIRD E.D. & IVERSEN L.L. (1974) Huntington's chorea: post-mortem measurement of glutamic acid decarboxylase, choline acetyltransferase and dopamine in basal ganglia. *Brain* **97**, 457–472.

BIRKMAYER W., DANIELCZYK W., NEUMAYER E. & RIEDERER P. (1972) The balance of biogenic amines as condition for normal behaviour. *J. Nerv. Trans.* **33**, 163–178.

BLACKWELL B. & MABBITT L.A. (1965) Tyramine in cheese related to hypertensive crises after monoamineoxidase inhibition. *Lancet* **1**, 938–940.

BLASBERG R. & LAJTHA A. (1966) Heterogeneity of the mediated transport systems of amino acid uptake in brain. *Brain Res.* **1**, 86–104.

BLASCHKO H. (1939) The specific action of L-DOPA decarboxylase. *J. Physiol.* **96**, 50P–51P.

BOUILLIN D.J., COLEMAN M., O'BRIEN R.A. & RIMLAND B. (1971) *J. Autism. Childh. Schiz.* **1**, 63–71.

BOUILLIN D.J. & O'BRIEN R.A. (1971) Abnormalities of 5-hydroxytryptamine uptake and binding by blood platelets from children with Down's syndrome. *J. Physiol.* **212**, 287–297.

BOURNE H.R., BUNNEY W.E., COLBURN R.W., DAVIS J.M., DAVIS J.N., SHAW D.M. & COPPEN A.J. (1968) Noradrenaline, 5-hydroxytryptamine and 5-hydroxyindoleacetic acid in hindbrains of suicidal patients. *Lancet* **2**, 805–808.

BOWERS M.B. (1970) Cerebrospinal fluid 5-hydroxyindoles and behaviour after L-tryptophan and pyridoxine administration to psychiatric patients. *Neuropharmac.* **9**, 599–604.

BOWERS M.B. (1972) Clinical measurements of central dopamine and 5-hydroxytryptamine metabolism: reliability and interpretation of cerebrospinal fluid acid monoamine metabolite measures. *Neuropharmac.* **11**, 101–111.

BREESE G.R., CHASE T.N. & KOPIN I.J. (1969) Metabolism of some phenylethylamines and their β-hydroxylated analogs in brain. *J. Pharmac.* **165**, 9–13.

BROWNSTEIN M., SAAVEDRA J.M. & PALKOVITS M. (1974) Norepinephrine and dopamine in the limbic system of the rat. *Brain Res.* **79**, 431–436.

BRUNE G.G. & HIMWICH H.E. (1962) Effects of methionine loading on the behaviour of schizophrenic patients. *J. nerv. ment. Dis.* **134**, 447–450.

BRUNE G.G. & HIMWICH H.E. (1963) Biogenic amines and behaviour in schizophrenic patients. *Recent Adv. biol. Psychiat.* **5**, 144–160.

BULAT M. & ZIVKOVIC B. (1971) Origin of 5-hydroxyindoleacetic acid in the spinal fluid. *Science N.Y.* **173**, 738–740.

BUNNEY W.E., BRODIE H.K., MURPHY D. & GOODWIN F.K. (1971) Studies of alpha-methyl-para-tyrosine, L-DOPA and L-tryptophan in depression and mania. *Am. J. Psychiat.* **127**, 872–881.

BUNNEY W.E. & DAVIS J.M. (1965) Norepinephrine in depressive reactions. A review. *Arch. Gen. Psychiat.* **13**, 483–494.

BUNNEY W.E., GOODWIN F.K., MURPHY D.L., HOUSE K.M. & GORDON E.K. (1972) The 'switch process' in manic depressive illness. II. Relationship to catecholamines, REM sleep and drugs. *Arch. Gen. Psychiat.* **27**, 304–309.

BURGERMEISTER J.J., DICK P., GARRONE G., GUGGISBERG M. & TISSOT M. (1963)

Urinary excretion of 5-hydroxyindoleacetic acid (5HIAA) in 150 patients with depressive syndrome and maniacal agitation (its modifications by 5-hydroxytryptophan loading and therapy in the depressive states) *Pr. Med.* **71**, 1116–1118.

CALNE D.B., CHASE T.N. & BARBEAU A. (Eds.) (1975) Dopaminergic mechanisms *Adv. Neurol.* **9**.

CALNE D.B., REID J.L. & VAKIL S.D. (1973) Parkinsonism treated with 3-0-methyldopa. *Clin. Pharm. Therap.* **14**, 386–389.

CALNE D.B., SPIERS A.S.D., STERN G.M., LAURENCE D.R. & ARMITAGE P. (1969a) L-DOPA in idiopathic parkinsonism. *Lancet* **ii**, 973–976.

CALNE D.B., STERN G.M., LAURENCE D.R., SHARKEY J.M. & ARMITAGE P. (1969b) L-DOPA in postencephalitic Parkinsonism. *Lancet* **i**, 744–746.

CAMPUS S., FABRIS F., RAPPELLI A., GASTALDI L., GAI V. & NATTERO G. (1967) Escrezione urinaria di acido 5-idrossiindolacetico durante crisi cefalalyica indotta da trinitro glicerina. *Boll. Soc. ital. biol. sper.* **43**, 1844–1847.

CARDON P.V. & MUELLER P.S. (1966) A possible mechanism: psychogenic fat mobilization. *Ann. N.Y. Acad. Sci.* **125**, 924–927.

CARLSSON A. & LINDQVIST M. (1963) Effect of chlorpromazine or haloperidol on formation of 3-methoxytyramine and normetanephrine in mouse brain. *Acta pharmacol et toxicol.* **20**, 140–144.

CARLSSON A., LINDQVIST M. & MAGNUSSON T. (1957) 3,4 Dihydroxyphenylalanine and 5-hydroxytryptophan as reserpine antagonists. *Nature* **180**, 1200.

CARNEY M.W.P., ROTH M. & GARSIDE R.F. (1965) The diagnosis of depressive syndromes and the prediction of E.C.T. response. *Brit. J. Psychiat.* **111**, 659–674.

CHASE T.N., GORDON E.K. & NG L.K.Y. (1973) Norepinephrine metabolism in the central nervous system of man: studies using 3-methoxy-4-hydroxyphenylethylene glycol levels in cerebrospinal fluid. *J. Neurochem.* **21**, 581–587.

CHASE T.N., & NG L.K.Y. (1972) Central monamine metabolism in Parkinson's disease. *Arch. Neurol.* **27**, 486–491.

CHASE T.N., NG L.K.Y. & WATANABE A.M. (1972) Parkinson's disease. Modification by 5-hydroxytryptophan. *Neurology* **22**, 479–484.

CHASE T.N., SCHNUR J.A. & GORDON E.K. (1970) Cerebrospinal fluid monamine catabolites in drug-induced extrapyramidal disorders. *Neuropharmac.* **9**, 265–268.

CIBA FOUNDATION SYMPOSIUM **22**. *Aromatic Amino Acids in the Brain* (1974) Amsterdam, Elsevier.

CLAVERIA L.E., CALNE D.B. & ALLEN J.G. (1973) 'On–Off' phenomena related to high plasma levodopa. *Brit. med. J.* **2**, 641–643.

CLAVERIA L.E., CURZON G., HARRISON M.J. & KANTAMANENI B.D. (1974) Amine metabolites in the cerebrospinal fluid of patients with disseminated sclerosis. *J. Neurol. Neurosurg. Psychiat.* **37**, 715–718.

COLEMAN M. (Ed.) (1973) *Serotonin in Down's Syndrome.* Amsterdam, North Holland.

CONNELL P.H. (1958) *Amphetamine Psychosis.* London, Chapman and Hall.

COPPEN A. (1973) Role of serotonin in affective disorders. In Barchas J. (Ed.) *Serotonin and Behaviour*, pp. 523–527. New York, Academic Press.

COPPEN A., ECCLESTON E.G. & PEET M. (1973) Total and free tryptophan in the plasma of depressive patients. *Lancet* **ii**, 60–63.

COPPEN A., METCALFE M., CARROLL J.D. & MORRIS J.G.L. (1972) Levodopa and L-tryptophan therapy in Parkinsonism. *Lancet* **i**, 654–658.

COPPEN A., SHAW D.M. & FARRELL J.P. (1963) Potentiation of the antidepressive effect of a monoamine-oxidase inhibitor by tryptophan. *Lancet* **i**, 79–81.

COSTA E., GESSA G.L. & SANDLER M. (1974a) *Serotonin: New Vistas. Histochemistry and Pharmacology.* New York, Raven Press.

COSTA E., GESSA G.L. & SANDLER M. (1974b) *Serotonin: New Vistas. Biochemistry and Behavioural and Clinical Studies*. New York, Raven Press.

COSTA E. & MEEK J.L. (1974) Regulation and biosynthesis of catecholamines and serotonin in the CNS. *Ann. Rev. Pharmac.* **14**, 491–511.

COTZIAS G.C., PAPAVASILIOU P.S. & GELLENE R. (1969) Modification of Parkinsonism-chronic treatment with L-DOPA. *New Engl. J. Med.* **280**, 337–345.

COTZIAS G.C., VAN WOERT M.H. & SCHIFFER L.M. (1967) Aromatic amino acids and modification of Parkinsonism. *New Engl. J. Med.* **276**, 374–379.

COYLE J.T. & SNYDER S.H. (1969) Antiparkinsonian drugs: inhibition of dopamine uptake in the corpus striatum as a possible mechanism of action. *Science, N.Y.* **166**, 899–901.

CRANE G.E. (1973) Persistent dyskinesia. *Brit. J. Psychiat.* **122**, 395–405.

CROW T.J. & GILLBE C. (1973) Dopamine antagonism and schizophrenic potency of neuroleptic drugs. *Nature, New Biol.* **245**, 27–28.

CURRAN D.A., HINTERBERGER H. & LANCE J.W. (1965) Total plasma serotonin 5-hydroxyindoleacetic acid and p-hydroxy-m-methoxymandelic acid excretion in normal and migrainous subjects. *Brain* **88**, 997–1007.

CURRAN D.A. & LANCE J.W. (1964) Clinical trial of methysergide and other preparations in the management of migraine. *J. Neurol. Neurosurg. Psychiat.* **27**, 463–469.

CURZON G. (1967) The biochemistry of dyskinesias. *Int. Rev. Neurobiol.* **10**, 323–370.

CURZON G. (1969) Tryptophan pyrrolase—a biochemical factor in depressive illness? *Brit. J. Psychiat.* **115**, 1367–1374.

CURZON G. (1972) Brain amine metabolism in some neurological and psychiatric disorders. In Cumings, J.N. (Ed.) *Biochemical Aspects of Nervous Diseases*, pp. 151–212. London, Plenum.

CURZON G. (1973) Involuntary movements other than Parkinsonism: biochemical aspects. *Proc. Roy. Soc. Med.* **66**, 873–876.

CURZON G. (1975) CSF Homovanillic acid: an index of dopaminergic activity? *Adv. Neurol.* **9**, 349–357.

CURZON G., BARRIE M. & WILKINSON M.I.P. (1969) Relationships between headache and amine changes after administration of reserpine to migrainous patients. *J. Neurol. Neurosurg. Psychiat.* **32**, 555–561.

CURZON G. & BRIDGES P. (1970) Tryptophan metabolites in depression. *J. Neurol. Neurosurg. Psychiat.* **33**, 698–704.

CURZON G., FRIEDEL J., KANTAMANENI B.D., GREENWOOD M.H. & LADER M.H. (1974) Unesterified fatty acids and the binding of tryptophan in human plasma. *Clin. Sci.* **47**, 415–424.

CURZON G., GUMPERT E.J.W. & SHARPE D.M. (1971) Amine metabolites in the lumbar cerebrospinal fluid of humans with restricted flow of cerebrospinal fluid. *Nature (New Biol.)* **231**, 189–191.

CURZON G., GUMPERT J. & SHARPE D. (1972) Amine metabolites in the cerebrospinal fluid in Huntington's Chorea. *J. Neurol. Neurosurg. Psychiat.* **35**, 514–519.

CURZON G. & KNOTT P.J. (1974) Fatty acids and the disposition of tryptophan. In *Aromatic Amino Acids in the Brain. Ciba Foundation Symposium* **22**, pp. 217–229. Amsterdam, Elsevier.

CURZON G., THEAKER P. & PHILLIPS B. (1966) Excretion of 5-hydroxyindolyl acetic acid (5HIAA) in migraine. *J. Neurol. Neurosurg. Psychiat.* **29**, 85–90.

DAIRMAN W., HORST W.D., MARCHELLE M.E. & BAUTZ D. (1975) The proportionate loss of L-3-4-dihydroxyphenylalanine and L-5-hydroxytryptophan decarboxylating activity in rat central nervous system following intracisternal administration of 5, 6-dihydroxytryptamine or 6-hydroxydopamine. *J. Neurochem.* **24**, 619–623.

DAMASIO A.R., CASTRO-CALDAS A. & LEVY A. (1973) The on–off effect. *Adv. Neurol.* **3**, 11–22.
DE LA TORRE J.C. (1972) Catecholamines in the human diencephalon; a histochemical fluorescence study. *Acta Neuropath. (Berl.)* **21**, 165–168.
DUNNER D.L., COHN C.K., WEINSHILBOUM R.M. & WYATT R.J. (1973) The activity of dopamine-beta-hydroxylase and methionine-activating enzyme in blood of schizophrenic patients. *Biol. Psychiat.* **6**, 215–220.
EASTHAM R.D. & JANCAR J. (1968) *Clinical Pathology in Mental Retardation*. Bristol, Wright.
ECCLESTON D., ASHCROFT G.W. & CRAWFORD T.B.B. (1965) 5-Hydroxyindole metabolism in rat. A study of intermediate metabolism using the technique of tryptophan loading. *J. Neurochem.* **12**, 493–503.
ECCLESTON D., MOIR A.T.B., READING H.W. & RITCHIE I.M. (1966) The formation of 5-hydroxytryptophol in brain *in vitro*. *Br. J. Pharmac.* **28**, 367–377.
EHRINGER H. & HORNYKIEWICZ O. (1960) Verteilung von Noradrenalin und Dopamin (3-Hydroxytyramin) in Gehirn des Menschen und ihr Verhalten bei Erkrankungen des Extrapyramiden Systems. *Klin. Wschr.* **38**, 1236–1239.
EVERETT G.M. & BORCHERDING J.W. (1970) L-DOPA: effect on concentration of dopamine, norepinephrine and serotonin in brains of mice. *Science N.Y.* **168**, 849–850.
FAHN S., LIBSCH L.R. & CUTLER L.W. (1971) Monoamines in the human neostriatum: topographic distribution in normals and in Parkinson's disease and their role in akinesia, rigidity, chorea and tremor. *J. Neurol. Sci.* **14**, 427–455.
FOG R. & PAKKENBERG H. (1970) Combined nitoman-pimozide treatment of Huntington's chorea and other hyperkinetic syndromes. *Acta. neurol. Scand.* **46**, 249–251.
FRAZER A., PANDEY N.G. & MENDELS J. (1973) Metabolism of tryptophan in depressive illness. *Arch. Gen. Psychiat.* **29**, 528–535.
FROHMAN C.E., WARNER K.A., BARRY C.T. & ARTHUR R.E. (1969) Amino acid transport and the plasma factor in schizophrenia. *Biol. Psychiat.* **1**, 201–208.
FUXE K., BUTCHER L.L. & ENGEL G. (1971) DL-5-Hydroxytryptophan induced changes in central monoamine neurons after peripheral decarboxylase inhibition. *J. Pharm. Pharmac.* **23**, 420–424.
FUXE K., GOLDSTEIN M. & LJUNGDAHL A. (1970a) Antiparkinsonian drugs and central dopamine neurons. *Life Sci.* **9**, Part 1, 811–824.
FUXE K., HOKFELT T., JONSSON G., LEVINE S., LIDBRINK P. & LOFSTROM A. (1973) Brain and pituitary-adrenal interactions. Studies on central monoamine neurons. In Brodish A. & Redgate E.S. (Eds.) *Brain–Pituitary–Adrenal Interrelationships*, pp. 239–269. Basle, Karger.
FUXE K., HOKFELT T. & UNGERSTEDT U. (1970b) Morphological and functional aspects of central monoamine neurons. *Int. Rev. Neurobiol.* **13**, 93–126.
FUXE K. & JONSSON G. (1974) Further mapping of 5-hydroxytryptamine neurons: studies with the neurotoxic dihydroxytryptamines *Adv. Biochem. Psychopharmac.* **10**, 1–12.
FYRO B., HELGODT S. & SEDVALL G. (1974) The effect of chlorpromazine on homovanillic acid levels in cerebrospinal fluid of schizophrenic patients. *Psychopharmacologia* **35**, 287–294.
GARELIS E., YOUNG S.N., LAL S. & SOURKES T.L. (1974) Monoamine metabolites in lumbar CSF: the question of their origin in relation to clinical studies. *Brain Res.* **79**, 1–8.
GESSA G.L. & TAGLIAMONTE A. (1974) Serum free tryptophan: control of brain concentrations of tryptophan and of synthesis of 5-hydroxytryptamine. In *Aromatic*

Amino Acids in the Brain. Ciba Foundation Symposium **22**, pp. 207–216. Amsterdam, Elsevier.

GIANUTSOS G., DRAWBAUGH R.B., HYNES M.D. & LAL H. (1974) Behavioural evidence for dopaminergic supersensitivity after chronic haloperidol. *Life Sci.* **14**, 887–898.

GLASSMAN A.H. & PLATMAN S.R. (1969) Potentiation of a monoamine-oxidase inhibitor by tryptophan. *J. Psychiat. Res.* **7**, 83–88.

GODWIN-AUSTEN R.B. KANTAMANENI B.D. & CURZON G. (1971) Comparison of benefit from L-DOPA in Parkinsonism with increase of amine metabolites in the CSF. *J. Neurol. Neurosurg. Psychiat.* **34**, 219–223.

GODWIN-AUSTEN R.B., TOMLINSON E.B., FREARS C.C. & KOK H.W.L. (1969) Effects of L-DOPA in Parkinsons' disease. *Lancet* ii, 165–168.

GOLDSTEIN M., ANAGOSTE B., BATTISTA A.F., OWEN W.S. & NAKATANI S. (1969) Studies of the amines in the striatum in monkeys with nigral lesions. *J. Neurochem.* **16**, 645–653.

GOPALAN C. & SRIKANTIA S.G. (1960) Leucine and Pellagra. *Lancet* i, 954–957.

GOTTFRIES C.G., GOTTFRIES I. & ROOS B.E. (1969) Homovanillic acid and 5-hydroxyindoleacetic acid in the cerebrospinal fluid of patients with senile dementia, presenile dementia and Parkinsonism. *J. Neurochem.* **16**, 1341–1345.

GRANERUS A.K., JAGENBURG R. & SVANBORG A. (1973) Intestinal decarboxylation of L-DOPA in relation to dose requirement in Parkinson's disease. *Arch. Pharmacol.* **280**, 429–439.

GREEN A.R. & GRAHAME-SMITH D.G. (1974) The role of brain dopamine in the hyperactivity syndrome produced by increased 5-hydroxytryptamine synthesis in rats. *Neuropharmac.* **13**, 949–959.

GREEN A.R. & CURZON G. (1975) Effects of hydrocortisone and immobilization on tryptophan metabolism in brain and liver of rats of different ages. *Biochem. Pharmac.* **24**, 713–716.

GREEN A.R., KOSLOW S.H. & COSTA E. (1973) Identification and quantitation of a new indole alkylamine in rat hypothalamus. *Brain Res.* **51**, 371–374.

GREENFIELD J.G. (1955) in Critchley M. (Ed.) *James Parkinson 1755–1824: A Bicentenary Volume of Papers Dealing with Parkinson's Disease*, pp. 219–243. New York, Macmillan.

GREENFIELD J.G. & BOSANQUET F.D. (1953) The brain-stem lesions in Parkinsonism. *J. Neurol. Neurosurg. Psychiat.* **16**, 213–226.

GUILDBERG H.C. (1969) Changes in amine metabolite concentrations in cerebrospinal fluid as an index of turnover. In Hooper G. (Ed.) *Metabolism of Amines in the Brain*, pp. 55–64. London, Macmillan.

GULDBERG H.C., TURNER J.W., HANIEH A., ASHCROFT G.W., CRAWFORD T.B.B., PERRY W.L.M. & GILLINGHAM F.J. (1967) On the occurrence of homovanillic acid and 5-hydroxyindol-3-ylacetic acid in the ventricular CSF of patients suffering from Parkinsonism. *Confinia neurol.* **29**, 73–77.

GULDBERG H.C. & YATES C.M. (1968) Some studies of the effects of chlorpromazine, reserpine and dihydroxyphenylalanine on the concentrations of homovanillic acid, 3,4-dihydroxyphenylacetic acid and 5-hydroxyindol-3-ylacetic acid in ventricular cerebrospinal fluid of the dog using the technique of serial sampling of the cerebrospinal fluid. *Br. J. Pharmac.* **33**, 457–471.

GUMPERT E.J.W., SHARPE D.M. & CURZON G. (1973) Amine metabolites in cerebrospinal fluid in Parkinson's disease and the response to L-DOPA. *J. Neurol. Sci.* **19**, 1–12.

HALL C.D., WEISS E.A., MORRIS C.E. & PRANGE J. (1972) Rapid deterioration in patients with Parkinsonism following tryptophan-pyridoxine administration. *Neurology* **22**, 231–237.

HANINGTON E. & HARPER A.M. (1968) The role of tyramine in the aetiology of migraine, and related studies on the cerebral and extracerebral circulations. *Headache* **8**, 84–97.

HANINGTON E., HORN M. & WILKINSON M. (1969) Further observations on the effects of tyramine. *Background to Migraine* **3**, 113–119.

HENDERSON W.R. & RASKIN N.H. (1972) 'Hot-dog' headache: individual susceptibility to nitrate. *Lancet* **ii**, 1162–1163.

HERRINGTON R.N., BRUCE A., JOHNSTONE E.C. & LADER M.H. (1974) Comparative trial of L-tryptophan and E.C.T. in severe depressive illness. *Lancet* **ii**, 731–734.

HILTON B.R. & CUMINGS J.N. (1972) 5-Hydroxytryptamine levels and platelet aggregation responses in subjects with acute migraine headache. *J. Neurol. Neurosurg. Psychiat.* **35**, 505–509.

HOCKADAY J.M., WILLIAMSON D.H. & WHITTY C.W.M. (1971) Blood glucose levels and fatty acid metabolism in migraine related to fasting. *Lancet* **i**, 1153–1159.

HOFFER A. (1972) Mechanism of action of nicotinic acid and nicotin-amide in the treatment of schizophrenia in Hawkins D. & Pauling D. (Eds.) *Orthomolecular Psychiatry*, pp. 202–262. San Francisco, Freeman.

HOKFELT T., FUXE M., GOLDSTEIN M. & JOHANSSON O. (1974a) Immunohistochemical evidence for the existence of adrenaline neurons in the rat brain. *Brain Res.* **66**, 235–251.

HOKFELT T., LJUNGDAHL A., FUXE K. & JOHANSSON O. (1974b) Dopamine nerve terminals in the rat limbic cortex: aspects of the dopamine hypothesis of schizophrenia. *Science, N.Y.* **184**, 177–179.

HOLE K. (1972) Reduced 5-hydroxyindole synthesis reduces postnatal brain growth in rats. *Eur. J. Pharmac.* **18**, 361–366.

HORNYKIEWICZ O. (1963) The topical localization and content of noradrenalin and dopamine (3-hydroxytyramine) in the substantia nigra of normal persons and patients with Parkinson's disease. *Wien. Klin. Wschr.* **75**, 309–312.

HORNYKIEWICZ O. (1964a) The distribution and metabolism of catecholamines and 5-hydroxytryptamine in human brain. In Richter D. (Ed.) *Comparative Neurochemistry*, pp. 379–386. Oxford, Pergamon.

HORNYKIEWICZ O. (1964b) The role of brain dopamine (3-hydroxytyramine) in Parkinsonism. *Biochemical and Neurophysiological Correlation of Centrally Acting Drugs (Proc. 2nd Int. Pharm. Meeting)*, pp. 57–68.

HSU L.L. & MANDELL A.J. (1974) Multiple N-methyltransferases for aromatic alkylamines in brain. *Adv. Biochemical Psychopharmac.* **11**, 75–84.

HUGHES R.C., POLGAR J.G., WEIGHTMAN D. & WALTON J.N. (1971) Levodopa in Parkinsonism: the effects of withdrawal of anticholinergic drugs. *Brit. med. J.* **2**, 487–491.

HUNTER K.R., LAURENCE D.R., SHAW K.M. & STERN G.M. (1973) Sustained levodopa therapy in Parkinsonism. *Lancet* **ii**, 929–931.

INOUYE A., SHIO H., SORIMACHI M. & KATOAKA K. (1970) Unsaturated fatty acids: platelet-serotonin releasers in tissue extracts. *Experientia* **26**, 308–309.

IVERSEN L.L. (1972) The uptake, storage, release and metabolism of GABA in inhibitory nerves. In Snyder S.J. (Ed.) *Perspectives in Neuropharmacology*, pp. 75–111. London, Oxford University Press.

IVERSEN L.L., HORN A.S. & MILLER R.J. (1975) Actions of dopaminergic agonists on cyclic AMP production in rat brain homogenates. *Adv. Neurol.* **9**, 197–212.

JANOWSKY D.S., EL-YOUSEF M., DAVIS J.M. & SEKERKE H.J. (1972) A cholinergic-adrenergic hypothesis of mania and depression. *Lancet* **ii**, 632–635.

JENNER F.A., GJESSING L.R., COX J.R., DAVIES-JONES A., HULLIN R.P. & HANNA S.M. (1967) A manic depressive psychotic with a persistent forty-eight hour cycle. *Brit. J. Psychiat.* **113**, 895–910.

JEQUIER E. & DUFRESNE J.J. (1972) Biochemical investigations in patients with Parkinson's disease treated with L-DOPA. *Neurology* **22**, 231–237.

JOHANSSON B. & ROOS B.E. (1974) 5-Hydroxyindoleacetic and homovanillic acid in cerebrospinal fluid of patients with neurological diseases. *Europ. Neurol.* **11**, 37–45.

KANAZAWA L., MIYATA Y., TOYOKURA Y. & OTSUKA M. (1973) The distribution of γ-aminobutyric acid (GABA) in the human substantia nigra. *Brain Res.* **51**, 363–365.

KAROUM F., WYATT R. & COSTA E. (1974) Estimation of the contribution of peripheral and central noradrenergic neurones to urinary 3-methoxy-4-hydroxy-phenylglycol in the rat. *Neuropharmac.* **13**, 165–176.

KAUFMAN S. (1974) Properties of the pterin-dependent aromatic amino acid hydroxylases. In *Aromatic Amino Acids in the Brain. Ciba Foundation Symposium* **22**, pp. 85–108. Amsterdam, Elsevier.

KAZAMATSURI H., CHIEN C.P. & COLE J. (1972a) Treatment of tardive dyskinesia. I. Clinical efficacy of a dopamine-depleting agent tetrabenazine. *Arch. Gen. Psychiat.* **27**, 95–99.

KAZAMATSURI H., CHIEN C.P. & COLE J. (1972b) Treatment of tardive dyskinesia. II. Short-term efficacy of dopamine-blocking agents haloperidol and thiopropazate. *Arch. Gen. Psychiat.* **27**, 100–103.

KEBABIAN J.W., PETZGOLD G.L. & GREENGARD P. (1972) Dopamine sensitive adenylate cyclase in the caudate nucleus of rat brain and its similarity to the 'dopamine receptor'. *Proc. Nat. Acad. Sci.* **69**, 2145–2149.

KEELE C.A. (1967) Polypeptides and other substances which may produce vascular headache. *Background to Migraine* **1**, 126–133.

KELLER H.H., BARTHOLINI G. & PLETSCHER A. (1974) Enhancement of noradrenaline turnover in rat brain by L-DOPA. *J. Pharm. Pharmac.* **26**, 649–651.

KENDELL R.E. (1968) *The Classification of Depressive Illnesses*, London, Oxford University Press.

KETY S. (1959) Biochemical theories of schizophrenia. *Science, N.Y.*, **129**, 1590–1596.

KIELY M. & SOURKES T.L. (1972) Transport of L-tryptophan into slices of rat cerebral cortex. *J. Neurochem.* **19**, 2863–2872.

KITS T.P. & VAN PRAAG H.M. (1970) A controlled study of the anti-depressant effect of p-chloro-N-methylamphetamine, a compound with a selective effect on the central 5-hydroxytryptamine metabolism. *Acta Psychiat. Scand.* **46**, 365–373.

KLAWANS H.L. (1972) Pathophysiology of schizophrenia and the striatum. *Dis. Nerv. Syst.* **33**, 711–719.

KLAWANS H.L., GOETZ C. & WEINER W.J. (1973a) Dopamine receptor site sensitivity in hyperthyroid guinea pigs: a possible model of hyperthyroid chorea. *J. Neurol. Trans.* **34**, 187–193.

KLAWANS H.L., MOSES H. & BEAULIEU D.M. (1973b) The influence of caffeine on d-amphetamine and apomorphine-induced stereotyped behaviour. *Life Sci.* **14**, 1493–1500.

KLAWANS H.L., PAULSON G.W., RINGEL S.P. & BARBEAU A. (1972) Use of L-DOPA in the detection of presymptomatic Huntington's chorea. *New Engl. J. Med.* **286**, 1332–1334.

KLAWANS H.L., RINGEL S.P. & SHENKER D.M. (1971) Failure of vitamin B_6 to reverse the L-DOPA effect in patients on a decarboxylase inhibitor. *J. Neurol. Neurosurg. Psychiat.* **34**, 682–686.

KLAWANS H.L. & RUBOVITS R. (1972) Central cholinergic-anticholinergic antagonism in Huntington's chorea. *Neurology*, **22**, 107–116.

KLAWANS H.L. & SHENKER D.M. (1972) Observations on the dopaminergic nature of hyperthyroid chorea. *J. Neurol. Trans.* **33**, 73–81.

KNOTT P.J. & CURZON G. (1972) Free tryptophan in plasma and brain tryptophan metabolism. *Nature, London,* **239**, 452–453.

KOPIN I.J. (1966) Biochemical aspects of release of norepinephrine and other amines from sympathetic nerve endings. *Pharmac. Rev.* **18**, 513–523.

KORF J. & VAN PRAAG H.M. (1971) Amine metabolism in the human brain: further evaluation of the probenecid test. *Brain Res.* **35**, 221–230.

KOSLOW S.H. (1974) 5-Methoxytryptamine: a possible central nervous system transmitter. *Adv. Biochem. Psychopharmac.* **11**, 95–100.

KRISHNASWAMY K. & RAGHURAM T.C. (1972) Effect of leucine and isoleucine on brain serotonin concentration in rats. *Life Sci.* **11, Part 2,** 1191–1197.

LANCE J.W. & ANTHONY M. (1971) Migrainous neuralgia or cluster headache? *J. Neurol. Sci.* **13**, 401–414.

LEHMANN J. (1972) Mental and neuromuscular symptoms in tryptophan deficiency. *Acta Psychiat Scand.* **Supp 237**, 1–28.

LEHMANN J. (1973) Tryptophan malabsorption in levodopa-treated Parkinsonian patients. Effect of tryptophan on mental disturbance. *Acta med. Scand.* **194**, 181–189.

LEWIS A. (1971) 'Endogenous' and 'exogenous': a useful dichotomy? *Psychol. Med.* **1**, 191–196.

LLOYD K.G., DAVIDSON L. & HORNYKIEWICZ O. (1973) Metabolism of levodopa in the human brain. *Adv. Neurol.* **3**, 173–188.

LLOYD K.G., FARLEY I.J., DECK J.H.N. & HORNYKIEWICZ O. (1974) Serotonin and 5-hydroxindoleacetic acid in discrete areas of the brainstem of suicide victims and control patients. *Adv. Biochem. Psychopharmac.* **11**, 387–397.

LORANGER A.W., GOODELL H., LEE J.E. & MCDOWELL F. (1972) Levodopa treatment of Parkinson's syndrome. Improved intellectual functioning. *Arch. Gen. Psychiat.* **26**, 163–168.

LOTT I.T., MURPHY D.L. & CHASE T.N. (1972) Down's syndrome. Central monoamine turnover in patients with diminished platelet serotonin. *Neurology.* **22**, 967–972.

MABRY P.D. & CAMPBELL B.A. (1973) Serotonergic inhibition of catecholamine induced behavioural arousal. *Brain Res.* **49**, 381–391.

MACLEAN R., NICHOLSON W.J., PARE C.M.B. & STACEY R.S. (1965) Effect of monoamine-oxidase inhibitors on the concentrations of 5-hydroxytryptamine in the human brain. *Lancet* **ii**, 205–209.

MACKENZIE D.Y. & WOOLF L.I. (1969) Maple syrup urine disease. *Brit. med. J.* **1**, 90–91.

MANDELL A.J. (Ed.) (1973) *New Concepts in Neurotransmitter Regulation.* New York and London, Plenum.

MAAS J.W. & LANDIS D.H. (1966) A technique for assaying the kinetics of norepinephrine metabolism in the central nervous system in vivo. *Psychosom. Med.* **28**, 247–256.

MARSDEN C.A. & CURZON G. (1974) Effects of lesions and drugs on brain tryptamine. *J. Neurochem.* 1171–1176.

MARSDEN C.D., DOLPHIN A., DUVOISIN R.C., JENNER P. & TARSY D. (1974) Role of noradrenaline in levodopa reversal of reserpine akinesia. *Brain Res.* **77**, 521–525.

MARSH G.G. & MARKHAM C.H. (1973) Does levodopa alter depression and psychopathology in Parkinsonism patients? *J. Neurol. Neurosurg. Psychiat.* **36**, 925–935.

MARSH G.G., MARKHAM C.H. & ANSEL R. (1971) Levodopa's awakening effect on patients with Parkinsonism. *J. Neurol. Neurosurg. Psychiat.* **34**, 209–218.

MCGEER P.L., MCGEER E.G. & FIBIGER H.C. (1973) Glutamic acid decarboxylase and choline acetylase in Huntington's chorea and Parkinson's disease. *Lancet* **ii**, 623–624.

MCGEER P.L., MCGEER E.G. & WADA J.A. (1971) Glutamic acid decarboxylase in Parkinson's disease and epilepsy. *Neurology* **21**, 1000–1007.

MCKEAN C.M., SCHANBERG S.M. & GIARMAN N.J. (1967) Aminoacidemias: effects on maze performance and cerebral serotonin. *Science, N.Y.* **157**, 213–215.

MCLENNAN D.L., CHALMERS R.J. & JOHNSON R.H. (1974) A double-blind trial of tetrabenazine, thiopropazate and placebo in patients with chorea. *Lancet* **i**, 104–107.

MEDICAL RESEARCH COUNCIL BRAIN METABOLISM UNIT (1972) Modified amine hypothesis for the aetiology of affective illness. *Lancet* **ii**, 573–577.

MEEK J.L. & NEFF N.H. (1973) Is cerebrospinal fluid the major avenue for the removal of 5-hydroxyindoleacetic acid from the brain? *Neuropharmac.* **12**, 497–499.

MENA I., COURT J., FUENZALIDA S., PAPAVASILIOU P.S. & COTZIAS G.C. (1970) Modification of chronic manganese poisoning. Treatment with L-DOPA or 5-OH tryptophan. *New Engl. J. Med.* **282**, 5–10.

MENA I., MARIN O. & FUENZALIDA S. (1967) Chronic manganese poisoning. *Neurology* **17**, 128–136.

MENDELS J. & FRAZER A. (1974) Brain biogenic amine depletion and mood. *Arch. Gen. Psychiat.* **30**, 447–451.

MENDELS J., FRAZER A., FITZGERALD R.G., RAMSEY T.A. & STOKES J.W. (1972) Biogenic amine metabolites in cerebrospinal fluid of depressed and manic patients. *Science, N.Y.* **175**, 1380–1382.

MILNE M.D., CRAWFORD M.A., GIRAO C.B. & LOUGHRIDGE L.W. (1960) The metabolic disorder in Hartnup disease. *Q. Jl. Med.* **29**, 407–421.

MOFFETT A., SWASH S. & SCOTT D.F. (1972) Effect of tyramine in migraine: a double-blind study. *J. Neurol. Neurosurg. Psychiat.* **35**, 496–499.

MORGAN M. & MANDELL A.J. (1969) Indole (ethyl) amine N-methyltransferase in the brain. *Science N.Y.* **166**, 492–493.

MUNKVAD I., PAKKENBERG H. & RANDRUP A. (1968) Aminergic systems in basal ganglia associated with stereotyped hyperactive behaviour and catalepsy. *Brain Behav. Evol.* **1**, 89–100.

MUSTAFA S.J. & CHANDRA S.V. (1971) Levels of 5-hydroxytryptamine, dopamine and norepinephrine in whole brain of rabbits in chronic manganese toxicity. *J. Neurochem.* **18**, 931–933.

NEFF N.H., BARRETT R.E. & COSTA E. (1969) Selective depletion of caudate nucleus dopamine and serotonin during chronic manganese dioxide administration to squirrel monkeys. *Experientia* **25**, 1140–1141.

NG K.Y., CHASE T.N., COLBURN R.W. & KOPIN I.J. (1970) L-DOPA-induced release of cerebral monoamines. *Science, N.Y.* **170**, 76–77.

NOBIN A. & BJORKLUND A. (1973) Topography of the monoamine neuron systems in the human brain as revealed in fetuses. *Acta Physiol. Scand.* **Supp. 388**.

NYBACK H. & SEDVALL G. (1969) Regional accumulation of catecholamines formed from tyrosine-14C in rat-brain: effect of Chlorpromazine. *Europ. J. Pharmac.* **5**, 245–252.

NYSTROM B., OLSON L. & UNGERSTEDT U. (1972) Noradrenaline nerve terminals in human cerebral cortices: first histochemical evidence. *Science N.Y.* **176**, 924–926.

O'KEEFE R., SHARMAN D.F. & VOGT M. (1970) Effect of drugs used in psychoses on cerebral dopamine metabolism. *Br. J. Pharmac.* **38**, 287–304.

OLSON L., NYSTROM B. & SEIGER A. (1973) Monoamine fluorescence histochemistry of human post mortem brain. *Brain Res.* **63**, 231–247.

PAPESCHI R., MOLINA-NEGRO P., SOURKES T.L., HARDY J. & BERTRAND C. (1970) Concentration of homovanillic acid in the ventricular fluid of patients with Parkinson's disease and other dyskinesias. *Neurology.* **20**, 991–995.

PARE C.M.B. & SANDLER M. (1959) A clinical and biochemical study of a trial of iproniazid in the treatment of depression. *J. Neurol. Neurosurg. Psychiat.* **22**, 247–251.

PARE C.M.B., SANDLER M. & STACEY R.S. (1959) The relationship between decreased 5-hydroxyindole metabolism and mental defect in phenylketonuria. *Archs. Dis. Childh.* **34**, 422–425.

PARE C.M.B., YEUNG D.P.H., PRICE K. & STACEY R.S. (1969). 5-Hydroxytryptamine, noradrenaline and dopamine in brainstem, hypothalamus and caudate nucleus of controls and of patients committing suicide by coal-gas poisoning. *Lancet* **ii**, 133–135.

PARKES J.D., MARSDEN C.D., REES J.E., CURZON G., KANTAMANENI B.D., KNILL-JONES R., AKBAR A., DAS S. & KATARIA M. (1974) Parkinson's disease, cerebral arteriosclerosis and senile dementia. *Q. Jl. Med.* **53**, 49–61.

PEREZ-REYES M. (1969) Differences in the capacity of the sympathetic and endocrine systems of depressed patients to react to a physiological stress. *Pharmacopsychiat. Neuropsychopharmakol.* **2**, 245–251.

PERRY T.L., HANSEN S. & KLOSTER B. (1973) Huntington's chorea: deficiency of γ-aminobutyric acid in brain. *New Engl. J. Med.* **288**, 337–342.

PERRY T.L., HANSEN S. & MACDOUGALL L. (1967) Amines of human whole brain. *J. Neurochem.* **14**, 775–782.

PFEIFFER R. & EBADI M. (1972) On the mechanism of the nullification of CNS effects of L-DOPA by pyridoxine in Parkinsonian patients. *J. Neurochem.* **19**, 2175–2181.

PLETSCHER A. (1968) Metabolism, transfer and storage of 5-hydroxytryptamine in blood platelets. *Brit. J. Pharmac.* **32**, 1–16.

POIRIER L.J., MCGEER E.G., LAROCHELLE L., MCGEER P., BEDARD P. & BOUCHER R. (1969) The effect of brain stem lesions on tyrosine and tryptophan hydroxylases in various structures of the telencephalon of the cat. *Brain Res.* **14**, 147–155.

POLLIN W., CARDON P. & KETY S. (1961) Effects of amino acid feeding in schizophrenic patients treated with iproniazid. *Science, N.Y.* **133**, 104–105.

POST R.M. & GOODWIN F.K. (1974) Effects of amitriptyline and imipramine on amine metabolites in the cerebrospinal fluid of depressed patients. *Arch. Gen. Psychiat.* **20**, 234–239.

POST R.M., GORDON E.K., GOODWIN F.K. & BUNNEY W.E. (1973) Central norepinephrine metabolism in affective illness: MHPG in the cerebrospinal fluid. *Science, N.Y.* **179**, 1002–1003.

POST R.M., KOTIN J., GOODWIN F.K. & GORDON E.H. (1973) Psychomotor activity and cerebrospinal fluid amine metabolites in affective illness. *Amer. J. Psychiat.* **130**, 67–72.

PRANGE A.J., WILSON I.C., LYNN C.W., ALLTOP L.B. & STIKELEATHER R.A. (1974) L-Tryptophan in mania. *Arch. Gen. Psychiat.* **30**, 56–62.

PULLAR I.A., WEDDELL J.M., AHMED R. & GILLINGHAM F.J. (1970) Phenolic acid concentrations in the lumbar cerebrospinal fluid of Parkinsonian patients treated with L-DOPA. *J. Neurol. Neurosurg. Psychiat.* **33**, 851–857.
RAMANAMURTHY P.S.V. & SRIKANTIA S.G. (1970) Effects of leucine on brain serotonin. *J. Neurochem.* **17**, 27–32.
RINNE U.K. & SONNINEN Y. (1972) Acid monoamine metabolites in the cerebrospinal fluid of patients with Parkinson's disease. *Neurology,* **22**, 62–67.
ROBERGE A.G. & POIRIER L.J. (1973) Effect of chronically administered L-DOPA on DOPA/5HTP decarboxylase and tyrosine and tryptophan hydroxylases in cat brain. *J. Neural. Trans.* **34**, 171–185.
ROBERTS H.J. (1967) Migraine and related vascular headaches due to diabetogenic hyperinsulism. *Headache.* **7**, 41–62.
ROBISON G.A., BUTCHER R.W. & SUTHERLAND E.W. (1971) *Cyclic AMP.* New York, Academic Press.
RODRIGUEZ R. (1972) Antagonism of tremorine induced tremor by serotoninergic agents in mice. Interactions with L-DOPA. *Life Sci.* **2, Part 1,** 535–544.
ROFFLER-TARLOV S., SHARMAN D.F. & TEGERDINE P. (1971) 3,4-Dihydroxyphenylacetic acid and 4-hydroxy-3-methoxyphenylacetic acid in the mouse striatum: A reflection of intra- and extra-neuronal metabolism of dopamine? *Br. J. Pharmac.* **42**, 343–351.
ROOS B.E. & STEG G. (1964) The effect of L-3,4-dihydroxyphenylalanine and DL-5-hydroxytryptophan on rigidity and tremor induced by reserpine, chlorpromazine and phenoxybenzamine. *Life Sci.* **3**, 351–360.
ROSECRANS J.A., LOVELL R.A. & FREEDMAN D.X. (1967) Effects of lysergic acid diethylamide on the metabolism of brain 5-hydroxytryptamine. *Biochem. Pharmac.* **26**, 2011–2021.
ROSNER F., ONG B.H., PAINE R.S. & MAHANAND D. (1965) Blood serotonin activity in trisomic and translocation Down's syndrome. *Lancet* **i**, 1191–1193.
ROTH M., GURNEY C., GARSIDE R.F., KERR T.A. & SCHAPIRA K. (1972) Studies in the classification of affective disorders. The relationship between anxiety states and depressive illness. *Brit, J. Psychiat.* **121**, 147–161.
SAAVEDRA J.M. & AXELROD J. (1972) A specific and sensitive enzymatic assay for tryptamine in tissue. *J. Pharmac.* **182**, 363–369.
SAAVEDRA J.M., COYLE J.T. & AXELROD J. (1973) The distribution and properties of the nonspecific N-methyltransferase in brain. *J. Neurochem.* **20**, 743–752.
SAAVEDRA J.M., PALKOVITS M., BROWNSTEIN M. & AXELROD J. (1974) Serotonin distribution in the rat hypothalamus and preoptic region. *Brain Res.* **77**, 157–165.
SACKS D. (1973) *Awakenings.* London, Duckworth.
SANDLER M., BONHAM-CARTER S., HUNTER K.R. & STERN G.M. (1973) Tetrahydroisoquinoline alkaloids: *in vivo* metabolites of L-DOPA. *Nature* **241**, 439–443.
SANDLER M. & YOUDIM M.B.H. (1974) Monoamine oxidases: the present status. *Int. Pharmacopsychiat.* **9**, 27–34.
SANDLER M., YOUDIM M.B.H. & HANINGTON E. (1974) A phenylethylamine oxidising defect in migraine. *Nature, London,* **250**, 335–337.
SANDLER M., YOUDIM M.B.H., SOUTHGATE J. & HANINGTON E. (1970) The role of tyramine in migraine: some biochemical mechanisms, *Background to Migraine* **3**, 103–112.
SANO I. (1972) L-5-Hydroxytryptophan-(L-5-HTP) therapie. *Folia Psychiat. Neurol. Japan* **26**, 9–17.
SANO I. & TANIGUCHI K. (1972) L-5-Hydroxytryptophan (L-5HTP)-Therapie des Morbus Parkinson. *Munch. Med. Woch.* **114**, 1717–1719.

SHARMAN D.F., POIRIER L.J., MURPHY G.F. & SOURKES T.L. (1967) Homovanillic acid and dihydroxyphenylacetic acid in the striatum of monkeys with brain lesions. *Canad. J. Physiol.* **45**, 57–62.

SHAW D.M., CAMPS F.E. & ECCLESTON E.G. (1967) 5-Hydroxytryptamine in the hindbrain of depressive suicides. *Brit. J. Psychiat.* **113**, 1407–1411.

SCHILDKRAUT J.J. (1971) Norepinephrine metabolism in affective disorders: recent clinical studies. *Excerpta Medica Int. Congress Series* **274**, 1040–1049.

SCHILDKRAUT J.J. (1974) Biochemical criteria for classifying depressive disorders and predicting responses to pharmacotherapy: preliminary findings from studies of norepinephrine metabolism. *Pharmakopsychiat. Neuropsychopharmak.* **7**, 98–107.

SHIELDS P.J. & ECCLESTON D. (1972) Effects of electrical stimulation of rat midbrain on 5-hydroxytryptamine synthesis as determined by a sensitive radioisotope method. *J. Neurochem.* **19**, 265–272.

SHIELDS P.J. & ECCLESTON D. (1973) Evidence for the synthesis and storage of 5-hydroxytryptamine in two separate pools in the brain. *J. Neurochem.* **20**, 881–888.

SHOPSHIN B., WILK S., SATHANANTHAN G., GERSHON S. & DAVIS K. (1974) Catecholamines and affective disorders revised: a critical assessment. *J. Nerv. Ment. Dis.* **158**, 369–383.

SICUTERI F. (1973) 5-Hydroxytryptamine supersensitivity as a new migraine theory. *Background to Migraine* **5**, 45–56.

SIMS K.L. & BLOOM F.E. (1973) Rat brain L-3,4-dihydroxyphenylalanine and L-5-hydroxytryptophan decarboxylase activities: differential effect of 6-hydroxydopamine. *Brain Res.* **48**, 165–175.

SJOSTROM R. (1972) Steady-state levels of probenecid and their relation to acid monoamine metabolites in human cerebrospinal fluid. *Psychopharmacologia* **25**, 96–100.

SNYDER S.H. (1973) Amphetamine psychosis: a 'model' schizophrenia mediated by catecholamines. *Am. J. Psychiat.* **130**, 61–67.

SNYDER S.H. & TAYLOR K.M. (1972) Histamine in the brain: a neuro-transmitter. In Snyder S.H. (Ed.) *Perspectives in Neuropharmacology: a Tribute to Julius Axelrod*, pp. 43–74. New York. Oxford University Press.

SOURKES T.L. (1973) On the origin of homovanillic acid (HVA) in the cerebrospinal fluid. *J. Neural. Trans.* **34**, 153–157.

STEIN L. (1971) Neurochemistry of reward and punishment: some implications for the etiology of schizophrenia. *J. psychiat. Res.* **8**, 345–361.

STEIN L. & WISE C.D. (1971) Possible etiology of schizophrenia: progressive damage to the noradrenergic reward system by 6-hydroxydopamine. *Science N.Y.* **171**, 1032–1036.

STROMBERG U. & WALDECK B. (1973) Behavioural and biochemical interaction between caffeine and L-DOPA. *J, Pharm. Pharmac.* **25**, 302–308.

TAGLIAMONTE A., TAGLIAMONTE P., PEREZ-CRUET J., STERN S. & GESSA G.L. (1971) Effect of psychotropic drugs on tryptophan concentration in the rat brain. *J. Pharmac.* **177**, 475–480.

TANIMUKAI H. & HIMWICH H.E. (1970) Biochemical changes in the schizophrenias with special reference to urinary products (indoleamines). In Price J.H. (Ed.) *Modern Trends in Psychological Medicine* **2**, pp. 78–101. London, Butterworth.

TAYLOR K.M. & SNYDER S.H. (1970) Amphetamine: differentiation by D and L isomers of behaviour involving brain norepinephrine or dopamine. *Science, N.Y.* **168**, 1487–1489.

THIERRY A.M., BLANC G., SOBEL A., STINUS L. & GLOWINSKI J. (1973) Dopaminergic terminals in the rat cortex. *Science N.Y.* **182**, 499–501.

TU J. & PARTINGTON M.W. (1972) 5-Hydroxyindole levels in the blood and CSF in

Down's Syndrome, phenylketonuria and severe mental retardation. *Develop. Med. Child. Neurol.* **14**, 457–466.
Tu J.B. & Zellwerger H. (1965) Blood serotonin deficiency in Down's syndrome. *Lancet* **i**, 715–717.
Ungerstedt U. (1971a) Stereotactic mapping of the monoamine pathways in the brain. *Acta Physiol. Scand.* **Supp. 367**.
Ungerstedt U. (1971b) Striatal dopamine release after amphetamine or nerve degeneration revealed by rotational behaviour. *Acta physiol. Scand.* **Suppl. 367**, 49–67.
Ungerstedt U., Avemo A., Ljungberg T. & Ranje C. (1973) Animal models of Parkinsonism. *Adv. Neurol.* **3**, 257–271.
Usdin E. & Snyder S.H. (1973) *Frontiers in Catecholamine Research.* New York, Pergamon.
Van Praag H.M. (1974) Therapy-resistant depressions. Biochemical and pharmacological considerations. *Pharmacopyschiat. Neuropsychopharmak.* **7**, 88–97.
Van Praag H.M. & Leijnse B. (1963) Die bedeutung der monoaminoxydase-hemmung als antidepressives Prinzip 1. *Psychopharmacologia* **4**, 1–14.
Van Praag H.M. & Leijnse B. (1966) Some aspects of the metabolism of glucose and non-esterified fatty acids in depressive patients. *Psychopharmacologia* **9**, 220–233.
Vogel W.H., Orfei V. & Century B. (1969) Activities of enzymes involved in the formation and destruction of biogenic amines in various areas of human brain. *J. Pharmac.* **165**, 196–203.
Vogt C. & Vogt O. (1937) Sitz und Wesen der Krankheiten im Lichte der topistischen Hirnforschung und des Variierens der Tiere. *J. Psychol. Neurol. (Lpz.)* **47**, 237–457.
Wurtman R.J., Larin F., Mostafapour S. & Fernstrom J.D. (1974) Brain catechol synthesis: control by brain tyrosine concentration. *Science, N.Y.* **185**, 183–184.
Wurtman R.J. & Romero J.A. (1972) Effects of levodopa on nondopaminergic brain neurons. *Neurology.* **22**, 72–81.
Weber R.B. & Reinmuth O.M. (1972) The treatment of migraine with propranolol. *Neurology* **22**, 366–369.
Wender P.H. (1972) The minimal brain dysfunction in children. *J. Nerv. Ment. Dis.* **155**, 55–71.
Wender P.H. (1974) Some speculations concerning a possible biochemical basis of minimal brain dysfunction. *Life Sci.* **14**, 1605–1621.
Wetterberg L., Gustavson K.H., Backstrom M., Ross S.B. & Froden O. (1972) Low dopamine-β-hydroxylase in Down's syndrome. *Clin. Genet.* **3**, 152–153.
Wise C.D. & Stein L. (1973) Dopamine-β-hydroxylase deficits in the brains of schizophrenic patients. *Science, N.Y.* **181**, 344–347.
Woolley D.W. & Shaw E. (1954) Some neurophysiological aspects of serotonin. *Brit. med. J.* **2**, 122–126.
Yamaguchi K., Shimoyama M. & Gholson R.K. (1967) Measurements of tryptophan pyrrolase *in vivo:* induction and feed back inhibition. *Biochim. biophys. Acta.* **146**, 102–110.
Yahr M.D. (Ed.) (1973) Treatment of Parkinson's disease—the use of DOPA decarboxylase inhibitors. *Adv. Neurol.* **2**, 1–303.
Youdim M.B.H., Carter S.B., Sandler M., Hanington E. & Wilkinson M. (1971) Conjugation defect in tyramine-sensitive migraine. *Nature, London,* **230**, 127–128.
Zis A.P., Fibiger H.C. & Phillips A.G. (1974) Reversal by L-DOPA of impaired learning due to destruction of the dopaminergic nigro-neostriatal projection. *Science, N.Y.* **185**, 960–962.

5 Biochemistry of coma

H. S. Bachelard

    I **Introduction**, 230

   II **Ischaemia**, 232

  III **Anoxia**, 239

  IV **Hypoxia**, 244

      A. Glycolysis and respiration, 244
      B. Glycolytic enzymes, 245
      C. Amines and amino acids, 247
      D. Kinetic requirements for oxygen, 249

   V **Hypoglycaemia**, 252

      A. Insulin-induced hypoglycaemia, 252
      B. Deoxyglucose-induced Hypoglycaemia, 254
      C. Glucose transport, 257
      D. Glucoreceptors, 259

  VI **Hyperglycaemia**, 259

      A. Hyperglycaemia and glucose efflux, 260

 VII **Hepatic coma**, 261

      A. Animal studies, 261

B Metabolic effects of hyperammonaemia, 262
C Glutamine formation, 266
D. α-Oxoglutaramate, 266
E. Ammonia and glycolytic enzymes, 266

VIII Coma and development, 267

A. Hypoglycaemia, 269
B Hypoxia, 271

I Introduction

Coma has been long regarded as the most serious manifestation of cerebral dysfunction, yet it is only relatively recently that the biochemical aspects of the comatose state have been seriously tackled. The clinical and physiological knowledge, with comprehensive treatment of the problems of diagnosis and of management, have been thoroughly reviewed by Fazekas & Alman (1962) and by Plum & Posner (1972).

In considering the biochemistry of coma, it seems necessary to be clear what is meant by coma. A short definition is a state of unconsciousness which cannot be reversed by normal external stimuli. This criterion, of failure to achieve arousal, distinguishes coma from the unconscious state of sleep. Coma is also distinct from sleep in that cerebral oxygen consumption is usually decreased in coma, whereas it is normal in sleep. The EEG may be the same in both states, especially in deep sleep, in which the EEG shows large slow waves. In deep coma the irregular large slow waves may disappear and the EEG may become isoelectric.

It is important to note that even in coma, many of the normal cerebral functions may be unimpaired. Consciousness seems to be associated with the thalamic region, but the association centres of the cerebral cortex are clearly affected in coma, as indicated by the stupor which precedes the comatose state (changes in orientation, attention and perception) and by the amnesia often observed after recovery from coma. Apart from amnesia, complete restoration of function often occurs during recovery, although this may take some time. Not only is there certainly selective vulnerability of brain regions, but also there seems likely to be selective sensitivity of specific areas of cerebral metabolism to the shocks which lead to coma.

In time, coma leads to structural damage of brain cells and this is usually associated with marked impairment of biochemical function, especially of energy metabolism. However, the behavioural symptoms and changes in EEG usually precede any detectable energy failure or cellular damage and there is increasing evidence that the functional deficits may not be due simply to energy failure.

The events leading to coma often include convulsions, although this may not always be the case. Studies on experimental animals often require the animals to be anaesthetized. The anaesthetics used, especially barbiturates, may be anticonvulsant and affect the EEG so any tendency for convulsions to occur may be masked. Certainly the energy failure which ultimately occurs can lead to convulsions, as is clear from the effects of vitamin deficiency or of convulsant drugs which interfere with energy metabolism (McIlwain & Bachelard, 1971). It seems that energy deprivation, from whatever cause, can trigger off convulsions.

The importance of glucose and oxygen to normal cerebral function is well known (McIlwain & Bachelard, 1971) and the biochemistry of many comatose states is clearly associated with a lack of these nutrients. The inability of intermediary metabolites (such as pyruvate, lactate and glutamate) or other hexoses (fructose and galactose) to arouse man or experimental animals from hypoglycaemic coma is one of the major pieces of evidence which indicate that these chemicals cannot normally replace glucose in cerebral metabolism. This can be due either to lack of adequate metabolic machinery within the brain or to an inadequate rate of entry of the substance from the bloodstream to the brain.

The frequency of comatose states is due to a wide variety of causes. These include cerebral tissue damage as a result of infection (meningitis, encephalitis), tumours, hamartoma, physical injury. Metabolic disorders which result in coma can be caused by vitamin deficiency, by abnormalities in electrolyte or water metabolism or by endocrine dysfunction. Many genetically-determined disorders of cerebral cellular metabolism, such as the lipidoses, lead to coma and death.

The most common metabolic causes of coma seem clearly to be due to failure of the supplies of glucose and oxygen reaching the brain and it is mainly these, shown in Table 5.1, which are to be discussed in this chapter. Attention will be

Table 5.1. Some metabolic causes of coma

Type	Metabolic cause	Remarks
1 Ischaemias (circulatory arrest a. General (cardiac arrest) b. Local	Lack of glucose, oxygen	Infarct, stroke. Structural damage.
2 Hypoxias a. Anoxic	Lack of oxygen	Heart fails first.
b. Hypoxic	Decreased arterial pO_2	
c. Anaemic	Decreased oxygen-carrying capacity of blood, due to abnormal haemoglobin; carbon monoxide poisoning	Normal arterial pO_2.
d. Oligaemic	Decreased cerebral blood flow	Normal arterial pO_2.
e. Histotoxic	Inhibited cytochrome oxidase by poisons, e.g. cyanide	
3 Hypoglycaemia	Decreased glucose (from insulin or from liver dysfunction)	
4 Hyperglycaemia	Increased glucose (diabetes)	
5 Hepatic coma	Increased ammonia? (liver disease)	

From Barcroft (1925), modified by Brierley (1975).

given to the sequence of metabolic events in relation to behavioural impairment and structural damage, to the question of energy failure and selective sensitivity of other specific biochemical mechanisms, and to possible reasons why the newborn mammalian brain seems to be less vulnerable to some of the causes of coma in the mature brain.

II Ischaemia

When the circulation to the brain is interrupted, cellular damage ultimately occurs. Knowledge of the biochemical, physiological and structural consequences of ischaemia is important in our understanding and potential methods of treating stroke, which may result from an interruption of the blood supply to a localized region of the brain.

Some of the definitive biochemical studies on the effects of ischaemia on the brain as a whole have been carried out in the laboratories of Professor Lowry in the United States (e.g. Lowry et al., 1964). It had been recognized for many years, since the studies of Kerr in 1935, that very rapid rates of intermediary metabolism were maintained *post mortem* in the brain, at least until endogenous energy supplies were exhausted. Lowry and his colleagues followed the time course of change in concentrations of intermediary metabolites in the brain in the first few seconds and minutes after the blood supply had been cut off. Mice were decapitated and, after various short time intervals, their heads were rapidly frozen for analysis of chemical constituents.

Fig. 5.1 shows the changes within the first minute of selected glycolytic intermediates and of high energy phosphate compounds. Creatine phosphate (CP) and adenosine triphosphate (ATP) concentrations fell very rapidly, whereas adenosine diphosphate (ADP) and adenosine monophosphate (AMP) rose. The total concentration of these adenine nucleotides (ATP, ADP and AMP) remained constant but their relative proportions had changed markedly. This is often assessed as the 'adenylate energy charge' or the 'energy charge potential' (ECP) from the ratio:

$$\frac{[ATP]+0.5[ADP]}{[ATP]+[ADP]+[AMP]}$$

This (Atkinson, 1968) was designed as an expression of the amount of high energy phosphate groups in the adenine nucleotide pool, but ignores the change in the high energy phosphate group of creatine phosphate. For this reason, the term 'adenylate energy charge' seems preferable to the 'energy charge potential'.

Thus, from Fig. 5.1, it can be seen that the 'adenylate energy charge' in the whole brain fell some 15% in the first 15 sec of ischaemia from 0·81 to 0·68. If the cerebral cortex alone is taken the value is originally higher, at 0·95, and falls at a greater rate. Fig. 5.1. also shows that in the first 15 sec the concentration of

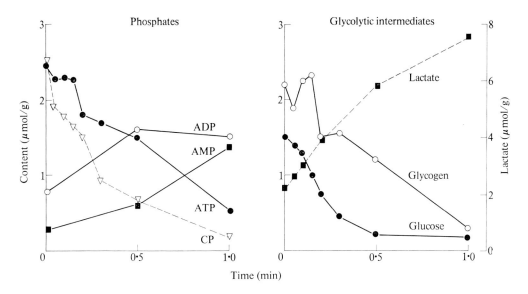

Fig. 5.1. Ischaemic changes in mouse brain. Mouse brains were frozen at various time intervals after decapitation. The zero time values were obtained from brains rapidly-frozen *in situ* (Lowry et al., 1964).

glucose fell by over 50% and lactate was increased to approximately twice its original level. These results indicate that glucose, and also glycogen, continue to be utilized by anaerobic glycolysis (see below) at the expense of the endogenous energy stores, ATP and creatine phosphate. With no replenishment of nutrients possible, due to the lack of the blood supply, energy failure soon ensues. Structural damage then follows, as illustrated in Fig. 5.2. This light micrograph clearly shows the vacuolation of the cellular cytoplasm which results from the anoxic-ischaemia produced by the 'Levine preparation'. One carotid artery of the rat is clamped (the other acts as a control) and the animal is subjected to intermittent periods of anoxia (Levine, 1960). Brown & Brierley (1966, 1968) modified the technique so the subsequent removal of the carotid clamp enabled symmetrical perfusion-fixation for microscopical examination to be performed. Fig. 5.2a shows the 'ischaemic cell damage' which occurred after 40 min of anoxic-ischaemia followed by a 30 min survival period in oxygen. Fig. 5.2b shows that the damage can still be observed in a similar preparation even after a 1 h survival period. The vacuoles were shown by electron microscopy (Fig. 5.3) to be due to swollen mitochondria which had lost most of their internal membranes, the cristae. Such cellular damage can be discerned by light microscopy within as short a time interval as 5 min and is described below.

Lowry and his colleagues also used the results of their studies on ischaemic mouse brain to assess the points at which metabolic regulation was being exerted under such conditions, and to assess the effects of anaesthesia on such changes.

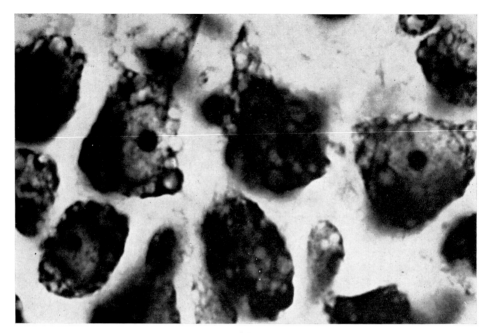

(a)

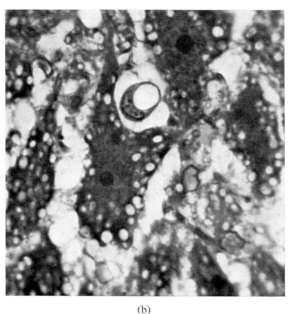

(b)

Fig. 5.2. Cellular damage caused by anoxic-ischaemia. Light microscopic pictures of the ipsilateral hippocampus (L_1) of a rat 'Levine preparation', showing microvacuoles in the cytoplasm of pyramidal neurones with normal nuclei. (a) Intermittent exposure to nitrogen for 40 min; followed by a survival time of 30 min. Paraffin-embedded, stained with cresyl fast violet. $\times 1,200$. (Brown & Brierley, 1968.) (b) Intermittent exposure to nitrogen for 40 min; followed by a survival time of 1 h. Epon section (1 μm), stained with toluidine blue. $\times 2,000$. (Brown, 1973.)

Photographs kindly donated by A. W. Brown, M.R.C. Laboratories, Carshalton, Surrey.

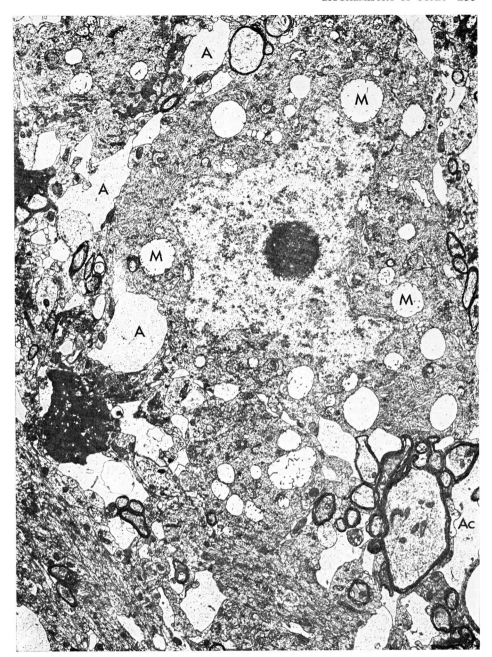

Fig. 5.3. Cellular damage caused by anoxic-ischaemia. Low-power electron micrograph of the hippocampal pyramidal cells of Fig. 5.2b. The microvacuoles of Fig. 5.2b are seen to be swollen mitochondria (M). Swollen astrocytic processes (A) surround, and slightly distort, the neuronal perikaryon. Note also the swollen perivascular astrocytic process (Ac) at bottom right. Perfusion-fixed with glutaraldehyde and post-osmicated. Stained with uranyl acetate, lead citrate. ×5,100.

The micrograph was kindly donated by A. W. Brown, M.R.C. Laboratories, Carshalton, Surrey.

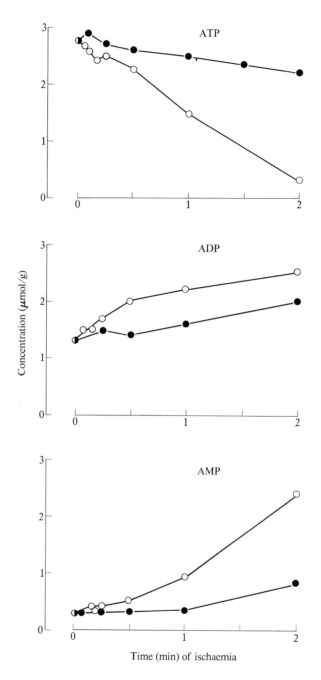

Fig. 5.4. Barbiturates and ischaemia. Effects of pentobarbital on ischaemic changes in 10-day-old mouse brain. The tissues were fixed as in Fig. 5.1. (Lowry et al., 1964.) ●: Anaesthetized. ○: Control.

Fig. 5.4 shows that the rates of post-mortem ischaemic change in adenine nucleotide concentrations were decreased in the presence of barbiturates. Detection of the control points operating in glycolysis is illustrated in Fig. 5.5 and is based on the relative changes observed in concentrations of metabolites. This technique cannot in itself be used to identify control points. However, if a substrate concentration changes in an appropriate direction at a stage known to be catalysed by a non-equilibrium enzyme, then it can be concluded that regulation is being exerted at that stage. It is essential to know if metabolism

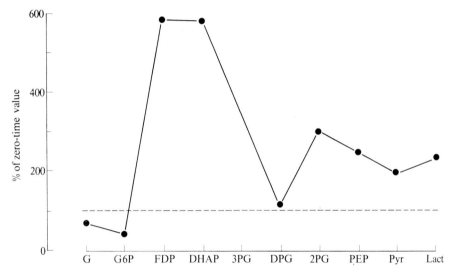

Fig. 5.5. Changes in concentrations of glycolytic substrates in ischaemia. Results from 10-day-old anaesthetised mice after 25 sec ischaemia. (Lowry et al., 1964.)

through the particular pathway is stimulated or inhibited under the experimental conditions so that the change in substrate concentration is clearly in a direction opposite to the change in the pathway as a whole (see Bachelard, 1975a, for a detailed discussion). In this case, metabolism through the pathway of glycolysis is known to be stimulated (from the increased lactate production shown in Fig. 5.1), and the substrate of a control point enzyme would therefore be expected to be decreased in concentration, provided that the enzyme normally catalyses a non-equilibrium reaction. Fig. 5.5 shows that in these experiments (Lowry et al., 1964) the intermediates of glycolysis between glucose and fructose 6-phosphate (F6P) were decreased and the subsequent intermediates were increased in concentration after a short period of ischaemia. The results were interpreted as regulation being exerted at phosphofructokinase, which converts F6P to fructose, 1,6-disphosphate (FDP), and possibly also at hexokinase, which produces glucose 6-phosphate (G6P) from glucose (see Fig. 5.6).

It is important to note that ischaemia represents a *mixture* of states, since

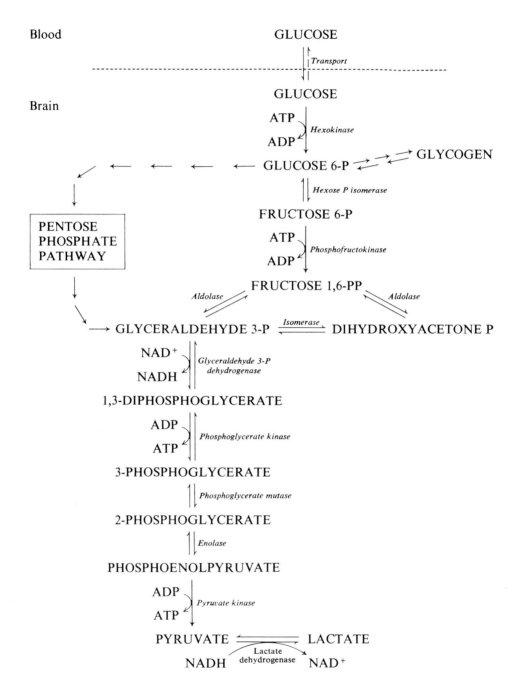

Fig. 5.6. Glycolysis.

supplies of both glucose and oxygen have been cut off, and the lack of blood circulation means also that metabolic products cannot be removed from the brain. The sections which follow describe the studies which have been performed under carefully controlled conditions for the separate states of anoxia, hypoxia and hypoglycaemia. The morphological and biochemical studies just described have been applied to these states. Due to the marked change in the 'adenylate energy charge' in ischaemia, it had long been assumed that the consequences of hypoxia and hypoglycaemia (convulsions, coma and subsequent death if not relieved) were also due to energy failure and that similar changes in adenine nucleotides were causative factors. We now know that the situation is more complex: the energy failure which occurs ultimately in hypoxia and hypoglycaemia may be just as much a consequence of the shock as is the cellular damage: both may be due to impairments of biochemical function not directly associated with the brain's energy metabolism.

III Anoxia

Complete removal of oxygen, usually by replacement with nitrogen, has been known to cause convulsions and coma since the results of Boyle's experiments on air deprivation were published in 1660. However, the effects of anoxia on the morphology of brain cells were essentially unknown until relatively recently. Van Harreveld showed that the electrical impedance of the cerebral cortex was increased by a few minutes of anoxia and concluded this to be due to changes in the extracellular spaces, which were decreased (Fig. 5.7) from a normal content of about 20% to less than 5% (van Harreveld et al., 1965; Van Harreveld & Malhotra, 1966, 1967). This was concluded to be largely due to cellular swelling, especially of glial cells.

Anoxia is also recognized as producing 'ischaemic cell damage' very quickly. The anoxic-ischaemic cell damage described above can be seen quite clearly by light microscopy within 5 min (Fig. 5.8). The earliest damage discernible is the formation of microvacuoles. Electron microscopical examination of these preparations (Fig. 5.9) confirmed that the microvacuoles are damaged mitochondria, with loss of cristae, and that swollen astrocytic processes also occur.

The loss of mitochondrial structure is clearly associated with changes in the energy state of the tissue in anoxia as indicated by the rapid loss of ATP and creatine phosphate which ultimately occurs (McIlwain & Bachelard, 1971). As in ischaemia, glycolysis in the anoxic brain is also known to be stimulated. Table 5.2 shows the effects of anoxia on glycolysis in the perfused dog brain (Drewes & Gilboe, 1973): a major increase in the rate of anaerobic lactate formation was detected within 1 min. The 'adenylate energy charge' was decreased by 50% after 10 min of anoxia. Glucose uptake was increased, from the arterio-venous differences of Table 5.2: studies specifically on glucose transport in this preparation led the investigators to conclude that the unidirectional

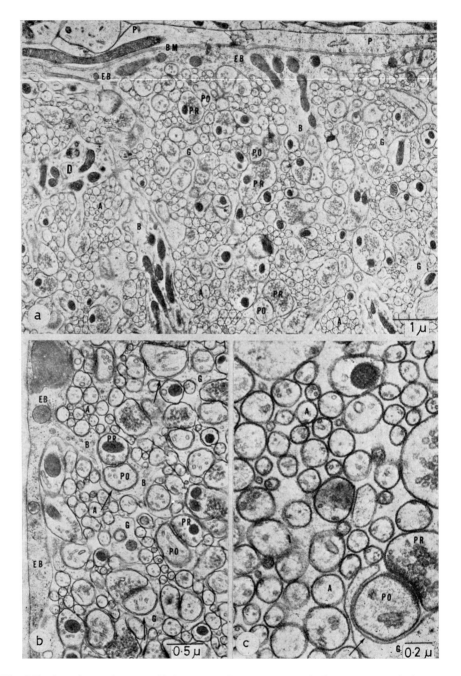

Fig. 5.7. Anoxia and extracellular space in mouse cerebellum. Most of the extracellular space (arrows) seen in a, b and c has disappeared after 8 min of circulatory arrest (d, e and f).

Abbreviations: A, axons; B, fibre of Bergmann; BM, basement membrane; D, den-

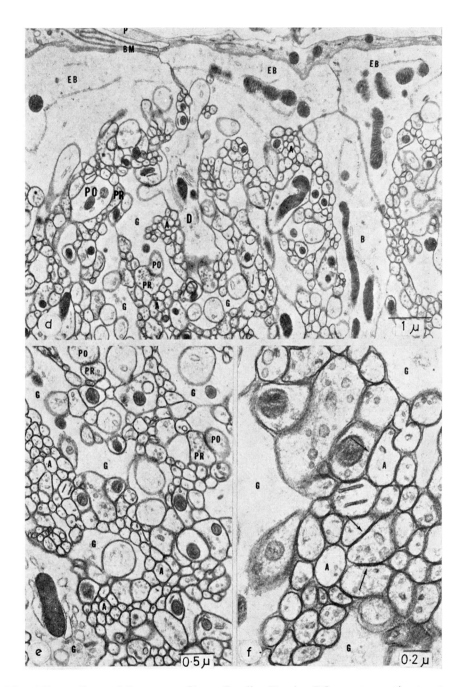

drite; EB, endfoot of Bergmann fibre; G, glia; P, pia; PO, postsynaptic structure; PR, presynaptic structure.

The electron micrographs were kindly donated by Professor A. Van Harreveld, California Institute of Technology. (Van Harreveld and Steiner, 1970.)

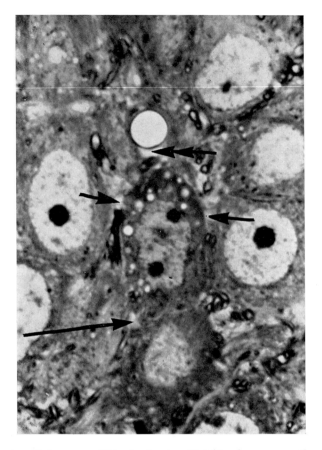

Fig. 5.8. Anoxic cell damage within 5 minutes. Rat 'Levine preparation': exposed to nitrogen intermittently for 5 min and killed immediately. Light microscopic picture of a pyramidal neurone of the ipsilateral hippocampus, showing microvacuoles in the cytoplasm of increased basophilia. Slight expansions of perineuronal processes around the damaged neurone (←) and of the pericapillary process (←←) can be seen. Epon section (15 μm), stained with toluidine blue. × 1,750.

The photograph was kindly donated by A. W. Brown, M.R.C. Laboratories, Carshalton, Surrey.

Table 5.2. Effects of anoxia on perfused dog brain

	Glucose used (μmole/g min)	Lactate produced (μmole/g min)	'Adenylate energy charge' After anoxia	After 15 min Recovery
Control	0.38	0.028	0.89	
Anoxia, 1 min	1.28 (+140%)	0.140 (+400%)		
Anoxia, 10 min	0.64 (+70%)	0.074 (+165%)	0.39	0.92

Data from Drewes & Gilboe, 1973; Drewes et al., 1973.

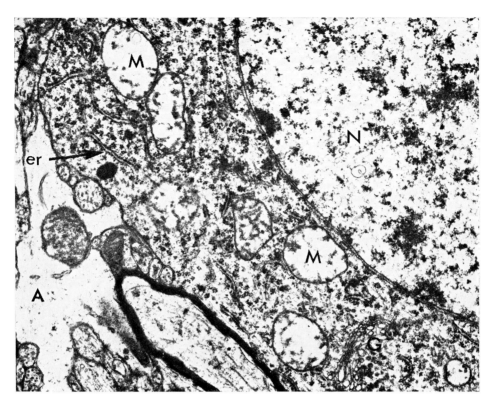

Fig. 5.9. Anoxic cell damage within 5 minutes. Electron micrograph of the neurone of Fig. 5.8. The nucleus (N), Golgi complex (G) and rough endoplasmic reticulum (er) appear normal. The microvacuoles of Fig. 5.8 are seen as damaged mitochondria (M) with loss of cristae. A swollen astrocytic process (A) is also present. The tissue was perfusion-fixed with glutaraldehyde, post-osmicated, and stained with uranyl-acetate, lead citrate. × 18,100. (Brown & Brierley, 1973.)

The micrograph was kindly donated by A. W. Brown, M.R.C. Laboratories, Carshalton, Surrey.

influx rate of glucose was unchanged, but that the increase in net uptake was due to decreased rates of efflux as the internal brain glucose became depleted (Betz et al., 1974).

So complete anoxia rapidly causes stimulation of glycolysis with a loss of energy and some structural damage indistinguishable from that caused by ischaemia. However, anoxia takes a considerable time to cause irreversible functional damage. The 50% decrease in 'adenylate energy charge' after 10 min anoxia could be reversed by 15 min oxygenation, and the effects of even 30 min anoxia on oxidative phosphorylation were found to be reversible (Drewes et al., 1973). It is worth noting that the 'adenylate energy charge' may not be the most sensitive criterion for functional sensitivity to oxygen deprivation. It has already been stated above that this concept ignores creatine phosphate. It represents a static situation (based on the concentrations of the adenine nucleotides) and gives no indication of the dynamic situation in terms of possible effects on the turnover rates of these substances. Some indication of the effects of anoxia on the dynamic metabolic state have emerged from the results of studies on brain slices incubated *in vitro*. Swanson (1969) found that anoxia caused impairment of cation transport *in vitro* even after a subsequent recovery period in oxygen, at stages when rates of synthesis of ATP and creatine phosphate appeared to be maintained. Cation pumping may therefore be a more sensitive metabolic criterion than the 'adenylate energy charge'; K^+ levels of incubated brain slices are routinely used to assess the metabolic integrity of the tissue (Bachelard et al., 1962, Bachelard, 1971a). Further effects of oxygen deprivation on the metabolism of biogenic amines are described subsequently.

The assumption that lack of oxygen causes cellular damage and coma simply as a result of energy failure has to be re-assessed from the results of controlled arterial hypoxia, described in the following section.

IV Hypoxia

A. GLYCOLYSIS AND RESPIRATION

In man, the oxygen content of the inspired air can usually be decreased from its normal value of around 20% to as low as 10% before any signs of stress, behavioural abnormality or appreciable decrease in the respiratory rate can be detected. A fall in the arterio-venous difference in oxygen under such conditions of mild arterial hypoxia is countered by an increase in the rate of blood flow through the brain, so the utilization of oxygen by the brain remains unchanged. However, changes in the glycolytic rate of the brain have been detected from the arterio-venous difference in lactate. This is more obvious below 10% oxygen: at 7% (Table 5.3) pronounced changes occur in the lactate output from the

Table 5.3. Hypoxia in man

		Normal	Hypoxia	% Change
CMR (O_2)	ml/min/100 g	3·02	3·13	+ 3
CMR(G)	μmol/min/100 g	24·9	31·9	+ 28
CMR(L)	μmol/min/100 g	2·31	10·73	+365
G oxid	as %	91·92	75·76	− 15·5
G → lactate	as %	4·49	18·86	+320
ATP synth. (calc.)	μmol/min/100 g	8·6	9·0	+ 5

Arterio-venous difference studies were performed in man, breathing approximately 7% oxygen, with $PaCO_2$ normal (Cohen, 1971).

human brain, with increased glucose input, under conditions where no alteration in the cerebral respiratory rate can be shown to occur. Normal pCO_2 was maintained in these studies (Cohen et al., 1967; Cohen, 1971). At high altitude man lives under conditions of constant mild hypoxia; arterio-venous difference studies there also gave evidence for increased rates both of cerebral blood flow and of anaerobic glycolysis (Severinghaus et al., 1966; Sørensen and Severinghaus, 1968).

The studies on man therefore provided strong indications of increased rates of glycolysis in the absence of any detectable change in respiration in mild arterial hypoxia, but possible effects on levels of creatine phosphate or adenine nucleotides could not be assessed, except by indirect calculation (Table 5.3). This was performed subsequently using experimental animals. Mild arterial hypoxia (6 to 7% oxygen) was given to the animals under rigorously-controlled physiological conditions of constant arterial pCO_2 and constant blood pressure, thus eliminating such complicating conditions as ischaemia or hypocapnia. The results of very short times of hypoxia confirmed the observations made in man of greatly increased rates of glycolysis without any change in rates of oxygen consumption by the brain (Duffy et al., 1972; MacMillan & Siesjö, 1972; Bachelard et al., 1974). Analysis of the cerebral energy intermediates showed that apart from a slight initial decrease in creatine phosphate, these remained at their normal concentrations. After only 5 min of mild arterial hypoxia, substrate analysis showed that the increased glycolytic rate was associated with activation of at least two of the control points of the pathway: hexokinase and pyruvate kinase (Fig. 5.10). Even shorter periods of 1 to 2 min provided evidence for activation also of phosphofructokinase (Norberg & Siesjö, 1975).

Thus the glycolytic responses to hypoxia, detectable within one minute, occur earlier than generalized ischaemic cell damage can be detected and before there is any perceptible energy failure. It seems that the energy failure which undoubtedly occurs subsequently is directly associated with structural damage, especially to mitochondria.

B. GLYCOLYTIC ENZYMES

From the data of Fig. 5.10 it seems obvious that the regulatory kinases of glycolysis are activated in hypoxia (seen also in ischaemia, see Fig. 5.5) but the molecular mechanisms of this enzymic activation are not clear. Phosphofructokinase is known to be very sensitive to changes in the balance of adenine nucleotides, since AMP, ADP and free excess ATP are inhibitory in a complex fashion (Lowry & Passonneau, 1966). This mechanism is unlikely to be involved in hypoxia in view of the lack of effect on the 'adenylate energy charge'. However, phosphofructokinase is activated by NH_4^+ (Lowry, 1966) as also is pyruvate kinase (Bachelard et al., 1975). Although NH_4^+ is elevated in ischaemia and anoxia (Table 5.4) it is increased only slightly in the whole brain under the

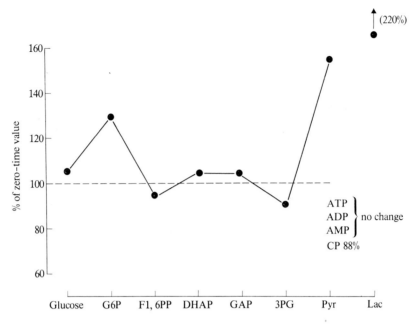

Fig. 5.10. Changes in glycolytic substrate concentrations after 5 minutes hypoxia. Rats were given 7% oxygen under nitrous oxide anaesthesia for 5 min and the brain was fixed by a rapid freezing technique. Each point represents the mean percentage change from the control value. (Bachelard et al., 1974.)

Table 5.4. GABA and NH_3 in hypoxic rabbit brain

	Time of hypoxia (min)			Ischaemia
	0	30	60	20
GABA	1·14	1·38	1·62	2·14
NH_3	0·27	0·62	0·95	0·95

Rabbits given 4% O_2, 5% CO_2 and 91% N_2 for the times shown. Results are μmol/g. frozen tissue (Thorn et al., 1973).

mildly hypoxic conditions which also produce greatly elevated rates of glycolysis (Bachelard et al., 1974). Nevertheless increased NH_4^+ may play a part in the activation which occurs in ischaemia and anoxia, and if highly localized, could also be involved in mild hypoxia. The effects of elevated NH_4^+ on glycolytic enzymes are discussed in more detail below under *Hepatic Coma*.

The third regulatory kinase, hexokinase, has also been implicated in the metabolic responses to hypoxia and ischaemia, although the results reported in the literature so far are conflicting. Knull et al. (1973) reported that total ischaemia in chick brain resulted in a redistribution of the enzyme between its normal sites of occurrence within the cell. Brain hexokinase occurs partly particulate (both in the mitochondria freely distributed within the cell and in the mitochondria within the nerve endings) and partly as a soluble cytoplasmic constituent (Bachelard, 1967). Knull et al. (1973) found that as a result of short periods of ischaemia, more of the chick brain enzyme appeared to be bound to the mitochondria and less was soluble, i.e., a movement apparently of cytoplasmic enzyme to the mitochondria. In contrast, Broniszewska-Ardelt and Miller (1974) reported that intermittent anoxia, a similar but not identical shock, produced in rat brain a decrease in the proportion of enzyme bound and in increase in the soluble form: i.e., a release of bound enzyme to the cytoplasm. If, as has been suggested (Bachelard, 1967; Bachelard & Goldfarb, 1969) it is the cytoplasmic form of the enzyme which is directly involved in cerebral glycolysis, increased cytoplasmic activity could contribute directly to the observed increase in rates of glycolysis.

Some unpublished studies on the activities of the regulatory glycolytic kinase enzymes in ischaemic mouse brain have also shown an increase in the activity of the particulate hexokinase. Over the first 2 min of ischaemia, the activity increased by 50%. At the same time there was a 30% decrease in the cytoplasmic hexokinase activity (Fig. 5.11). Slight decreases in activity of phosphofructokinase and of pyruvate kinase also occurred. At 30 sec or 1 min, the increase in particulate hexokinase activity of 25 to 30% seems unlikely to make a significant contribution in itself to the dramatically-increased overall rates of glycolysis described above.

C. AMINES AND AMINO ACIDS

The acceleration of cerebral glycolytic rate appears to be more marked than the changes observed in the other areas of cerebral metabolism that have been examined to date. Conditions of hypoxia which are harsher than those described above produced only relatively slight changes in biologically active amino acids. Thorn et al. (1973) found only slightly increased concentrations of γ-aminobutyrate in the brains of rabbits exposed at 4% O_2, 5% CO_2 and 91% N_2 for 30 min (Table 5.4). The changes in cerebral ammonia were also less than those which resulted from ischaemia in the same experimental studies. The concentration of glutamate, a key excitant amino acid in the brain and the precursor of γ-aminobutyrate (Fig. 5.12), is apparently not significantly affected by hypoxia (Macmillan & Siesjö, 1972).

Some changes in the turnover rates of aromatic biogenic amines have been detected in hypoxia and warrant particular attention. Davis and Carlsson (1973a) found that changes in the metabolism of dopamine, noradrenaline and 5-hydroxytryptamine (serotonin) could not be demonstrated after exposure of the animals to 5·6% O_2, unless inhibitors of appropriate enzymes were present.

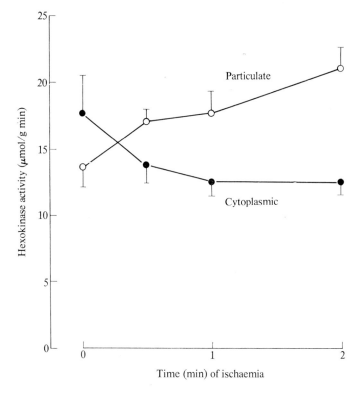

Fig. 5.11. Changes in glycolytic enzyme activities during ischaemia. Mice were decapitated and their brains frozen subsequently. Zero times are for the brains rapidly frozen *in situ*. Each point is the mean of four animals and the vertical bars are S.D. values. (Data from our unpublished experiments.)

Table 5.5. Hypoxia and amines

Brain content* (% of original)	Time (min) of hypoxia			
	0	30	60	120
DOPA	100	85	70	45
5HTry	100	50	55	70
Turnover rate** (calculated as % of control)	Dopamine	82		
	Noradrenaline	87		
	5HT	74		

Rats were exposed to 5·6% O_2.
*With an inhibitor of aromatic amino acid decarboxylase.
**With pargyline, an inhibitor of monamine oxidase.
Davis & Carlsson (1973).

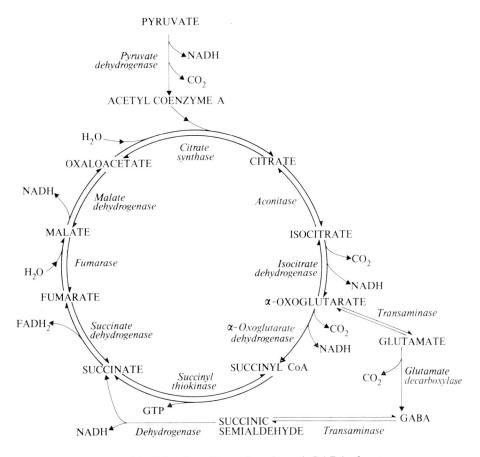

Fig. 5.12. Tricarboxylic acid cycle and GABA shunt.

If the aromatic amino acid decarboxylase activity (Fig. 5.13) was inhibited, 30 min of hypoxia produced a 15% decrease in DOPA and a 50% decrease in 5-hydroxytryptophan (Table 5.5). Use of pargyline (an inhibitor of monoamine oxidase) led to a slight but significant decrease in calculated rates of turnover of the three amines, dopamine, noradrenaline and serotonin (see also Carlsson, 1974).

D. KINETIC REQUIREMENTS FOR OXYGEN

Changes in the metabolism of these amines, or of neurotransmitters generally, would not be expected to show any effect on the total energy state of the brain, because they involve only a minute fraction of the energy used elsewhere, particularly for cation transport (Table 5.6). However, while the effects of hypoxia on the metabolism of these amine neurotransmitters are quantitatively much less than the effects on glycolysis, there is a reason to believe that the amines

Table 5.6. Proportion of available energy consumed in transmitter synthesis

Process	Concentration in brain[1] (nmol/g)	Turnover rate (in vivo)[2] (nmol/g min)	Calculated energy consumption as ATP[2] (nmol/g min)
Synthesis of ATP	3,000	12,000	—
Transport of Na$^+$	55,000	5,000	2,500
Synthesis of ACh	10–35	0·5–2·0	0·5–2·0
Synthesis of amines	2–6	0·02–0·2	0·1–1·5

[1] The range of concentration is due to variations in different species (McIlwain & Bachelard, 1971; Bachelard, 1974).
[2] For details of the calculations see Bachelard, 1975b.

may be particularly vulnerable to a decrease in available oxygen. Oxygen is required for two major metabolic functions in the brain: for mitochondrial respiration, where molecular oxygen is directly consumed in the final stage (cytochrome oxidase) of the electron transport system, and in the aromatic amine hydroxylation reactions of Fig. 5.13, also as molecular oxygen. The relevant hydroxylase activities have been shown to be decreased after hypoxia (Davis & Carlsson, 1973b).

Comparison of these two processes shows clearly that the hydroxylation reactions are far more sensitive to low oxygen than is the respiratory chain. The K_m value for oxygen in cerebral mitochondrial respiration is approximately 0·1 μM (Clark et al., 1975), slightly lower than that reported for liver mitochondria (0·5 μM, Degn & Wohlrab, 1971; Peterson et al., 1974). This is 1,000 times smaller than the K_m values for the brain hydroxylases: 1 mM for tryptophan hydroxylase (Kaufman, 1974) and 0·5 mM for tyrosine hydroxylase (Fisher & Kaufman, 1972). The values for these brain and liver processes are summarized in Table 5.7. Although the K_m value cannot be used as a quantitative inverse index of the affinity of a process for its substrate, it does give a good

Table 5.7. K_m Values for oxygen

Process	Tissue	Approx. K_m (mM–O_2)	Reference
Respiration	Brain	0·0001	Clark et al., 1975.
	Liver	0·0005	Peterson et al., 1974.
Tyrosine hydroxylase	Brain	0·5	Fisher & Kaufman, 1972.
Tryptophan hydroxylase	Brain	1·0	Kaufman, 1974.
Phenylalanine hydroxylase	Liver	0·2	Fisher & Kaufman, 1972.

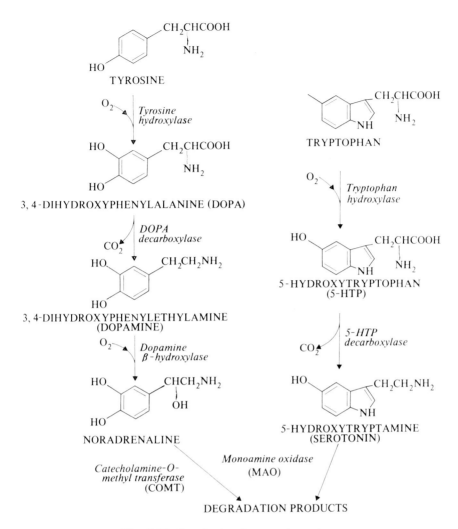

Fig. 5.13. Synthesis of aromatic amines.

indication of the *relative* affinity; the hydroxylases obviously show a far lower affinity for oxygen than does the respiratory process. The hydroxylases are therefore much more likely to be affected by low oxygen than is respiration, bearing in mind that the normal tissue concentration of oxygen has been calculated to be around 2 mM (Kaufman, 1974).

The amines may also react to hypoxia by another mechanism. Mild hypoxia causes behavioural disturbances, studied in rats by measuring locomotor activity and a conditioned avoidance response. The disturbance in behaviour caused by 8·65% oxygen could be reversed by DOPA, if an inhibitor of peripheral DOPA decarboxylase (see Fig. 5.13) was present. The workers who performed these studies (Brown & Engel, 1973; Brown *et al.*, 1973) have suggested that mild hypoxia may affect dopamine receptors which were argued to be even more sensitive to low oxygen than the hydroxylases concerned with dopamine synthesis.

The pronounced effects of hypoxia on cerebral glycolytic rates still remain to be explained; there arises the possibility of some mechanism for sensitive control of glycolysis in response to neurotransmitters or to synaptic events. The first and most sensitive functional response to hypoxia may therefore be dopaminergic neurotransmission and the studies performed to date may have exposed a hitherto unexpected mechanism of metabolic regulation.

V Hypoglycaemia

A simple lack of glucose has been recognized for many years as causing changes in behaviour and in EEG, with normal pO_2 and pCO_2 in the bloodstream. In time, convulsions, coma and death can ensue. Again, as described for hypoxia (above), the first symptoms of hypoglycaemia, the first indications of impairment of brain function, are manifested before any change in the energy status of the tissue can be demonstrated.

A. INSULIN-INDUCED HYPOGLYCAEMIA

Some of the earliest definitive studies on hypoglycaemia (as on hypoxia) were performed by Himwich nearly 40 years ago. Table 5.8 shows the results of his studies on the effects of insulin hypoglycaemia on the carbohydrate metabolism of brain and muscle. Dogs were fasted (mild hypoglycaemia) or fasted and then injected with insulin (intense hypoglycaemia) and arteriovenous differences in glucose and oxygen were measured. With mild hypoglycaemia (at a blood glucose concentration of below 2 mM) utilization of glucose by the brain was normal

Table 5.8. Insulin hypoglycaemia in dogs

	Conditions	Arterial blood glucose (mM)	Arterio-venous difference	
			Glucose (mM)	Oxygen (mM)
Brain	Control	5·0	0·73	4·15
	Mild hypoglycaemia	1·67	0·70	3·60
	Intense hypoglycaemia	0·67	0·17	1·70
Muscle	Control	5·0	0·42	3·1
	Hypoglycaemia	1·1	0·10	2·7

Data from Himwich & Fazekas, 1937.

and there was no marked effect on respiration. However, under intense hypoglycaemia (which would produce coma) the consumption of both glucose and oxygen was seriously impaired. In contrast to the effect on the brain, a drastic decrease in glucose utilization by hypoglycaemic muscle was not reflected in the rate of consumption of oxygen which remained normal. Clearly the muscle was able to maintain its respiration from alternative substrates to glucose, but the brain could not.

However, closer examinations of the relationship between glucose consumption and respiration in insulin-induced hypoglycaemia have recently shown that the brain may be capable of using endogenous substrates. The normal respiration rate can be maintained even after glucose consumption has started to fall. No energy failure is detectable at that stage, so it seems as if the brain is able to utilize endogenous materials for maintenance of its oxidative energy production (Lewis et al., 1974; see also Tews et al., 1965).

In time, the lack of glucose causes irreversible structural damage. This was shown originally by chemical analysis as a loss of membrane components (proteins and phospholipids) in Geiger's experiments on the perfused brain (see Geiger, 1958). More recently 'ischaemic cell damage' has been detected by light microscopy within the first few minutes of insulin-induced hypoglycaemia (Fig. 5.14). The structural damage is identical to that seen after ischaemia or anoxia, described previously in this chapter. Then, as in hypoxia, the damage can be correlated with energy failure, but behavioural symptoms have been shown to precede both energy failure and cell damage.

Rates of utilization of glucose in monkeys rendered hypoglycaemic with insulin were compared with the EEG by Meldrum et al. (1971). They found that the cerebral consumption of glucose remained constant or increased slightly over the first hour after the injection of insulin and that the changes in EEG only occurred when the rate of glucose utilization had decreased. Cellular damage was not detected until later. In their experiments there was no hypoxia and no change in intracranial pressure.

This sequence of events has been compared with cerebral concentrations of metabolic intermediates in smaller laboratory animals. Studies on insulin-hypoglycaemic mice (Ferrendelli & Chang, 1973; Ferrendelli, 1975) also showed that changes in EEG and behaviour preceded any detectable change in the concentrations of creatine phosphate or of ATP, not only in the cerebral cortex but in all of the other brain regions that were studied. Similar results were reported from studies on the effects of insulin hypoglycaemia in rats: the EEG was observed to change only after the intracellular concentrations of glucose in the brain had been exhausted and convulsions occurred subsequently (Lewis et al., 1974). The depletion of glucose and the change in EEG occurred at a stage where, again, no change in the concentrations of ATP or of creatine phosphate were observed, although ammonia had accumulated. These experiments were most rigorously controlled to exclude ischaemia and hypoxia and the results

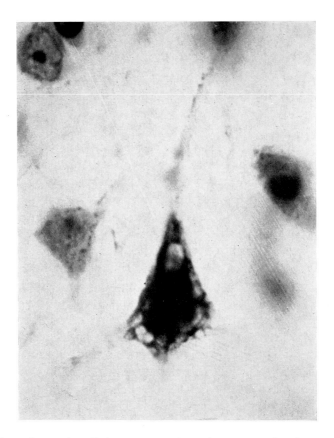

Fig. 5.14. Hypoglycaemic cell damage. Monkeys (*Macaca mulatta*) were made hypoglycaemic with insulin; blood sugar concentrations were below 1·1 mM. The light microscopic picture is of a pyramidal neurone showing microvacuoles similar to those observed in ischaemia and anoxia (*q.v.*). The brain was fixed by perfusion with FAM (40% formaldehyde–glacial acetic acid–methanol; 1:1:8) and tissues embedded in Paraplast. Paraffin sections were stained with Luxol fast blue and cresyl fast violet. ×1,600. (Brierley *et al.*, 1971.)

The photograph was kindly supplied by A. W. Brown, M.R.C. Laboratories, Carshalton, Surrey.

can reasonably be discussed in terms of pure hypoglycaemia. So from these insulin hypoglycaemic studies a clear message emerges: behavioural and EEG abnormalities can be demonstrated to occur as a consequence of hypoglycaemia without any detectable impairment of energy metabolism.

B. DEOXYGLUCOSE-INDUCED HYPOGLYCAEMIA

Concurrently with the use of insulin to produce hypoglycaemia an alternative approach was being pursued. This exploited the ability of a glucose analogue,

2-deoxyglucose, to inhibit the rate of glucose entry to the brain. If propranolol is given to counter the adrenergic response, deoxyglucose produces cellular hypoglycaemia in the presence of normal blood glucose concentrations and under normoxic conditions. In 1958, Landau and Lubs had described the use of deoxyglucose in cats to produce hypoglycaemic symptoms (which included convulsions) and had argued that deoxyglucose-induced hypoglycaemia was indistinguishable from insulin hypoglycaemia. These authors were the first to suggest that deoxyglucose could be used to produce 'cellular hypoglycaemia'.

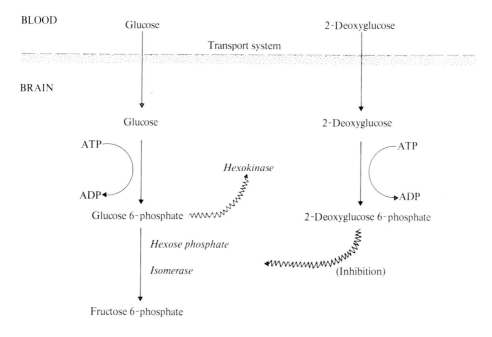

Fig. 5.15. Metabolism of deoxyglucose in the brain.

Deoxyglucose shares with glucose the cerebral mechanisms for transport and phosphorylation (Fig. 5.15) but it is not metabolized further. Relatively low concentrations (2 mM) of the final product, deoxyglucose 6-phosphate, seem to have no effect on cerebral metabolism, but higher concentrations (20 mM) exert strong inhibitory action on hexose phosphate isomerase, see also Fig. 5.6 (Horton et al., 1973). At times when the main effect of deoxyglucose is on inhibition of glucose transport (i.e. cellular hypoglycaemia), rather than on the isomerase, changes in behaviour and EEG were observed without any detectable change in creatine phosphate or ATP (Horton et al., 1973). Effects on energy metabolism only become apparent at a later stage when sufficient deoxyglucose 6-phosphate had accumulated to inhibit the isomerase (Table 5.9).

Table 5.9. Deoxyglucose-induced hypoglycaemia

Time (min)	Whole brain concentrations								Effect on behaviour	Effect on glucose metabolism[1]
	(mM)		(% of control)							
	DOG	DOG 6P	G	G6P	Lactate	ATP	CP			
0	0	0	100	100	100	100	100		Change in EEG	Glucose uptake inhibited by 60%.
5	1·09	2·30	62[2]	137	104	103	106		Tonic clonic jerks	Glycolysis inhibited by 70%.
10	5·82	3·12	137[2]	189[2]	69[2]					
20	6·33	3·82	300[2]	219[2]	67[2]	110			Failing respiration	
30	6·74	5·45	494[2]	191[2]	68[2]		49[2]			

[1] From radioactive labelling studies.   [2] Significantly different from zero-time values.
Mice were injected with deoxyglucose (3 mg/g body weight) with body temperature maintained, and killed at the times shown (Horton et al., 1973).

C. GLUCOSE TRANSPORT

Two quite independent approaches have produced the same conclusion: the first symptoms of hypoglycaemia are not due to energy failure and we must therefore look elsewhere for the metabolic cause. It is clear from studies on the kinetics of glucose transport to the brain that the hypoglycaemia produced by insulin or by deoxyglucose is a consequence of limitations in the rate at which glucose can enter the brain from the bloodstream. The kinetic data for cerebral

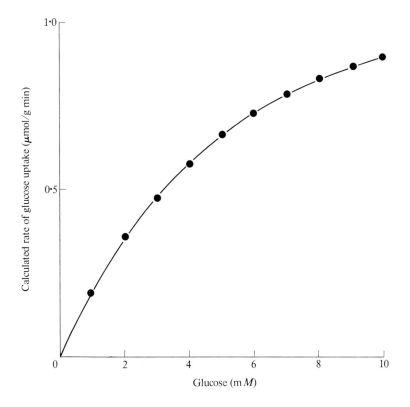

Fig. 5.16. Calculated rate of glucose entry to the brain.

glucose transport systems are summarized in Table 10. Using a mean K_m value of 6·5 mM and a maximum velocity of 1·5 μmol/g min, the uptake curve of Fig. 5.16 has been calculated from the Michaelis–Menten equation, $v = V/(1 + K_m/S)$. At blood glucose concentrations below 30 mg/100 ml (1·7 mM) such as are caused by insulin, there is not sufficient glucose calculated to be entering the brain to maintain the normal glucose consumption rate of 0·3 μmol/g min (Bachelard, 1971, 1975b).

With deoxyglucose present at normal blood glucose concentrations, the amount of glucose entering the brain is limited by competitive inhibition.

Table 5.10. Cerebral glucose transport: kinetic properties

	Technique	Species	K_m (mM)	V_{max} (μmol/g min)	Reference
Brain, *in vitro*	Glucose analogues	Guinea pig	5	1·3	Bachelard, 1971a.
Brain, *in vivo*	Glucose analogues	Rat	7	1·5	Buschiazzo et al., 1970.
	Glucose analogues	Mouse	6	2·0	Growdon et al., 1971.
	Glucose, infusion	Rat	7	1·25	Bachelard et al., 1973.
	Glucose, indicator dilution	Dog	8	1·75	Betz et al., 1973.
	Glucose infusion	Sheep	6	2·6	Pappenheimer & Setchell, 1973.
Brain Synaptosomes	Glucose analogues	Guinea pig	0·2		Heaton & Bachelard, 1973.
	Glucose analogues	Rat	0·24		Diamond & Fishman, 1973.
Choroid plexus					
(*in vivo*)	Perfusion	Cat	16		Brønsted, 1970.
(*in vitro*)		Frog	20		Prather & Wright, 1970.

D. GLUCORECEPTORS

If, as appears to be the case, a slight limitation in available glucose causes hypoglycaemic symptoms without any detectable decrease in metabolically-utlizable energy stores, then these symptoms may not be due simply to a lack of glucose as a nutrient for energy production. Ferrendelli (1975), from the studies on insulin hypoglycaemia noted above, argued that the neurological consequences of such relatively slight limitations in available glucose could not be due to energy failure and suggested that they might be due to an alteration in the concentration of some metabolite derived from glucose other than one of the energy phosphates. Alternative substrates for energy metabolism, such as ketone bodies or amino acids, do not seem to be directly implicated (Lewis et al., 1974; Ferrendelli, 1975).

Another possible cause of the behavioural changes in hypoglycaemia is that glucose itself might have a function which is independent of its role as the major source of energy. A process or receptor with a relatively low affinity for glucose could be highly sensitive to a relatively small decrease in the glucose reaching it. Potential types of glucose receptor have already been described. Lesions, produced electrically or by using gold thioglucose, in the ventromedial nucleus of the hypothalamus, are known to bring about remarkable effects on feeding: the animals (mice or rats) never feel satiated and eat until they resemble balloons. The chemically-produced lesions are concluded to be more specific than the electrolytic lesions because gold thioglucose elicits feeding responses only. Specificity is also shown by the gold thioglucose molecule itself: replacement of the glucose moiety by other hexoses gives a biologically-inactive molecule (Mayer & Arees, 1968). The feeding response elicited by gold thioglucose can be reversed by glucose, and the decrease in cellular firing rate in the ventromedial nucleus after lesioning can be reversed by glucose infusions (Cross, 1964). These studies clearly indicate some special sensitivity of specific hypothalamic cells to variations in the blood glucose concentration.

Growth hormone synthesis in incubated rat anterior pituitaries has been shown to be affected by inhibitors of glucose transport in a fashion which seems to be consistent with their binding to a glucoreceptor (Betteridge & Wallis, 1974).

Thus reaction with receptors in the hypothalamic-pituitary system, specifically sensitive to the glucose in the circulation, could result in behavioural changes which are not directly associated with the classical role of glucose as a fuel. Although the glucoreceptors indicated to be present seem to be involved primarily with growth and feeding, these or others so far undetected may react to low glucose in eliciting the first detectable symptoms of hypoglycaemia.

VI Hyperglycaemia

Diabetic coma is always associated with elevated blood sugar concentrations and usually with acidosis. Some hyperglycaemic patients die from acute cerebral oedema, even if the blood glucose concentration has been caused to decrease. Under such circumstances the brain and cerebrospinal fluid concentrations of glucose, fructose and sorbitol are increased. Hyperglycaemic acidosis and oedema have been reproduced in dogs, and similarly increased concentrations of glucose, fructose and sorbitol, and also of inositol, were found. The results were interpreted (Prockop, 1971) in terms of increased conversion of glucose to fructose, sorbitol and inositol which, together with the acidosis, caused an increase in osomotic pressure leading to the increased intracranial pressure and oedema observed in hyperglycaemia in man.

In diabetic coma, cerebral oxygen uptake was found to be decreased by up to 50% and cannot be accounted for simply in terms of reduced blood flow, which was decreased by only 15% (Kety et al., 1948). There appeared to be some correlation between the blood concentration of ketone bodies and the fall in respiration. However, the acidosis associated with ketosis does not seem to be the cause of the coma because correction of blood pH or of the water balance does not reverse the coma (Fazekas & Alman, 1962; Plum & Posner, 1972). Also cases of non-ketotic diabetic coma have been observed in which the plasma concentration may be above 400 mg/100 ml (over 22 mM) but without apparent cerebral oedema. The consequences of non-ketotic hyperglycaemia (severe encephalopathy, focal neurological dysfunction, and sometimes, seizures) have been attributed to generalized hyperosmolarity rather than to an increase in concentration of any particular chemical, and a metabolic basis of nonketotic hyperosmolar coma is far from clear (Arieff and Carrol, 1972; Maccario, 1968; Wright and Gann, 1963; see also the editorial in *Lancet*, ii, 1071, 1072). The accumulation specifically of glucose itself might cause a metabolic deficit, from the speculative arguments which follow.

A. HYPERGLYCAEMIA AND GLUCOSE EFFLUX

There is indirect evidence of active efflux of glucose from the brain under normal conditions and especially in hyperglycaemia (Crone, 1965; Cutler & Sipe, 1971; Bachelard et al., 1973), but the mechanism or the kinetic properties of this efflux do not appear to have been studied at all, unless one regards sugar transport across the choroid plexus as a type of efflux. This has been studied directly (Brønsted, 1970; Prather & Wright, 1970) and the kinetic constants of the choroid plexus transport system are included in the data of Table 5.10. This shows a higher K_m value (16 to 20 mM) than the influx process (5 to 8 mM).

The choroid plexus system also shows evidence for a direct energy requirement (Brønsted, 1970) in contrast to influx, which does not (see Bachelard, 1975b, for discussion of this). Under normal circumstances, glucose influx to the brain is 'downhill', i.e., from a relatively high glucose concentration in the blood to a lower concentration in the brain. Its properties of 'facilitated diffusion' are compatible with this. On the other hand, while efflux could be due to 'counterflow', it could be argued to be an active energy-requiring transport process since it occurs 'uphill' against a concentration gradient. This will be more so in hyperglycaemia with very high blood glucose concentrations. If, and there is no direct evidence for this, glucose efflux from the brain is an energy-consuming process, then under conditions of severe hyperglycaemia it is possible that the energy consumed in pumping glucose out of the brain might result in a local depletion of ATP, and energy failure. This will remain speculative until we obtain evidence that glucose efflux is an energy-dependent process and until we have some idea of the quantitative kinetic data, not only for glucose efflux, but also for the ATP utilized in the process. Energy failure could therefore be a contributory factor to the coma which results from hyperglycaemia; it may be highly localized, as in stroke. It must be stressed that there is no evidence one way or the other available at present. Generalized energy failure (assessed from concentrations of ATP and creatine phosphate) was not observed in rats and mice rendered hypernatraemic by NaCl injection, in the presence of a 25% decrease in cerebral metabolic rate, but there was no evidence that the animals were hyperglycaemic (Lockwood, 1975).

VII Hepatic coma

Gross liver damage, arising from chronic alcoholism, organic solvent poisoning or from infective hepatitis, can often result in gross neurological and psychiatric disturbance, with convulsions leading to coma and death. The condition, which is characterized by high concentrations of ammonia in the circulation, can sometimes be ameliorated by haemodialysis and by a low-protein diet. It is the high circulating ammonia which is generally believed to cause the cerebral dysfunction, although the mechanism of the toxicity is not clear. In man and experimental animals the severity of the neurological disturbance seems to be a function of both the concentration of the ammonia and the time of exposure.

Post-mortem examination of the brains in hepatic coma shows that morphological damage has occurred in glial cells, in particular in the astrocytes (Adams & Foley, 1953). The characteristic changes appear in the astrocyte nuclei, which are swollen and lobulated, and resemble the 'Alzheimer Type II' astrocytes observed earlier in Wilson's disease (von Hösslin & Alzheimer, 1912).

It appeared therefore that the high ammonia concentrations in the brain caused degeneration specifically of astroyctes rather than of neurons, and the condition has been simulated in various ways in laboratory animals, in order that the morphological and biochemical changes can be monitored.

A. ANIMAL STUDIES

Three types of technique have been employed in attempts to develop animal models of hepatic coma.

(1) *Portocaval anastomosis.* The portal vein is tied off and cut high up in the portal fissure; the other end is attached to the anterior wall of the inferior vena cava above the right renal vein (Kyu & Cavanagh, 1970; Cavanagh & Kyu, 1971). This portocaval shunt technique causes the blood to by-pass the liver, so the normal major mechanism for ammonia detoxification (the liver urea cycle) no longer operates. A relatively long period (8 to 12 weeks) is required after the operation in rats before neurological disturbance and gross morphological change becomes pronounced and the animals do not normally become comatose. However, swelling of astrocytic end-feet can be seen within 1 week, changes in astrocyte nuclei have been discerned after 5 weeks, and the cells after 8 to 12 weeks closely resemble the Alzheimer type II cells of clinical hepatic coma (Fig. 5.17). The biochemical changes in the brain (below) are also similar to those found using other techniques.

The shunt technique has been performed also on dogs which went into coma and died within 2 days (Norenberg et al., 1972), although the blood pressure and the arterial concentrations of glucose, O_2, CO_2 and electrolytes were all normal. Examination of the brain showed the Alzheimer type II astrocytes and large glycogen granules were present in astrocyte processes.

Coma can also be produced in shunted rats if NH_4^+ levels are increased by NH_4^+ injection, or by gavage feeding of Dowex 50W-X8, an ammonium ion exchange resin (Norenberg et al., 1974).

(2) *Injection of urease.* Hydrolysis of tissue urea following single or multiple intraperitoneal injections of purified jackbean urease has been used to produce considerable increases in blood ammonia concentrations. The animals resemble the clinical human condition insofar as convulsions and coma result. This treatment also produces Alzheimer type II astrocytes (Prior et al., 1970; Prior & Visek, 1972; Gibson et al., 1974).

(3) *Injections of ammonium salts.* Single or multiple injections of NH_4^+ are toxic in the adult rats at dose levels above 80 mmol/100 g body weight; death is preceded by convulsions (Hindfelt & Siesjö, 1971). Biochemical studies on acute ammonia intoxication have largely been performed using sub-toxic doses.

B. METABOLIC EFFECTS OF HYPERAMMONAEMIA

Ammonia has long been known to be toxic to the brain and also to produce convulsions. One of the major changes which precede the onset of convulsions in animals given methionine sulphoximine (which inhibits glutamine synthetase, *q.v.*) is increased cerebral ammonia (Folbergrova et al., 1969; see also Richter & Dawson, 1948). The three experimental methods of producing hyperammonaemia, described above, also give increased cerebral concentrations of ammonia and of glutamine (Table 5.11).

Hepatic coma in man is associated with decreased cerebral utilization of

Table 5.11. Concentrations of intermediary metabolites in experimental hyperammonaemia

Technique		Blood (mM)		Cerebrum (μmol/g)			
		Ammonia	Glutamine	Ammonia	Glutamine	Glucose	Glycogen
Portocaval shunt	Control	0·04[c]; 0·07[a]	0·8[a]	0·3[c]	10·2[a]; 3·8[d]	1·1[d]	1·7[d]
(3 to 6 weeks)	Treated	0·13[c]; 0·32[a]	0·9[a]	0·9[c]	24·1[a]; 13·1[d]	0·7[d]	1·3[d]
Urease (2 h)	Control	0·02[b]					
	Treated	0·20[b]					
NH_4^+ injection	Control	0·04[c]		0·4[c]			
(7·8 μmol/g; 30 min)	Treated	0·46[c]		3·6[c]			

Note: The glucose of the blood, and the glutamate and aspartate of the brain, were unchanged. Data from: [a]Williams *et al.* (1972); [b]Gibson *et al.* (1974); [c]Hindefelt & Siesjö (1971); Holmin & Siesjö (1974); [d]Cremer & Cavanagh (personal communication of unpublished results).

oxygen (Fazekas *et al.*, 1956) and increased rates of glycolysis, as was found in hypoxic coma (*q.v.*). Similar metabolic events have been observed in the studies on experimental hyperammonaemia in laboratory animals: glycolysis is increased at stages where no change in the energy status of the brain can be demonstrated. Increased cerebral arterio-venous differences in glucose have indicated a 30 to 35% increase in its rate of consumption and lactate was found

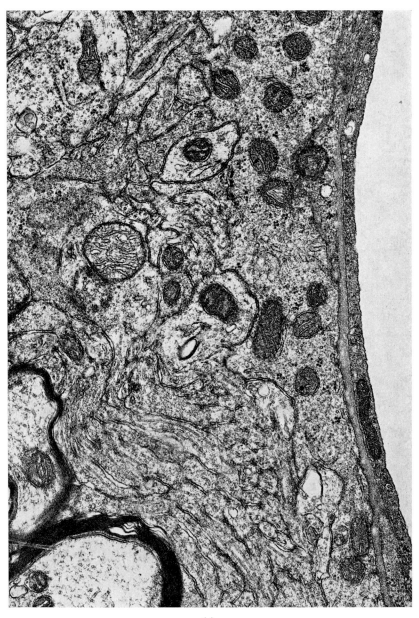

(a)

to accumulate in the brain. There was no detectable effect on adenine nucleotide levels and arterial pH, pO_2 and pCO_2 were all normal (Hawkins et al., 1973; Hindfelt & Siesjö, 1971; Holmin & Siesjö 1974; Prior & Visek, 1972). The toxic effects of NH_4^+ are decreased by barbiturates and more so by hypothermia; hyperthermia gives increased sensitivity to NH_4^+ toxicity (Schenker & Warren, 1962; Schenker et al., 1967).

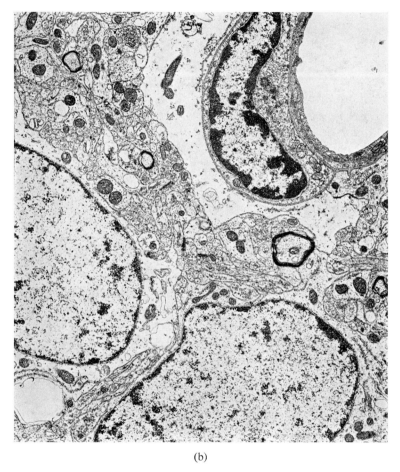

(b)

Fig. 5.17. Cell damage after porto-caval anastomosis in the rat. Sections of the rat thalamus were fixed by perfusion *in situ* with glutaraldehyde (which avoids postmortem fixation artefacts) and stained with uranyl acetate, lead citrate (see Zamora et al., 1973).

(a) Control tissue, showing a normal astrocytic end foot against a capillary wall, with dense cytoplasm and organelles in contrast to b.

(b) Tissue, 2 weeks after porto-caval anastomosis, showing watery swelling of astrocyte cytoplasm and vacuolation of the endoplasmic reticulum.

The micrographs were kindly donated by Professor J. B. Cavanagh, Institute of Neurology, London.

C. GLUTAMINE FORMATION

The major cerebral mechanism for removing ammonia is not urea formation but by formation of glutamine (McIlwain & Bachelard, 1971). NH_4^+ is removed by two enzymic stages of glutamate metabolism: formation of glutamate from α-oxoglutarate catalysed by glutamate dehydrogenase, and formation of glutamine from glutamate, catalysed by glutamine synthase (Fig. 5.18). Both of these reactions are energy consuming: glutamine synthase uses ATP directly and glutamate dehydrogenase consumes energy indirectly as NADH. The observation of elevated glutamine in hyperammonaemia is therefore to be expected, but it seems puzzling that the apparent increase in glutamine synthesis is not reflected by energy loss. Changes in glucose and glycogen are relatively slight (Table 5.11) and cannot account for the amount of energy which should theoretically be lost.

It is feasible that glutamine accumulation could be due, not only to increased rates of formation, but also to decreased rates of its removal. Glutaminase, a very active enzyme in the mammalian brain (Svenneby, 1970), is known to be inhibited under certain conditions by lipid metabolites such as stearyl-CoA and palmityl-CoA (Kvamme & Torgner, 1974), although it is not clear if these are elevated in the circulation as a result of liver damage.

D. α-OXOGLUTARAMATE

The accumulated glutamine, by whatever the mechanism, is partly converted to α-oxoglutaramate (Fig. 5.18), a metabolite not normally found in appreciable amounts. Cerebrospinal fluid concentrations of α-oxoglutaramate were found to be elevated, from a normal concentration of about 3 μM, to nearly 50 μM in patients with hepatic coma (Vergara et al., 1974). This enormous increase in α-oxoglutaramate was far more pronounced than the increases in the cerebrospinal fluids of NH_4^+ (2-fold), glutamate (2-fold) or glutamine (5-fold). Whether α-oxoglutaramate is a key toxic metabolite in the clinical condition has yet to be determined but toxic effects have been observed in rats after infusion (in concentrations of above 10 mM) into the cerebral lateral ventricle (Vergara et al., 1974).

E. AMMONIA AND GLYCOLYTIC ENZYMES

The stimulation of cerebral glycolysis in hyperammonaemia may be interpreted in terms of ammonium ion activation of at least two glycolytic enzymes, phosphofructokinase and pyruvate kinase. NH_4^+ activates cerebral phosphofructokinase, synergistically with other effectors such as inorganic phosphate and AMP (Lowry & Passonneau, 1966). Alone, 3 mM-NH_4^+ increased enzymic activity some 3-fold at low substrate concentrations by affecting the maximum velocity

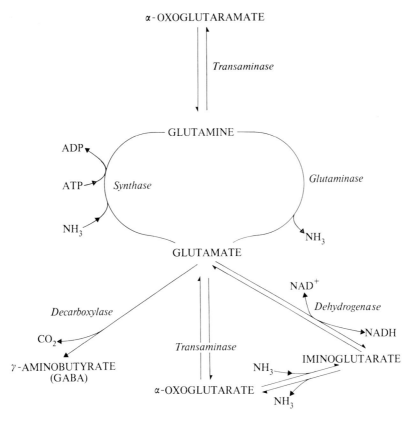

Fig. 5.18. Glutamate metabolism in the brain. Note: α-Oxoglutaramate also occurs as a degradation product of histidine in mammalian tissues.

rather than by lowering the K_m value of the enzyme for ATP. At higher (inhibitory) concentrations of ATP, the activating effect of NH_4^+ was stronger.

Concentrations of NH_4^+ of 3 mM or more have been found in the brain in hyperammonaemia, and lower concentrations of about 1 mM have been reported in hypoxia and ischaemia (Table 5.4). In the intense hypoglycaemia produced by insulin in rats, the cerebral ammonia rose from 0·3 mM in the untreated animals to approximately 3 mM when the blood glucose concentration was down to 1 mM (Lewis et al., 1974a). Cerebral pyruvate kinase activity shows its greatest sensitivity to NH_4^+ over this concentration range (Fig. 5.19). The potential activity of the enzyme is increased 10- to 30-fold by an increase in NH_4^+ concentration from 0·2 to 3 mM in the absence of K^+, and some 2- to 3-fold in the presence of normal concentrations of K^+ (Bachelard et al., 1975).

VIII Coma and development

The preceding discussion in this Chapter has emphasized the apparent anomaly in the underlying biochemistry of many comatose states: an impairment of function (EEG, behaviour) is clearly seen before cell damage and energy failure can be detected. The first metabolic change discernible is often a stimulation of glycolysis. It is possible that energy failure may be highly localized in a small region or in relatively few of the cells in the brain, and any change in energy status may be masked by the majority of cells or regions where no change has occurred. This seems unlikely since the cellular damage which ultimately follows does not appear to be so highly localized, and that detectable generalized energy failure occurs concomitantly with the structural damage. It seems reasonable therefore to search for some involvement with sensitive processes other than the energy-producing oxidative processes of the brain. Such sensitive processes might be metabolic or structural. Some guidance as to the sensitive processes likely to be involved may emerge from a study of the relevant biochemical and morphological changes which take place during development, because the newborn mammalian brain seems less vulnerable, at least to hypoglycaemia and hypoxia, than the adult brain. In considering changes during early brain development it must be noted that the various stages of development occur at different times according to the species. Thus whereas myelination is virtually complete at birth in the guinea pig, it continues for years or weeks after birth in man and the rat.

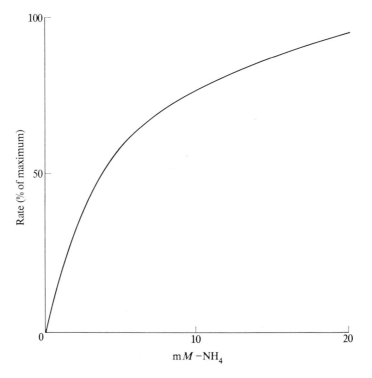

Fig. 5.19. NH_4^+ and cerebral pyruvate kinase activity. (Bachelard et al., 1975.)

A. HYPOGLYCAEMIA

If mice are rendered hypoglycaemic with insulin, great differences in survival between newborn and adults are apparent. Mayman and Tijerina (1971) found that whereas 30% of the adult mice died within 4 h, all of the newborn survived. However, although the newborn are less sensitive to hypoglycaemia, prolonged neonatal hypoglycaemia can lead to brain damage. In newborn babies, this is known to lead to mental retardation, to cerebral palsy or to death. At birth, man has a blood glucose concentration of 2 to 3 mM which normally rises to the adult value of 6 to 7 mM within one week. Structural damage to the newborn human brain may occur as a result of prolonged hypoglycaemia with blood glucose levels below 1 mM. This damage consists of acute degeneration of neurones and glial cells throughout the brain (Anderson et al., 1966; Jones & Smith, 1971). Transient neonatal hypoglycaemia may yield no immediate behavioural symptoms but, as with neonatal hypoxia, there is some reason to suspect that febrile convulsions and epilepsy, with accompanying lesions in the limbic system, may occur subsequently (Falconer, 1970, 1974).

Nevertheless, although neonatal hypoglycaemia, especially in premature babies, is an important cause of mental and neurological disorder, it is clear that the young are less sensitive than the adults, especially in species (mouse, rat and man) where significant brain development occurs in the post-natal period of growth. Working with mice, Mayman and Tijerina (1971) showed that the newborn have higher cerebral stores than the adult of glucose and of glycogen, which are less depleted in hypoglycaemia (Table 5.12). In these experiments there were no changes in ATP or creatine phosphate. Studies on glycolytic rates also indicated that rates of glucose utilization are not only slower in the newborn but also that the proportion of anaerobic glycolysis to oxidative respiration is higher in the newborn. Less glucose is utilized, even less oxygen is consumed, and of the glucose which is utilized at birth, a higher proportion is

Table 5.12. Hypoglycaemia in Mice

		Serum G	Brain G	$\dfrac{\text{Brain}}{\text{Serum}}$	Glycogen
Newborn (1–2 day)	Control	9·35	3·39	0·35	5·5
	Insulin	3·21	0·82	0·27	5·05
Adult	Control	9·91	1·53	0·15	2·2
	Insulin	2·84	0·37	0·14	1·0

The mice were given insulin (25 μg/kg) and killed after 1 hr. Results are μmoles/g frozen tissue. ATP and CP were unchanged throughout (Mayman & Tijerina, 1971).

metabolized via the NADPH-producing hexose monophosphate shunt. Less energy is consumed in the newborn, presumably because there is less electrical activity and less cation transport.

Fig. 5.20 shows the relative time course during development of changes in terms of glycolysis, respiration and formation of myelin and nerve endings. From this, it appears that the major growth period for synaptic formation (to judge both from the very approximate number of synapses and from the content of N-acetylneuraminic acid which is a component of synaptic membranes) occurs before the full formation of mitochondria. The development of oxidative metabolism and of myelin also would appear to occur later than the development of synapses. However, the evidence from Fig. 5.20 on the formation of synaptic structures is based on visual (microscopic) assessment and on chemical analysis, but it does not tell us much about synaptic function. Indeed other evidence suggests that the full electrical activity of the brain does not occur

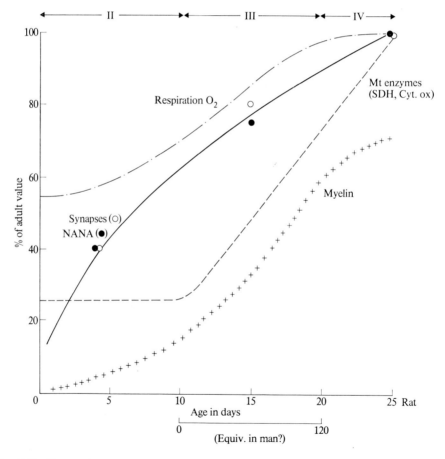

Fig. 5.20. Some changes in the developing brain. Data from: Davison & Dobbing, 1968; McIlwain & Bachelard, 1971; Cragg, 1974.

until later, i.e. until myelination has progressed considerably. For example, the respiratory or glycolytic responses to electrical stimulation (McIlwain & Bachelard, 1971) are not apparent until myelination is virtually complete.

Some conclusion on the sensitivity of synaptic function to hypoglycaemia may be drawn from considerations of the properties of glucose transport. The much lower K_m value for glucose transport across the synaptic membrane (Table 5.10) renders very unlikely a peculiar sensitivity of the synaptic region to limitations in available glucose. It may be that the newborn brain is protected against low glucose simply as a result of its lower metabolic requirement for glucose. There is also the possibility that the glucoreceptors of the hypothalamus (*q.v.*) may form in the later rather than in the earlier stages of development.

B. HYPOXIA

The sensitivity of newborn mammals to anoxia also varies with the species. Whereas most adult mammals go into anoxic coma within 3 min, their newborn remain conscious for much longer periods, which vary according to their stage of development at birth, e.g. rat (50 min.), cat (25 min), dog (23 min), rabbit (17 min), and guinea pig (7 min). As with hypoglycaemia, protective features include higher carbohydrate stores, a slower metabolic rate and a higher proportion of anaerobic metabolism (Himwich *et al.*, 1942).

Whereas in hypoglycaemia, a peculiarly sensitive metabolic or structural process is not clearly apparent (apart from speculations about the importance of glucoreceptors), there is a possibility that the synaptic region may be involved in the behavioural response to hypoxia.

Certain aspects of synaptic metabolism (synthesis of aromatic amine transmitters) have been argued above to be particularly sensitive to low oxygen, from the reported K_m values for oxygen, and there is some evidence (noted above) that dopamine receptors may also be particularly vulnerable to oxygen deprivation.

In conclusion, our current state of knowledge of the biochemistry of coma strongly suggests that it is not simply due to energy failure, as was believed in the past, but perhaps to sensitivity of specialized cerebral processes which may not be directly involved in energy metabolism. The newborn seem to be protected essentially because of their slower rates of metabolism of both glucose and oxygen, and it may be that their higher endogenous reserves ensure that the specialized non-energy requiring processes still have sufficient of these molecules available for maintenance of their specific functions. It could also be partly due to later development of these functions. In the adult, there is some evidence to suggest that amine metabolism in the synaptic region may be particularly sensitive to oxygen deprivation. Sensitivity to low glucose may reside in specific glucoreceptors in the hypothalamic-pituitary axis.

There is obviously much to be learned about specialized roles of glucose and oxygen in the maintenance of cerebral function.

The experimental work described in this chapter was supported by a grant (NS 07918) from the U.S. Public Health Service.

References

ADAMS R.D. & FOLEY J.M. (1953) The neurological disorder associated with liver disease. *Proc. Assoc. Res. Nerv. Ment. Dis.* **32**, 198–326.

ANDERSON J.M., MILNER R.D.G. & STRICH S.J. (1966) Pathological changes in the nervous system in severe neonatal hypoglycaemia. *Lancet*, **ii**, 372–375.

ARIEFF A.I. & CARROL H.J. (1972) Nonketotic hyperosmolar coma with hyperglycemia: clinical features, pathophysiology, renal function, acid-base balance, plasma-cerebrospinal fluid equilibria and the effects of therapy in 37 cases. *Medicine (Baltimore)*, **51**, 73–94.

ATKINSON D.E. (1968) The energy charge of the adenylate pool as a regulatory parameter. *Biochemistry, Easton*, **7**, 4030–4034.

BACHELARD H.S. (1967) The subcellular distribution and properties of hexokinases in the guinea-pig cerebral cortex. *Biochem. J.* **104**, 286–292.

BACHELARD H.S. (1971a) Specificity and kinetic properties of monosaccharide uptake into guinea-pig cerebral cortex *in vitro*. *J. Neurochem.* **13**, 213–222.

BACHELARD H.S. (1971b) Glucose transport and phosphorylation in the control of carbohydrate metabolism in the brain. In Brierley J.B. & Meldrum B.S. (Eds.) *Brain Hypoxia*, pp. 251–260. London, Heinemann.

BACHELARD H.S. (1975a) Carbohydrate metabolism in the brain. *Handbook of Clinical Neurology*, **29**, Chapter 4.

BACHELARD H.S. (1975b) How does glucose enter brain cells? In Ingvar D. & Lassen D. (Eds.) *Alfred Benzon Symposium VIII: The Working Brain*. Copenhagen, Munksgaard.

BACHELARD H.S., CAMPBELL W.J. & McILWAIN H. (1962) The sodium and other ions of mammalian cerebral tissues, maintained and electrically-stimulated *in vitro*. *Biochem. J.* **84**, 225–232.

BACHELARD H.S., DANIEL P.M., LOVE E.R. & PRATT O.E. (1973) The transport of glucose into the brain of the rat *in vivo*. *Proc. Roy. Soc. B.* **183**, 71–82.

BACHELARD H.S., FISHER S.K. & NICHOLAS P.C. (1975) Monovalent cations and cerebral pyruvate kinase activity. *Biochem. J.* In press.

BACHELARD H.S. & GOLDFARB P.S.G. (1969) Adenine nucleotides and magnesium ions in relation to control of mammalian cerebral-cortex hexokinase. *Biochem. J.* **112**, 579–586.

BACHELARD H.S., LEWIS L.D., PONTÉN U. & SIESJÖ B.K. (1974) Mechanisms activating glycolysis in the brain in arterial hypoxia. *J. Neurochem.* **22**, 395–401.

BARCROFT R. (1975) *The Respiratory Function of the Blood*. Cambridge Univ. Press.

BETTERIDGE A. & WALLIS M. (1974) Control of growth hormone synthesis *in vitro* by glucose. *Biochem. Soc. Trans.* **2**, 956–957.

BETZ A.L., GILBOE D.D., YUDILEVICH D.L. & DREWES L.R. (1973). Kinetics of unidirectional glucose transport into the isolated dog brain. *Am. J. Physiol.* **225**, 586–592.

BETZ A.L., GILBOE D.D. & DREWES L.R. (1974) Effects of anoxia on the net uptake and unidirectional transport of glucose into the isolated dog brain. *Brain Res.* **67**, 307–316.

BOYLE R. (1660) *New physico-mechanical Experiments touching the Spirit of Air, and its Effects*. Oxford.
BRIERLEY J.B. (1975) Personal communication.
BRIERLEY J.B., BROWN A.W. & MELDRUM B.S. (1971). The nature and time course of the neuronal alterations resulting from oligaemia and hypoglycaemia in the brain of *Macaca Mulatta*. *Brain Res.* **25**, 483–499.
BRONISZEWSKA-ARDELT B. & MILLER A.T. (1974) Hypoxic changes in brain hexokinase distribution: phylogenetic and developmental considerations. *Comp. Biochem. Physiol.* **49B**, 151–156.
BRØNSTED H.E. (1970). Ouabain-sensitive carrier-mediated transport of glucose from the cerebral ventricles to surrounding tissues in the cat. *J. Physiol. (Lond.)*, **208**, 187–201.
BROWN A.W. (1973) The morphological characteristics and evolution of anoxic-ischaemic brain damage in different species. *M. Phil. Thesis*. Council for National Academic Awards.
BROWN A.W. & BRIERLEY J.B. (1966). Evidence for early anoxic-ischaemic cell damage in the rat brain. *Experientia*, **22**, 546–560.
BROWN A.W. & BRIERLEY J.B. (1968) The nature, distribution and earliest stages of anoxic-ischaemic nerve cell damage in the rat brain as defined by the optical microscope. *Brit. J. Exp. Path.* **49**, 87–106.
BROWN A.W. & BRIERLEY J.B. (1973) The earliest alterations in rat neurones and astrocytes after anoxia-ischaemia. *Acta Neuropath. (Berlin)*, **23**, 9–22.
BROWN R., DAVIS J.N. & CARLSSON A. (1973). DOPA reversal of hypoxia-induced disruption of the conditioned avoidance response. *J. Pharm. Pharmac.* **25**, 412–414.
BROWN R. & ENGEL J. (1973) Evidence for catecholamine involvement in the suppression of locomotor activity due to hypoxia. *J. Pharm. Pharmac.* **25**, 815–819.
BUSCHIAZZO P.M., TERRELL E.B. & REGEN D.M. (1970). Sugar transport across the blood–brain barrier. *Am. J. Physiol.* **219**, 1505–1513.
CARLSSON A. (1974) The *in vivo* estimation of rates of tryptophan and tyrosine hydroxylation: effects of alterations in enzyme environment and neuronal activity. In *Aromatic Amino Acids in the Brain*, CIBA Foundation Symposium No. 22, Amsterdam, Elsevier, pp. 117–134.
CAVANAGH J.B. & KYU M.H. (1971). Type II Alzeheimer change experimentally produced in astrocytes in the rat. *J. Neurol. Sci.* **12**, 63–75.
CLARK J.B., DEGN H. & NICHOLLS P. (1975) Apparent K_m for oxygen of rat brain mitochondrial respiration. *J. Neurochem.* In press.
COHEN P.J. (1971). Energy metabolism of the human brain. In Siesjö B.K. & Sørensen S.C. (Eds.) *Ion Homeostasis of the Brain*. Copenhagen, Munksgaard, pp. 417–423.
COHEN P.J., ALEXANDER S.C., SMITH T.C., REIVICH M. & WOLLMAN H. (1967) Effects of hypoxia and normocarbia on cerebral blood flow and metabolism in conscious man. *J. Appl. Physiol.* **23**, 183–189.
CRAGG B.G. (1974) Plasticity of synapses. *Brit. Med. Bull.* **30**, 141–144.
CRONE C. (1965) Facilitated transfer of glucose from blood into brain tissue. *J. Physiol. (Lond.)*, **181**, 103–113.
CROSS B.A. (1964) The hypothalamus in mammalian homeostasis. *Sympos. Soc. Exp. Biol.* **18**, 157–193.
CUTLER R.W.P. & SIPE J.C. (1971) Mediated transport of glucose between blood and brain in the cat. *Am. J. Physiol.* **220**, 1182–1186.
DAVIS J.N. & CARLSSON A. (1973a) The effect of hypoxia on monoamine synthesis, levels and metabolism in rat brain. *J. Neurochem.* **21**, 783–790.

Davis J.N. & Carlsson A. (1973b) Effect of hypoxia on tyrosine and tryptophan hydroxylation in unanaesthetized rat brain. *J. Neurochem.* **20**, 913–915.

Davison A.N. & Dobbing J. (1968) *Applied Neurochemistry.* Oxford, Blackwell Scientific Publications.

Degn H. & Wohlrab H. (1971) Measurement of steady-state values of respiration rate and oxidation levels of respiratory pigments at low oxygen tensions. A new technique. *Biochim. Biophys. Acta,* **245**, 347–355.

Diamond I. & Fishman R.A. (1973) High affinity transport and phosphorylation of 2-deoxy-D-glucose in synaptosomes. *J. Neurochem.* **20**, 1533–1542,

Drewes L.R. & Gilboe D.D. (1973a) Cerebral metabolite and adenylate energy charge recovery following 10 min of anoxia. *Biochem. Biophys. Acta,* **320**, 701–707.

Drewes L.R. & Gilboe D.D. (1973b) Glycolysis and the permeation of glucose and lactate in the isolated, perfused dog brain during anoxia and post-anoxic recovery. *J. Biol. Chem.* **218**, 2489–2496.

Drewes L.R., Gilboe D.D. & Betz A.L. (1973) Metabolic alterations in brain during anoxic-anoxia and subsequent recovery. *Arch. Neurol.* **29**, 385–390.

Duffy T.E., Nelson S.R. & Lowry O.H. (1972) Cerebral carbohydrate metabolism during acute hypoxia and recovery. *J. Neurochem.* **19**, 959–977.

Falconer M.A. (1970) Significance of surgery for temporal lobe epilepsy in childhood and adolescence. *J. Neurosurg.* **33**, 233–252.

Falconer M.A. (1974) Mesial temporal (Ammon's Horn) sclerosis as a common cause of epilepsy. Aetiology, treatment and prevention. *Lancet,* **ii**, 767–770.

Fazekas J.F. & Alman R.W. (1962) *Coma: Biochemistry, Physiology and Therapeutic Principles.* Springfield, Illinois, Thomas.

Fazekas J.F., Ticktin H.E., Ehrmantraut W.R. & Alman R.W. (1956) Cerebral metabolism in hepatic insufficiency. *Am. J. Med.* **21**, 843–849.

Ferrendelli J.A. (1975) The role of CNS energy metabolism in cerebral dysfunction produced by hypoglycaemia. In Ingvar D. & Lassen D. (Eds.) *The Working Brain.* Copenhagen, Munksgaard.

Ferrendelli J.A. & Chang M-M. (1973) Brain metabolism during hypoglycaemia. *Arch. Neurol.* **28**, 173–177.

Fisher D.B. & Kaufman S. (1972) The inhibition of phenylalanine and tyrosine hydroxylase by high oxygen levels. *J. Neurochem.* **19**, 1359–1365.

Folbergrova J., Passonneau J.V., Lowry O.H. & Schulz D.W. (1969) Glycogen, ammonia and related metabolites in the brain during seizures evoked by methionine sulphoximine. *J. Neurochem.* **16**, 191–203.

Geiger A. (1958) Correlation of brain metabolism and function by the use of a brain perfusion method *in situ. Physiol. Revs.* **38**, 1–20.

Gibson G.E., Zimber A., Krook L., Richardson E.P. & Visek W.J. (1974) Brain histology and behaviour of mice injected with urease. *J. Neuropath. Exp. Neurol.* **33**, 201–211.

Growdon W.A., Bratton T.S., Houston M.C., Tarpley H.L. & Regen D.M. (1971) Brain glucose metabolism in the intact mouse. *Am. J. Physiol.* **221**, 1738–1745.

Hawkins R.A., Miller A.L. & Veech R.L. (1973) The acute action of ammonia on rat brain metabolism *in vivo. Abstracts 4th I.S.N. Meeting, Tokyo,* 82–83.

Heaton G.M. & Bachelard H.S. (1973) The kinetic properties of hexose transport into synaptosomes from guinea pig cerebral cortex. *J. Neurochem.* **21**, 1099–1108.

Himwich H.E., Bernstein A.O., Herrlich H., Chesler A. & Fazekas J.F. (1942) Mechanisms for the maintenance of life in the newborn during anoxia. *Am. J. Physiol.* **135**, 387–391.

HIMWICH H.E. & FAZEKAS J.F. (1937) The effect of hypoglycaemia on the metabolism of the brain. *Endocrinology*, **21**, 800–805.

HINDFELT B. & SIESJÖ B.K. (1971a) Cerebral effects of acute ammonia intoxication: the influence on intracellular and extracellular acid-base parameters. *Scand. J. Clin. Lab. Invest.* **28**, 353–364.

HINDFELT B. & SIESJÖ B.K. (1971b) Cerebral effects of acute ammonia intoxication: the effect upon energy metabolism. *Scand. J. Clin. Lab. Invest.* **28**, 365–374.

HOLMIN T. & SIESJÖ B.K. (1974) The effect of porta-caval anastomosis upon the energy state and upon acid-base parameters of the rat brain. *J. Neurochem.* **22**, 403–412.

HORTON R.W., MELDRUM B.S. & BACHELARD H.S. (1973) Enzymic and cerebral metabolic effects of 2-deoxy-D-glucose. *J. Neurochem.* **21**, 507–520.

HÖSSLIN C. VON & ALZHEIMER A. (1912) Ein Beitrag zur Klinik, und pathologischen Anatomie der Westphal-Strümpellschen Pseudosklerose. *Z. Neurol. Psychiat.* **8**, 183–209.

JONES E.L. & SMITH W.T. (1971) Hypoglycaemic brain damage in the neonatal rat. In Brierley J.B. & Meldrum B.S. (Eds.) *Brain Hypoxia*, pp. 231–241. London, Heinemann.

KERR S.E. (1935) Studies on the phosphorus compounds of brain. I. Phosphocreatine. *J. Biol. Chem.* **110**, 625–635.

KERR S.E. (1936) The carbohydrate metabolism of brain. I. The determination of glycogen in nervous tissue. *J. Biol. Chem.* **116**, 1–8.

KETY S.S. & SCHMIDT C.F. (1948) The quantitative determination of cerebral blood flow in man; theory, procedure and normal values. *J. Clin. Invest.* **27**, 476–483.

KETY S.S. WOODFORD R.B., HARMEL M.H., FREYHAN F.A., APPEL K.E. & SCHMIDT C.F. (1948) Cerebral blood flow and metabolism in schizophrenia; effects of barbiturate semi-narcosis, insulin-coma and electroshock. *Am. J. Psychiat.* **104**, 765–770.

KAUFMAN S. (1974) *Aromatic Amino Acids in the Brain. CIBA Foundation Symposium No. 22*, pp. 85–108. Amsterdam, Elsevier.

KNULL H.R., TAYLOR W.F. & WELLS W.W. (1973) Effects of energy metabolism on *in vivo* distribution of hexokinase in brain. *J. Biol. Chem.* **248**, 5414–5417.

KVAMME E. & TORGNER I.A. (1974) Phosphate-dependent effects of palmityl-CoA and stearyl-CoA on phosphate-activated pig brain and pig kidney glutaminase. *FEBS Letters*, **47**, 244–247.

KYU M.H. & CAVANAGH J.B. (1970) Some effects of porto-caval anastomosis in the male rat. *Brit. J. exp. Path.* **51**, 217–227.

LANDAU B.R. & LUBS H.A. (1958) Animal responses to 2-deoxy-D-glucose administration. *Proc. Soc. Exp. Biol. Med.* **99**, 124–127.

LEVINE S. (1960) Anoxic-ischaemic encephalopathy in rats. *Am. J. Path.* **36**, 1–17.

LEWIS L.D., LJUNGGREN B., NORBERG K. & SIESJÖ B.K. (1974a) Changes in carbohydrate substrates, amino acids and ammonia in the brain during insulin-induced hypoglycaemia. *J. Neurochem.* **23**, 659–671.

LEWIS L.D., LJUNGGREN B., RATCHESON R.A. & SIESJÖ B.K. (1974b) Cerebral energy state in insulin-induced hypoglycaemia related to blood glucose and to EEG. *J. Neurochem.* **23**, 673–679.

LOCKWOOD A.H. (1975) Acute and chronic hyperosmolality. Effects on cerebral amino acids and energy metabolism. *Arch. Neurol.* **32**, 62–64.

LOWRY O.H. (1966) Energy metabolism of the nerve cell. In Rodahl K. & Issekutz B. (Eds.) *Nerve as a Tissue*, pp. 163–174. New York, Harper and Row.

LOWRY O.H. & PASSONNEAU J.V. (1966) Kinetic evidence for multiple binding sites on phosphofructokinase. *J. Biol. Chem.* **241**, 2268–2279.

LOWRY O.H., PASSONNEAU J.V., HASSELBERGER F.X. & SCHULZ D.W. (1964) Effect of ischaemia on known substrates and cofactors of the glycolytic pathway in brain. *J. Biol. Chem.* **239**, 18–30.

MCILWAIN H. & BACHELARD H.S. (1971) *Biochemistry and the Central Nervous System.* 4th ed., Churchill, London.

MACCARIO M. (1968) Neurological dysfunction associated with non-ketotic hyperglycemia. *Arch. Neurol.* **19**, 525–534.

MACMILLAN V. & SIESJÖ B.K. (1972) Brain energy metabolism in hypoxemia. *Scand. J. Clin. Lab. Invest.* **30**, 127–136.

MAYER J. & AREES E.A. (1968). Ventromedial glucoreceptor system. *Fed. Proc.* **27**, 1345–1348.

MAYMAN C.I. & TIJERINA M.L. (1971) The effect of hypoglycaemia on energy reserves in adult and newborn brain. In Brierley J.B. & Meldrum B.S. (Eds.) *Brain Hypoxia*, pp. 242–249. London, Heinemann.

MELDRUM B.S., HORTON R.W. & BRIERLEY J.B. (1971) Insulin-induced hypoglycaemia in the primate: relationship between physiological changes and neuropathology. In Brierley J.B. & Meldrum B.S. (Eds.) *Brain Hypoxia*, pp. 207–224. London, Heinemann.

NORBERG K. & SIESJÖ B.K. (1975) Personal communication of unpublished results.

NORENBERG M.D., LAPHAM L.W., BENSON R.W. & MAY A.G. (1972) Ultrastructural observations in protoplasmic astrocytes in experimental acute hepatic coma. *J. Neuropath. Exp. Neurol.* **31**, 184–185.

NORENBERG M.D., LAPHAM L.W., NICHOLS F.A. & MAY A.G. (1974). An experimental model for the study of hepatic encephalopathy. *Arch. Neurol.* **31**, 106–109.

PAPPENHEIMER J.R. & SETCHELL B.P. (1973) Cerebral glucose transport and oxygen consumption in sheep and rabbits. *J. Physiol. (Lond.)*, **233**, 529–551.

PETERSON L.C., NICHOLLS P. & DEGN H. (1974) The effect of energization on the apparent Michaelis–Menten constant for oxygen in mitochondrial respiration. *Biochem. J.* **142**, 247–252.

PLUM F. & POSNER J.B. (1972) *The Diagnosis of Stupor and Coma.* Philadelphia, Davies.

PROCKOP L.D. (1971) Hyperglycaemia, polyol accumulation and increased intracranial pressure. *Arch. Neurol.* **25**, 125–140.

PRATHER J.W. & WRIGHT E.M. (1970). Molecular and kinetic parameters of sugar transport across the frog choroid plexus. *J. Membrane Biol.* **2**, 150–172.

PRIOR R.L., CLIFFORD A.J. & VISEK W.J. (1970) Tissue amino acid concentrations in rats during acute ammonium intoxication. *Am. J. Physiol.* **219**, 1680–1683.

PRIOR R.L. & VISEK W.J. (1972) Effects of urea hydrolysis on tissue metabolite concentrations in rats. *Am. J. Physiol.* **223**, 1143–1149.

RICHTER D. & DAWSON R.M.C. (1948) The ammonia and glutamine content of the brain. *J. Biol. Chem.* **176**, 1199–1210.

SCHENKER S., MCCANDLESS D.W., BROPHY E. & LEWIS M.S. (1967). Studies on the intracerebral toxicity of ammonia. *J. Clin. Invest.* **46**, 838–848.

SCHENKER S. & WARREN K.S. (1962) Effect of temperature variation on toxicity and metabolism of ammonia in mice. *J. Lab. Clin. Invest.* **60**, 291–301.

SEVERINGHAUS J.W., CHIODI H., EGER E.I., BRANDSTATER B. & HORNBEIN T.F. (1966). Cerebral blood flow in man at high altitude. *Circ. Res.* **19**, 274–282.

SØRENSEN S.C. & SEVERINGHAUS J.W. (1968). Irreversible respiratory insensitivity to acute hypoxia in man born at high altitude. *J. Appl. Physiol.* **25**, 217–220.

SVENNEBY G. (1970). Pig brain glutaminase; purification and identification of different enzyme forms. *J. Neurochem.* **17**, 1591–1599.

SWANSON P.D. (1969). The effects of oxygen deprivation on electrically stimulated cerebral cortex slices. *J. Neurochem.* **16**, 35–45.

TEWS J.K., CARTER S.H. & STONE W.E. (1965). Chemical changes in the brain during insulin hypoglycaemia and recovery. *J. Neurochem.* **12**, 679–693.

THORN W., GRIESHABER, TH. & JUNGE H. (1973). γ-Aminobuttersäure- und NH_3-Gehalt in Kaninchengehirn nach Hypoxie und Ischämie. *Pflügers Arch.* **345**, 347–351.

VAN HARREVELD A., CROWELL J. & MALHOTRA S.K. (1965). A study of extracellular space in central nervous tissue by freeze substitution. *J. cell. Biol.* **25**, 117–137.

VAN HARREVELD A. & MALHOTRA S.K. (1966). Demonstration of extracellular space by freeze-drying in the cerebellar molecular layer. *J. cell. Sci.* **1**, 223–228.

VAN HARREVELD A. & MALHOTRA S.K. (1967). Extracellular space in the cerebral cortex of the mouse. *J. Anat.* **101**, 197–207.

VAN HARREVELD A. & STEINER J. (1970). Extracellular space in frozen and ethanol substituted central nervous tissue. *Anat. Rec.* **116**, 117–130.

VERGARA F., PLUM F. & DUFFY D.E. (1974). α-Ketoglutaramate: increased concentrations in the cerebrospinal fluid of patients in hepatic coma. *Science*, **183**, 81–83.

WILLIAMS A.H., KYU M.H., FENTON J.C.B. & Cavanagh J.B. (1972). The glutamate and glutamine content of rat brain after portocaval anastomosis. *J. Neurochem.* **19**, 1073–1077.

WRIGHT H.K. & GANN D.S. (1963) Hyperglycemic hyponatremia in nondiabetic patients. *Arch. Intern. Med.* **112**, 344–346.

ZAMORA A.J., CAVANAGH J.B. & KYU M.H. (1973). Ultrastructural responses of the astrocyte to porto-caval anastomosis in the rat. *J. Neurol. Sci.* **18**, 25–45.

6 Brain specific antigens: biochemical role in selective pathogenesis

E. J. Thompson

I Introduction, 281

II Neuronal antigens: unclassified, 281

    A. Nerve growth factor, 281
    B. 14-3-2 Protein or antigen alpha, 283
    C. (Na^+, K^+) ATPase, 284
    D. Neuronin, 284
    E. Gangliosides, 284
        1. Monosialoganglioside GM_1, 284
        2. Disialoganglioside GD_3, 284
        3. Disialoganglioside GD_{1b}, 285
        4. Trisialoganglioside GT_1, 285

III Neuronal antigens: synaptic, 285

    A. Acetylcholine receptor protein (dicotinic) 285
    B. Acetylcholinesterase, 288
    C. Choline acetyl transferase (choline acetylase), 288
    D. Glutamic acid decarboxylase, 288
    E. Dopamine beta hydroxylase or chromomembranin A, 288
    F. Tyrosine hydroxylase, 289
    G. Brain tropomyosin, 289
    H. GP350 Glycoprotein, 290
    I. Chromogranin A, 290
    J. Neurophysins, 290
    K. Vesiculin, 290

IV Neuronal antigens: axonal, 290

    A. Actin, 290
    B. Myosin, 292
    C. Tubulin, 293
    D. Filarin, 293
    E. Dynein, 295
    F. AP22 Protein, 295

V Glial antigens: unclassified, 295

    A. Glial fibrillary acidic protein, 295
    B. S-100 protein, 295
    C. Alpha 2 glycoprotein, 297
    D. Beta trace protein, 297
    E. 10 B Glycoproteins, 299

VI Glial antigens: myelin, 299

    A. Myelin basic protein (encephalitogenic), 299
    B. CNS Myelin glycoprotein, 300
    C. PNS Myelin glycoprotein, 300
    D. Wolfgram protein, 301
    E. Proteolipid proteins, 301
    F. 2', 3', Cyclic phosphodiesterase, 301
    G. Cerebrosides, 301

VII Appendix: techniques, 302

  A. Protein characterization, 302
   1. Separation of monomers in the presence of sodium dodecyl sulphate, 302
   2. Other potentially specific antigens, 304
   3. Peptide mapping, 305
  B. Immunization, 305

VIII Conclusion, 306

I Introduction

Specialization of tissue function is well demonstrated by the brain. Different cell types provide one basis for this specialization. However, even the membranes of a given cell type often show striking segregation of function in different regions of intercellular contact. Hence, the differential localization of specific markers for individual kinds of neural cells or for discrete membrane areas is of considerable significance.

There are now several well-characterized antigens specific to the nervous system. In this chapter, their biochemical similarities and differences will be discussed as well as their role in the pathogenesis of specific diseases. In addition, specific antibodies have given much useful information regarding the normal structure and function of the nervous system.

Specific antigens will be considered under the headings of 'Neuronal' (including 'Synaptic' and 'Axonal') and of 'Glial' (including 'Myelin'). The bulk of the information will be presented in tabular form. The text will be used rather to illustrate major points relating to each specific antigen. Detailed biochemical methods relating to the analysis and characterization of these antigenic classes are given in an appendix.

Mixed antigens derived from brain subcellular fractions or different cell types will not be considered here in detail, as it is assumed that eventually the specific antigens within these preparations will be isolated.

Different tumours of the nervous system synthesize various brain specific proteins. As this is potentially useful in the diagnosis and treatment of such tumours, reference will be made to appropriate examples. In this case, immunization with such tumours has proved useful in the study of the normal nervous system.

II Neuronal antigens: unclassified (Table 6.1)

A. NERVE GROWTH FACTOR

This protein was found to allow outgrowth of nerve fibres from sympathetic ganglia in culture, and when antiserum to this protein was given to young animals it was found to produce an immunological sympathectomy (Levi-Montalcini & Angeletti, 1968).

The protein is composed of three subunits: alpha (acidic), which may have a protective action during dissociation of ganglia, since it gives higher total cell yields; beta (basic), which maintains the normal outgrowth of neurites; and gamma (neutral), which has esterase activity (Shooter & Roboz-Einstein 1971). Although the richest source has been male mouse salivary glands, it has also been found to be released from glial cells (Longo & Penhoet, 1974) and from fibroblasts (Oger et al., 1974). Since the protein has a similar amino acid sequence to insulin (Frazier et al., 1972) it was of interest to note that the NGF

Table 6.1. Comparative data on specific brain antigens: neuronal antigens: unclassified

Antigen	Biochemical			
	Monomeric molecular weight ($\times 10^{-3}$ Daltons)	Structure		Amount
		Primary	Quaternary	
Nerve growth factor	12	α-Acidic β-Basic γ-Acidic (slightly)	$(\alpha\beta\gamma)_2$	2×10^4 per neurone
14-3-2 or Alpha	49	Acidic	Dimers	1% Soluble protein
Na^+, K^+ ATPase	105			5% Membrane protein
Neuronin	47	Acidic		1% Soluble protein
Ganglioside GM_1	1	Acidic	Micelles	Total Ganglioside values: 25 μmoles per gm of Nerve ending protein
GD_3	1	Acidic	Micelles	
GD_{1b}	1	Acidic	Micelles	
GT_1	2	Acidic	Micelles	

receptor did not compete with insulin (Banerjee et al., 1973). Both receptors are located on the cell surface, and full response is obtained if the protein has been coupled to Sepharose beads (Frazier et al., 1973). The protein may also initiate polymerization of tubulin (Calissano & Cozzari 1974).

Labelled nerve growth factor (or fragments thereof) have been shown to be transported from the axon terminals in the iris of the eye back to the cell bodies in the superior cervical ganglion. It is thus considered as possibly having a 'reverse trophic' effect of end organ upon the innervating neurons (Hendry et al., 1974). This protein also aids in the regrowth of nerve into muscle when the muscle is transplanted into the brain stem (Bjerre et al., 1974). It is also considered to have more general trophic function such as a stimulus for the proliferation of blood vessels.

Cell Type	Anatomical		Physiological	Pathological
	Membrane locus	Expression in tissue culture	Function	Associated diseases
Sympathetic neurones	Surface	Glioma (C-6) Sympathetic Neurones	α-Salvage β-Maintain γ-Esterase	Immuno-Sympathectomy
Neurones	Intra-cellular	Neuroblastoma		Optic Nerve Section
Neurones	Axons	Neuroblastoma (N2a) Glioma (C-6)	Maintain Membrane Potential	
	Soluble			Dementia
Neurones Fibroblasts	Surface	Fibroblasts (L)	Surface Negative Charge	Cholera
Neurones Muscle	Surface		Surface Negative Charge	Strychnine
Neurones 'T' lymphocytes	Surface	Neuroblastoma (N2a)	Surface Negative Charge	Tetanus
Neurones	Surface	Neuroblastoma (N2a)	Surface Negative Charge	Tetanus

B. 14-3-2 PROTEIN OR ANTIGEN ALPHA

These two proteins have recently been shown to be identical (Bennett, 1974). The 14-3-2 protein was originally noted as part of the investigation which described the S-100 protein (Moore & McGregor, 1965). The 14-3-2 protein was separated by DEAE column followed by electrophoresis. Antigen alpha was found by rather similar means. However, in this preparation the electrophoresis was performed before the amino cellulose column fractionation (Bennett & Edelman, 1968).

The 14-3-2 protein had been considered as a marker for neurones based in part upon its synthesis by some, but not all classes of neuroblastoma (Herschman & Lerner, 1973). This disjunction has also been noted in ethylnitrose urea induced nerve cell lines (Schubert *et al.*, 1974). Unlike S-100, 14-3-2 production seems not to vary during the cell cycle (Kolber *et al.*, 1974). The function of this protein is not clear, but it decreased 10-fold when the optic nerve was cut (Perez *et al.*, 1970). The antigen alpha protein has also been noted to aggregate easily (Bennett & Edelman 1968). Like the S-100 protein, the 14-3-2 protein has been found to constitute about 1% of the water soluble cellular proteins (Perez & Moore, 1970).

C. (NA^+, K^+) ATPase

This enzyme is a major constituent of axonal membranes in fish (Grefarth & Reynolds, 1973), where it has a molecular weight of 105,000 Daltons. This subunit is phosphorylated during hydrolysis of the ATP when stimulated by cations. In electric eel the enzyme has a smaller subunit of 58,000 Daltons (Jean *et al.*, 1974, Askari, 1974). Both portions are antigenic and some antibody preparations against either subunit inhibit the activity of the enzyme. However, it has also been noted (Kyte, 1974) that other antibody preparations do not block the enzyme; using the latter antibody preparation, the enzyme was localized on the cytoplasmic side of the plasma membrane (canine kidney medulla).

D. NEURONIN

This protein has been shown to be decreased in senile brains (see Chapter 1 for full details).

E. GANGLIOSIDES

This group of glycolipids has the characteristic sugar *N*-acetyl neuraminic acid, (sialic acid) which is in part responsible for the net negative charge on surface membranes. Many viruses which attach to cell membranes contain an enzyme to hydrolyse this sugar (neuraminidase, sialidase, or receptor destroying enzyme). Gangliosides are enriched in neurones, specifically in synaptic areas (Breckenridge *et al.*, 1972, Hamberger & Svennerholm, 1971).

1. MONOSIALOGANGLIOSIDE GM_1

This ganglioside is not restricted to brain, but is also normally found in gut and skin. It has been shown to be the receptor for cholera toxin (Cuatrecasas, 1973). The toxin acts after binding to the receptor by increasing intracellular levels of cyclic AMP. In the gut this leads eventually to the loss of fluids.

2. DISIALOGANGLIOSIDE GD_3

This ganglioside is normally found not only in brain but also in muscle. It is the receptor for various amine neurotransmitters such as serotonin (5-HT), tryptamine, and for lysergic acid diethylamine (LSD), strychnine and related drugs (van Heyningen, 1974).

3. DISIALOGANGLIOSIDE GD_{1b}

This ganglioside is found on neurones (at synapses) as well as on T lymphocytes (Esselman & Miller, 1974). It is a receptor for tetanus toxin, which blocks synaptic transmission (a) in spinal cord at inhibitory synapsis (b) at the neurotransmitter junction (van Heyningen, 1974).

4. TRISIALOGANGLIOSIDE GT_1

This ganglioside is found mainly on neurones. Being closely allied in chemical structure to GD_{1b}, it is also a receptor for tetanus toxin.

III Neuronal Antigens: synaptic (Table 6.2.)

A. ACETYLCHOLINE RECEPTOR PROTEIN (NICOTINIC)

Recent appreciation of the affinity and specificity of the curare-like snake toxins has allowed swift progress in the isolation of this receptor protein. Investigation of ligand binding has become a useful general model for studying the mechanisms of hormone action, by analogy with insulin and nerve growth factor. Antibody to this exposed cell surface protein produces a myasthenic syndrome in rabbits. This experimental condition responds to the action of eserine, an inhibitor of the acetylcholinesterase, which acts by presumably raising the concentration of free acetylcholine to compete with the antibody. The cholinergic receptor is located on muscle and neuronal cell surfaces and is accessible to antibody. The antibody not only blocks the access of carbamylcholine (Sugiyama et al., 1973) but the binding constant demonstrates high affinity since it is quite difficult to wash off the antibody.

Patients with myasthenia gravis have been shown to have decreased numbers of binding sites for α-bungarotoxin (Fambrough et al., 1973). Since this disease has been shown to fit into the general category of autoimmune diseases (Feltkamp et al., 1974), it is relevant to ask what portion of this decreased number of sites for the toxin is due to antibody being present on the receptor (Almon et al., 1974).

The receptor protein is normally restricted to endplate areas of muscle (Miledi & Potter, 1971) as well as the surface of sympathetic neurones (Greene et al., 1973) and is a useful marker for such cell types in culture. It is also found on neurones in synaptic areas of the retina (Vogel et al., 1974).

Table 6.2. Comparative data on specific brain antigens: neuronal antigens: synaptic

Antigen	Monomeric molecular weight ($\times 10^{-3}$ Daltons)	Biochemical		Amount
		Structure Primary	Quaternary	
Acetylcholine receptor	54 45	Acidic	2 heavy chains 2 light chains	3×10^5 per sympathetic neurone
Acetylcholine esterase	90 65	Acidic sialo-glyco protein	2 heavy chains 2 light chains	10^7 per muscle synapse
Choline acetyl transferase	64			
Glutamic acid decarboxylase	58			
Dopamine beta hydroxylase	77	Acidic glycoprotein	4 chains	
Tyrosine hydroxylase				
Brain tropomyosin	30			
GP 350 Glycoprotein	12	Acidic sialo-glyco-protein		
Chromogranin A	40		Catechols and dopamine beta hydroxylase	
Neurophysins (I, II & III)	I 9 II 14 III 9	Acidic	I Oxytocin II Vasopressin	
Vesiculin	10			

	Anatomical		Physiological	Pathological
Cell type	Membrane locus	Expression in tissue culture	Function	Associated diseases
Neurone muscle	Surface (synapse)	Neuroblastoma (NG 108) Neurones (symp.) Muscle (L-6)	Bind acetylcholine	Mysathenia gravis
Neurone muscle	Surface (synapse)	Neuroblastoma (N-18) Neurones Muscle (L-6)	Hydrolyse acetylcholine	Poison (Nerve gas)
Cholinergic neurones	Intracellular	Neuroblastoma (NS20Y)	Synthesize acetylcholine	Huntington's chorea
GABA-ergic neurones	Intracellular		Synthesize GABA	Dementia
Adrenergic neurones	Intracellular	Neuroblastoma (N115)	Synthesize noradrenaline	Hypertension; affective disorders
Adrenergic neurones	Intracellular	Neuroblastoma (N1115)	Synthesize DOPA	
Neurones	Synaptic vesicle and Plasma membrane		Transmitter release?	
Neurones	Soluble and nerve ending membranes			
Neurones	Soluble from synaptic vesicle			
Hypothalamic neurones	I Para-ventricular II Supraoptic nuclei III Minor amount		Bind neurotrans-mitter	
Neurones	Soluble from synaptic vesicle			

B. ACETYLCHOLINESTERASE

This surface membrane protein is easily separated from the acetylcholine receptor protein. It has a different amino acid composition from the latter (Klett et al., 1973). Antibody to the esterase has been demonstrated with specificity using the innervated versus non-innervated sides of the electroplax organ (Benda et al., 1970). This protein is also restricted in muscle to endplates (Rogers et al., 1969). The esterase is a glycoprotein and hence estimates of molecular weight on gels containing sodium dodecyl sulphate must be regarded as tentative (Dudai & Silman, 1974) due to possible atypical binding of the detergent to sugar containing proteins (Grefarth & Reynolds, 1974). There may be two pairs of subunits (with different molecular weights) per molecule of enzyme.

C. CHOLINE ACETYL TRANSFERASE (CHOLINE ACETYLASE)

This enzyme is a marker for cholinergic neurones. Raising specific antiserum has proved difficult (as with actin or tubulin) because it is a highly conserved protein (see Section IV C and Appendix VII B) (Rossier et al., 1973). The enzyme is intracellular and antibody will react with motor neurones in the anterior horns of spinal cord (Eng et al., 1974). The protein has a molecular weight of 60,000–67,000 Daltons, and is one of the proteins transported from soma to nerve endings as is seen by blockade of axonal flow due to nerve ligation. The protein is decreased in patients with Huntington's chorea (Hiley & Bird, 1974).

D. GLUTAMIC ACID DECARBOXYLASE

This enzyme is a marker for the neurones which synthesize the inhibitory neurotransmitter gamma amino butyric acid (GABA). Antibody against the protein (Molecular weight 58,000 Daltons) has been shown to react mainly with the outermost layers (I–III) of the dorsal horn in the spinal cord of rat (McLaughlin et al., 1974). In the cerebellum the main reaction was with synaptic input to nucleus interpositus and the Purkinje cells, consistent with the enzyme containing cells being Purkinje cells, basket cells, Golgi II cells and satellite cells (Saito et al., 1974).

E. DOPAMINE BETA HYDROXYLASE OR CHROMOMEMBRANIN A

This enzyme is a marker for those neurones which synthesize the adrenergic neurotransmitters, noradrenalin or adrenalin (see Fig. 6.1). It is of particular interest in the peripheral nervous system since it is released along with the neurotransmitter and is hence monitored in hypertensive patients, as an index of adrenal sympathetic activity (Rush et al., 1974). A portion of the membrane surrounding the vesicles appears to be released with neurotransmitter (Smith,

1971); it is thus of interest that chromomembranin A (being just such a protein component of the vesicle) has been shown to be identical with dopamine beta hydroxylase (Hörtnagl *et al.*, 1972). It is also proposed that chromogranin A constitutes two of the four chairs (Aunis *et al.*, 1975).

Antibody made against this protein has been shown to give very similar results for localization of neurones in the central nervous system as does localization by the formaldehyde condensation method: namely, in the locus coeruleus in the pontine tegmentum (Swanson & Hartman, 1974). Similar results were obtained in the peripheral nervous system. As with choline acetylase (Section III C), the protein is probably transported towards the nerve ending as it also accumulates on the proximal side following nerve ligation (Nagatsu *et al.*, 1974).

F. TYROSINE HYDROXYLASE

This enzyme is a more general marker for neurones which synthesize catechols, since it occurs two enzymatic steps before dopamine beta hydroxylase in the synthetic pathway to noradrenaline (see Fig. 6.1). Antibody to this protein

Tyrosine $\xrightarrow{A}$ DOPA $\xrightarrow{B}$ DOPAmine $\xrightarrow{C}$ Noradrenaline $\xrightarrow{D}$ Adrenaline

    A Tyrosine Hydroxylase
    B DOPA Decarboxylase
    C DOPAmine Beta Hydroxylase
    D Phenyl N-Methyl Transferase

Fig. 6.1. Sequence of synthesis of adrenergic neurotransmitters.

(Hoeldtke *et al.*, 1974) has been shown to detect dopaminergic neurones of the CNS in addition to those visualized by the antibody to dopamine beta hydroxylase (see section IIIE). In the peripheral nervous system, neurones of sympathetic ganglia are positive to tyrosine hydroxylase antibody (Pickel *et al.*, 1974).

There are several other enzymes which are concerned with neurotransmitter metabolism. Some of the adrenergic enzymes, notably DOPA decarboxylase, are considered in Chapter 4.

G. BRAIN TROPOMYOSIN

Brain tropomyosin (different from muscle tropomyosin) has been isolated from cultured neurones (Fine *et al.*, 1973), synaptic plasma membranes and synaptic vesicles, while being absent from the soluble cytoplasmic fraction of the nerve (Blitz & Fine, 1974). It is also possible that the same protein may be found in preparations of axons (Grefarth & Reynolds, 1973).

H. GP350 GLYCOPROTEIN

This protein is said to be specific for intact synaptosomes. It is also found in the soluble cytoplasmic fraction. Although the monomeric molecular weight is apparently 11,600 Daltons, the protein is excluded from G-200 columns, i.e. molecular weight in excess of 200,000 Daltons. Each mole of glycoprotein is presumed to contain one mole of sialic acid, two of galactosamine, as well as galactose, glucosamine and possibly other sugars on two carbohydrate chains (van Nieuw Amerongen & Roukema, 1974). The protein has been localized in sections, on Purkinje cells of cerebellum, as well as pyramidal and stellate neurones in the cerebral cortex (van Nieuw Amerongen *et al.*, 1974).

I. CHROMOGRANIN A

This protein is released during sympathetic nerve stimulation, together with catecholamine and dopamine beta hydroxylase (Smith, 1971). It has been proposed as part of the dopamine beta hydroxylase protein (see section III E). As with nerve transmitter synthesizing enzymes, this protein is transported down the axon, as it accumulates, with nerve ligation (Geffen *et al.*, 1969). Although it has a higher molecular weight (Kirshner & Kirshner 1969) it has been thought to be analogous to vesiculin (section III K).

J. NEUROPHYSINS

These acidic proteins (I, II, III) bind to the basic polypeptide neurohormones oxytocin and vasopressin (Uttenthal & Hope, 1970). Neurophysin II has been localized by fluoresceinated antibody technique to the supraoptic nuclei of the hypothalamus and is released along with the hormone (vasopressin) by nervous activity (Livett *et al.*, 1971). Neurophysin I binds to oxytocin and Neurophysin III is present in much smaller amounts than I and II.

K. VESICULIN

This protein was isolated from the soluble fraction of synaptic vesicles. Mixed antibodies against synaptosomes (nerve ending particles) do not cross-react with vesiculin (Ulmar & Whittaker, 1974). It is not yet clear that its localization to synaptic vesicles is specific (Whittaker *et al.*, 1974).

IV Neuronal Antigens: axonal (Table 6.3)

A. ACTIN

Although traditionally thought of as a muscle protein, actin has been found in brain in amounts comparable with muscle itself (Fine & Bray, 1971). Since neurones only constitute a fraction of brain volume and since astrocytes and

Table 6.3. Comparative data on specific brain antigens: neuronal antigens: axonal

Antigen	Biochemical				Anatomical			Physiological	Pathological
	Monomeric molecular weight ($\times 10^{-3}$ Daltons)	Structure Primary	Structure Quaternary	Amount	Cell type	Membrane locus	Expression in tissue culture	Function	Associated diseases
Actin	43	Acidic	Myosin	20% Soluble protein	Neurones wide distribution	Intracellular microfilaments (50 Å)	Wide distribution	Growth Cone Cell movement	Dementia (Alzheimer)
Myosin	200 170	Acidic	Actin	2% Plasma membrane protein	Neurones wide distribution	Intracellular microfilaments (50 Å)	Wide distribution	Cell Movement	? above
Tubulin	56 55	Acidic	Dimers & polymers	20% Soluble protein	Neurones wide distribution	Intracellular microtubules (230 Å)	Neurones Neuroblastoma (N-18)	Cell support Axonal flow	Vinca Alkaloids
Filarin	53	Acidic	Polymers	30% Axonal protein	Neurones	Intracellular neurofilaments (100 Å)		? Argentophilia	? Senile dementia
Dynein	>300		Tubulin		? Neurones	Axons		? Axonal flow	
AP 22	42	Acidic			Neurones	Soluble			

oligodendrocytes also extend processes, it is almost certain to find actin in these and other cell types in brain. Brain actin has also been termed neurin (Berl *et al.*, 1973). It has been shown to be ubiquitous in other cultured cells (Bray, 1972) and is felt to be specifically related to cell movement, for example, in the rapidly growing axons of nerve cells (Wessels *et al.*, 1971). Actin is a cytoplasmic protein and specific intracellular loci have been seen in fibroblasts using immunofluorescence (Lazarides & Weber, 1974). Production of antibody has been achieved after complete denaturation in sodium dodecyl sulphate (see Appendix VII B).

The morphological location of this protein is probably on the microfilaments (Yamada *et al.*, 1971), since heavy meromyosin fragments will 'decorate' these filaments in a characteristic way (Chang & Goldman, 1973). Microfilaments are found in a loose network at the growing tips of axons (Bray, 1973). Actin has also been found in the cytoplasm of nerve endings, as well as associated with the membranes, namely, the synaptic plasma membrane and the synaptic vesicles (Blitz & Fine, 1974). See Fig. 6.2 for diagrammatic representation of microfilament (actin) location in nerve axon and ending.

An actin-relaxing protein complex has been isolated from rat brain (Puszkin & Kochwa, 1974) by affinity with muscle myosin on a polystyrene support. The brain actin-relaxing protein complex reacted with synaptic vesicles to release 14C-glutamate following preloading with the labelled dicarboxylic acid.

B. MYOSIN

Myosin-like molecules have been found in brain (Berl *et al.*, 1973) and neurones (Blitz & Fine, 1974). Brain myosin has also been termed stenin, as it changes viscosity in response to ATP and hydrolyses the substrate with stimulation by Ca^{++} and Mg^{++} (Puszkin *et al.*, 1972). The heavy chain doublet has a molecular weight of about 200,000 Daltons. The three light chains (about 20,000 Daltons) have characteristic distributions in different types of muscle: red, white and heart (Sarkar *et al.*, 1971, Weeds *et al.*, 1974). Muscle myosin has been successfully used as immunogen (Lowey & Steiner, 1972).

Like actin, myosin has been found in many cells where it is probably associated with production of movement, namely, the microfilament network. Using antibody to visualize the actin or myosin on these intracellular structures, the myosin appears to show 'striations', i.e. interruptions along the length of the filaments, whereas actin is continuous in its distribution (Weber & Groeschel-Stewart, 1974). Myosin is certainly associated with the plasma membrane and it can be detected with antibody on the external surface of cells, probably because it spans the thickness of the membrane (Willingham *et al.*, 1974). Myosin-like molecules were reported to be enriched in synaptic vesicles (Berl *et al.*, 1973). The specific molecule seems to be the 170,000 Daltons chain, rather than both heavy chains, as in the synaptic plasma membrane (Blitz & Fine, 1974).

C. TUBULIN

Two closely related tubulin proteins (Fine, 1971) constitute the monomeres of the polymeric neurotubules or microtubules seen by electron microscopy (Olmstead *et al.*, 1970). Tubulin also constitutes a major protein of the mitotic spindle or the cilia. Neurotubules probably serve a directing role analogous to a cellular skeleton. They have also been presumed to be the channels along which materials are transported within neurones, i.e., the so-called axonal flow (Schmitt, 1968). An intriguing possibility in this regard is that the tubulin-dynein (see section IV E) may possibly help to move small particles along the inside of the axon.

Tubulin has been purified using vinca alkaloids such as colchine or vinblastine. Although these drugs are used clinically to block mitosis (to disrupt the mitotic spindle), they have well-known side effects, notably neuropathy. Since other proteins may co-purify with the presumptively pure antigen, antibody specificity must be rigorously checked.

Aside from questions of purity of antigen, it has proved difficult to prepare antibody against tubulin. As with other highly conserved proteins (in the evolutionary sense that they are widely utilized across many species) it has been found advantageous to give the rabbit a long interval between boosts (see section III C and Appendix VII B).

An attempt to characterize the fibrous protein of astrofilaments has been made by removing the neuronal cell bodies of the optic nerve (enucleation of rabbits) and allowing astrocytes (and other cells) to infiltrate. Much later, nerves are removed and preparation of a fraction which shows principally 80 Å filaments is isolated. The major protein appears to be closely allied to tubulin dimers, i.e. only a few differences of peptide mapping (Johnson & Sinex, 1974).

A recent finding of interest is that solubilization of post-synaptic densities with detergents reveals that a major constituent protein appears to be very similar to tubulin monomers (Banker *et al.*, 1974). Tubulin-like molecules also constitute about 20% of the membrane proteins of synaptic vesicles and synaptosomal plasma membranes (Blitz & Fine, 1974), i.e. that portion of the cytoplasm which is directly under the synaptic membrane and has high affinity for OsO_4 as seen in electron microscopy.

D. FILARIN

This protein is thought to be the structural counterpart of the neurofilaments and was initially studied in squid axons (Huneeus & Davison, 1970). In this species the protein has atypical interactions with the detergent sodium dodecyl sulphate, but the molecular weight has been estimated to be about 74,000 Daltons.

In mammalian brain, presumptive neurofilaments have been isolated and the major protein of bovine white matter has been found to have a molecular weight of 51,000 Daltons (Shelanski *et al.*, 1971; Iqbal *et al.*, 1974). Davison's recent work has given an estimate of 54,000 Daltons for calf brain (Davison & Winslow, 1974). The protein is also probably associated with the membranes of synaptic vesicles and/or myelin with perhaps trapped axoplasm (Blitz & Fine, 1974).

Table 6.4

Name	Diameter (Å)	Cross linkage	MW ($\times 10^{-3}$ Daltons)	Protein	Presumed function
Microtubules	200–260	13 per tube	55, 56	Tubulin	Intracellular support Axonal flow
Astrofilaments	70–80*	Bundles	55, 56	Tubulin Analogue	Not known
Microfilaments	40–60	Single or Bundles	43	Actin	Cell movement (Growth cone)
Neurofilaments	90–100	Single	53	Filarin	? Axonal flow ? Stain with Ag^+

*Normal astrofilaments { Protoplasmic 60–80 Å ; Fibrous 80–90 Å

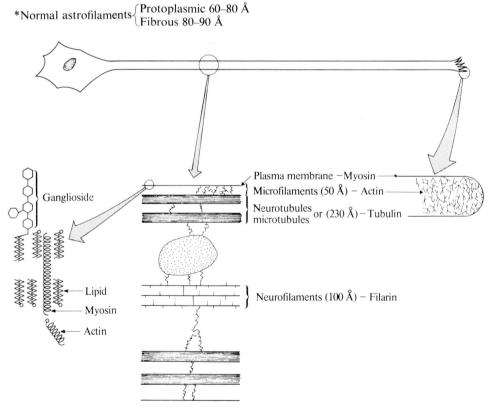

Fig. 6.2. Cellular localization of specific antigens.

The 'twisted tubule' fraction from dementia brain has a molecular weight of 50,000 Daltons (Iqbal et al., 1974). Table 6.4 summarizes the results found for several proteins which are associated with various cellular filaments. Their morphological locations are represented in Fig. 6.2.

E. DYNEIN

This protein, molecular weight over 300,000 Daltons, has been found to co-purify with tubulin (Gaskin et al., 1974) (see also section IV C). As the R_f of dynein is very close to the origin of the polyacrylamide gel, it is important to run the protein well into the gel and to ensure that the degree of polymerization is not different at the top as opposed to the rest of the gel. It may be that tubulindynein could be in part responsible for some of the axonal flow.

F. AP22 PROTEIN

This protein, molecular weight approximately 42,000 Daltons, is distinctive in that it stains pink with Commassie blue and is not denaturated by boiling water. It also shows inter-conversion between two different charged forms, but in an unpredictable fashion (Lim & Goedken, 1973). It has been localized in axons of stria terminalis and stria medularis. In cerebellum it was found in axons of basket cells and mossy fibres (Hartman & Lim, 1974).

V Glial antigens: unclassified (Table 6.5)

A. GLIAL FIBRILLARY ACIDIC PROTEIN

This protein was originally isolated from the dense plaques that occur in those areas of brain which have lost their oligoedendrocytes due to multiple sclerosis or other diseases. These sclerotic plaques are due to an infiltration of cells, including astrocytes, into the space left by the destroyed oligodendroglia (Eng et al., 1971). The protein appears in the protoplasmic astrocytes which grow in the cortex after cellular injury (Bignami & Dahl, 1974). It is also synthesized by glioma cells in culture (Bissell et al., 1974). It is found along the distribution of Bergmann's astrocytes in cerebellum (Bignami & Dahl, 1973). This is of particular interest since astrocytes may guide the growing nerve fibre in cerebellum (Rakic, 1971). This protein is also found in platelets and growing fibroblasts (menningioma) and its relationship to actin would be of interest (Dahl & Bignami, 1973). Both also have similar molecular weights and are precipitated by vinblastin. However, rigorous separation on polyacrylamide gels reveals a principal doublet with a third minor component (Dahl & Bignami, 1974). It is possible that there is interconversion of the various forms.

B. S-100 PROTEIN

This highly acidic soluble protein was found to be enriched in brain as compared with liver (Moore, 1965). Water soluble proteins were fractionated on the basis

Table 6.5. Comparative data on specific brain antigens: glial antigens: unclassified

Antigen	Biochemical				Anatomical			Physiological	Pathological
	Monomeric molecular weight ($\times 10^{-3}$ Daltons)	Structure Primary	Quaternary	Amount	Cell type	Membrane locus	Expression in tissue culture	Function	Associated diseases
Glial fibrillary acidic protein	44 43	Acidic	? Dimers	2% Normal white matter 16% M.S. plaque	Astrocytes	Intracellular	Glioma (C-6)	C.N.S. scar tissue	Multiple sclerosis
S-100	7	Quite acidic	$2 \times A$ $1 \times T$	0.2% Soluble protein	Glial	Intracellular	Glioma (C-6)	Ca^{++} binding	Wallerian degeneration
Alpha 2 glycoprotein	47	Acidic glycoprotein			Glial	Soluble and neuropil cellular fraction			
Beta trace protein	31	Acidic glycoprotein			Glial				Brain degeneration
10B glycoproteins (I & II)		Acidic glycoprotein			Glial				

of charge differences. After an initial ammonium sulphate cut, soluble proteins were fractionated on amino cellulose columns and further separated by electrophoresis (Moore & McGregor, 1965). The protein was designated S-100 as it proved 100% soluble in saturated ammonium sulphate solution. Although Moore was not the first to attempt such a separation of brain proteins, the fact that he produced an antiserum allowed quantitative analysis, as well as cellular and subcellular visualization.

Most of the protein is found in water-soluble extracts, but some remains membrane bound. Both forms were quantitated in investigations of different human tumours. (Haglid et al., 1973 and Stravou et al., 1971). The protein, which constitutes about 0·2% of soluble cell proteins, has been localized intracellularly in glial cells and on nuclei of some neurones (Hyden & McEwan, 1966). It is synthesized by glial cell lines in tissue culture (Pfeiffer et al., 1971). Its function is not clear, but it appears to bind significant amounts of calcium (Calissano et al., 1969).

It has been found to decrease 10-fold when the sciatic nerve is cut (Perez & Moore, 1968) and to increase 0·5-fold when the optic nerve is cut (Perez et al., 1970). Aggregation is a very significant problem in its handling (Dannies & Levine, 1971) and hence multiple forms have been detected (Filipowicz et al., 1968). Since this protein has a wide evolutionary distribution (i.e. is a highly conserved protein) antibody formation was enhanced by mixing the immunogen with methylated bovine serum albumin (see Appendix VII B).

C. ALPHA 2 GLYCOPROTEIN

This protein fraction is said to be specific for glial cell cytoplasm. It contains two lines by immunoprecipitation, although one was present in higher concentration. It will be of interest to follow the resolution of these proteins. They have been demonstrated to be more concentrated in the neuropil fraction (presumptive glia) compared with the neuronalperikaryal fraction of human brain. (Warecka et al., 1972). On sodium dodecyl sulphate gels, the protein fraction showed one major and two minor glycoproteins (PAS stain) and five bands when stained for protein. The apparent molecular weight of the major glycoprotein is 47,000 Daltons (Brunngraber et al., 1974).

D. BETA TRACE PROTEIN

This protein was initially found in cerebrospinal fluid. It has since been localized in glia of the normal nervous system, as well as in glial tumours, and thus parallels the distribution of the S-100 protein (section V B). However, immunodiffusion shows no relation to S-100, 14-3-2 or alpha$_2$ glycoproteins. It is found

Table 6.6. Comparative data on specific brain antigens: glial antigens: myelin

Antigen	Monomeric molecular weight ($\times 10^{-3}$ Daltons)	Biochemical			Anatomical			Physiological	Pathological
		Structure		Amount	Cell type	Membrane locus	Expression in tissue culture	Function	Associated diseases
		Primary	Quaternary						
Basic protein	19	Basic	Lipids	30% Myelin proteins	Oligodendroglia Schwann cell	Within membrane (non-surface of myelin) Surface	Schwannoma (RN-2)	Structural protein of myelin sheath	Experimental allergic encephalomyelitis, multiple sclerosis
CNS Myelin glycoprotein	110	Acidic glycoprotein		See Table 6.7	Oligodendroglia	Myelin			
PNS Myelin glycoprotein	28	Basic glycoprotein		See Table 6.7	Schwann cell	Myelin			Experimental allergic neuritis
Wolfgram protein	52 48	Acidic		See Table 6.7	Oligodendroglia Schwann cell	Myelin			
Proteolipid protein	25	Hydrophobic	Lipid	See Table 6.7	Oligodendroglia Schwann cell	Myelin	Schwannoma (RN-2)		
2', 3' Cyclic phosphodiesterase					Oligodendroglia Schwann cells	Myelin			
Cerebroside	0·8	Neutral	Mixed micelles	20% Myelin lipids	Oligodendroglia Schwann cells	Myelin		Structural lipid of myelin	Experimental allergic encephalomyelitis, multiple sclerosis

to be increased in cerebrospinal fluid after brain injury (stroke or multiple sclerosis during exacerbation) but not with gliomas (Olsson et al., 1974).

E. 10 B GLYCOPROTEINS

These soluble proteins, 10 BI, 10 BII, were removed from a DEAE column with $1N$ phosphoric acid. They have been proposed as being involved with learning in pigeons (Bogoch, 1970).

VI Glial antigens: myelin (Table 6.6)

A. MYELIN BASIC PROTEIN (ENCEPHALITOGENIC)

This protein has attracted much attention over the years, not only because of its possible relevance to human demyelinating diseases, but also because it is responsible for the best characterized model of an autoimmune disease. Although antibody can be directed against at least three regions of the protein (Driscoll et al., 1974) and the encephalitis induced by at least two regions (Eylar et al., 1972), the precise pathogenesis of experimental allergic encephalomyelitis is still not clear.

Basic protein can be demonstrated as a major constituent (30% of total protein) of myelin (Reynolds & Green, 1973). However, it is not normally exposed on the surface of the membrane. Attempts at chemical labelling with lactoperoxidase or pyridoxal phosphate (Golds & Braun, 1974) have shown that the basic protein is only labelled if the myelin is initially damaged. Likewise, labelling with antibody against the protein which has been coupled to ferritin to permit visualization under electron microscopy, has shown that the external surface does not bind antibody (Henderson et al., 1973). Antibody is bound only along the interperiod line. The disease can be transferred by white cells rather than antibody (Paterson, 1966) and the demyelination seen in tissue culture does not appear to be antibody directed against basic protein (Kies et al., 1973). Thus, antibody and T cells may recognize different determinants.

The rise in antibody is apparently linked to the re-establishment of tolerance in the animal (Lennon & Dunkley, 1974). Obviously, quantitation of T cell function, i.e. cell-mediated immunity during the course of the disease will be most instructive (Greaves et al., 1973). The role of basic protein in the electrostatic binding of acidic lipids has been partially characterized (Palmer & Dawson 1969, Banik & Davison, 1974). These considerations are important for understanding the sequence of events in the destruction of myelin both *in vitro* and *in vivo* (Wood & Dawson, 1974).

It is important to note differences between the basic proteins found in the peripheral nervous system versus the central nervous system. These are given in Table 6.7. It will be seen that there is a second basic protein (of lower molecular

Table 6.7. *Percentage of individual proteins in myelin from the CNS (Oligodendrocyte) and the PNS (Schwann cell)*

Protein	Molecular weight of monomers ($\times 10^{-3}$ Daltons)	CNS (Oligodendrocyte)	PNS (Schwann cell)
Glycoprotein (CNS)	110	2%	—
Wolfgram protein	52, 48	15–21%	2%
Glycoprotein (PNS)	30	—	50–60%
Proteolipid protein	25	24–28%	5–10%
Basic protein (BP_1)	19	26–34%	2–15%
PNS BP_2	12	*	2–15%

*Rat CNS $\begin{cases} BP_1\ 10\% \\ BP_2\ 14\% \end{cases}$ (Different molecular weight: 16,000 Daltons)

weight) found in the CNS myelin from rat. This smaller basic protein (BP_2) is capable of producing the same disease as BP_1. Two proteins are found in PNS myelin of many species. The larger basic protein (BP_1) is probably the same in central and peripheral myelin (Brostoff et al., 1972). However, the PNS myelin BP_2 is quite different from the CNS myelin BP_2 (and BP_1). It has a different amino acid composition and a smaller molecular weight. CNS myelin BP_2 is 16,000 Daltons; PNS myelin BP_2 is 12,000 Daltons.

B. CNS MYELIN GLYCOPROTEIN

In terms of total amounts of protein, this is a minor constituent of myelin prepared from the central nervous system (Quarles et al., 1973). One of the exposed sugars is mannose and it binds to concanavalin A. The protein has an apparent molecular weight of 110,000 Daltons and is probably found on the external surface (Matthieu et al., 1974) and possibly the interperiod line of myelin (Matus et al., 1973).

C. PNS MYELIN GLYCOPROTEIN

This glycoprotein is the major constituent protein of myelin, prepared from the peripheral nervous system (Everley et al., 1973; Brostoff et al., 1972). It has an apparent molecular weight of 28,000 Daltons. The major sugars appear to be mannose and galactose in equimolecular amounts. There may possibly be neuraminic acid (Wood & Dawson, 1974a & b). When animals were immunized with this protein, they developed experimental allergic neuritis, analogous to that using PNS basic protein, although demyelination was less severe with the glycoprotein.

D. WOLFGRAM PROTEIN

This protein is a doublet with molecular weight of 48,000 and 52,000 Daltons, and is found in myelin prepared from both the central and peripheral nervous system. However, there is proportionally more in CNS myelin than PNS myelin (Wood & Dawson, 1973). It is still possible that this protein is of axonal origin, due to its loss with hypotonic washing of myelin. However, it is also argued that it is part of the 'precursor' myelin-like membrane fraction (Waehneldt & Neuhoff, 1974).

E. PROTEOLIPID PROTEINS

This group of proteins (Reynolds & Green, 1973) was originally named because of solubility in chloroform-methanol. The major protein has been found principally in myelin from both CNS and PNS and has an apparent molecular weight of 25,000 Daltons. Being so hydrophobic, it is important to suspect abnormal migration in the presence of sodium dodecyl sulphate (see Appendix VII A, 1).

The relative amounts of these myelin proteins vary to some extent between species, and between brain and spinal cord in a given species. The main differences are between CNS and PNS myelin originating from oligodendrocyte (CNS) (Morell et al., 1973) and Schwann cell (PNS) (Greenfield et al., 1973). Some of these differences are compared in Table 6.7.

F. 2′, 3′, CYCLIC PHOSPHODIESTERASE

This enzyme is a marker for myelin (Prohaska et al., 1973). It has been shown to be enriched not only in normal CNS myelin but also in Schwann cell tumours (Trams, 1973). The protein has not yet been purified sufficiently for immunization. It would be of interest to know if antisera against other myelin components will block the enzymatic activity.

G. CEREBROSIDES

The galactose containing form of this glycolipid constitutes approximately 25% of the lipids of the myelin sheath. On membranes elsewhere, the predominant cerebroside is the glucose containing form. The galactose form thus constitutes a marker lipid for myelin.

Antibodies to galacto-cerebroside are found in experimental allergic encephalomyelitis. It was noted that antibody to basic protein does not produce demyelination in myelinated cerebellar explants (section VI A), whereas antibody to whole white matter does. Absorption of the latter with cerebroside removed the demyelinating antibody activity. Hence, in experimental allergic encephalomyelitis there are two major antigens derived from whole white matter, namely basic protein and galacto-cerebroside. It is the latter which is important, in the demyelination seen in tissue culture. It has also been found that antibody specifically raised against galacto-cerebroside demyelinates and inhibits myelination in culture (Fry *et al.*, 1974).

Cellular locations of selected antigens are represented in Fig. 6.3.

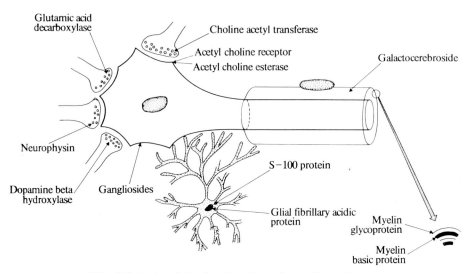

Fig. 6.3. Subcellular localization of specific antigens.

VII Appendix: techniques

A. PROTEIN CHARACTERIZATION

In this section it can be seen that well-established methods for the separation and identification of proteins have been supplemented by two procedures which have proved of special value in the characterization of specific proteins.

1. SEPARATION OF MONOMERS IN THE PRESENCE OF SODIUM DODECYL SULPHATE

This detergent has gained wide usage for two reasons: (a) it solubilizes membranes, as well as, if not better than, other agents (Fairbanks *et al.*, 1971); (b) because most proteins are uniformly coated with detergent and attracted by the electric field, they are effectively sieved through the polyacrylamide gel, hence the distance travelled in the gel is proportional to the logarithm of the molecular weight. The discovery of this so-called '1·4 law' was first made by Pitt-Rivers (Pitt-Rivers & Impiombato, 1968). They found that proteins, *when fully reduced*, bind 1·4 g of detergent per 1·0 g of protein. This discovery was amplified by Reynolds & Tanford (1970), who also showed the intermediate binding (the '0·4' law) and this explained the empirical finding of Maizel. Maizel's finding of R_f proportional to log of molecular weight in the presence of the detergent was published after, but without knowledge of, the Pitt-Rivers '1·4 law'. Reynolds & Tanford stressed the importance of full reduction of proteins, as well as the free monomer concentration of the detergent which is in equilibrium with the micellar form of the detergent. This equilibrium between monomeric and micellar forms is a function of ionic strength and pH (Reynolds & Tanford, 1970). Full consideration must therefore be given to the precise experimental conditions used if anomalies are to be avoided (Campagnoni & Magno, 1974; Fairbanks et al., 1971).

Deviation from the '1·4 law' is quite marked with proteins containing many sugars (Grefarth & Reynolds, 1974). Hence, molecular weights derived from proteins which are PAS positive should be regarded as tentative (see Section III B). The same tentative assessment of molecular weights for glycoproteins should be noted for values obtained from guanidine hydrochloride columns (Trayer et al., 1971). The use of different concentrations of polyacrylamide for glycoproteins does not answer the question: how much detergent is bound to a given glycoprotein? However, this procedure will show an anomaly of detergent binding, if such is present. For the optimal separation of closely related proteins, inclusion of urea can be quite useful (Fine, 1971). Another method depends on forming a continuous gradient of increasing concentrations of polyacrylamide, using the applied electrophoretic force to allow proteins to reach equilibrium; separation is thereby effected on the basis of molecular weight (Willard et al., 1974).

Proteins with a large complement of hydrophobic amino acids can give anomalous mobilities in the gel (Sierra & Tzagoloff, 1973), where they run ahead of the tracking dye, as if with the free detergent. Because of hydrodynamic considerations, i.e. the shape of the detergent coated molecule, molecular weights below 15,000 Daltons are not so reliable. The proteins behave as different-sized spheres rather than different-length cylinders. It should be noted that gangliosides can form mixed micelles with SDS and give an apparent molecular weight of either 10,000 or 30,000 Daltons, depending upon their manipulation (Dutton & Barondes, 1972).

Different methods are available for staining proteins. Although Commassie blue is most sensitive, it does not stain glycoproteins which have substantial amounts of sugar (Fairbanks et al., 1971). This problem can be overcome by performing the staining reaction at 60°C (Righetti et al., 1973). Fast green is used for semi-quantitative results (Gorovsky et al., 1970), but binds primarily to basic proteins. Naphthol blue black (Amido Schwartz) can give a metachromatic reaction with the basic encephalitogenic protein. All three stains rely upon electrostatic attraction for the formation of a coloured complex. Hence they are well known to vary from run to run and to depend upon the amount of destaining, i.e. amount of methanol, and to vary with the endogenous charge on the individual proteins. Many of these variables are overcome if quantitative results are required, by using a covalent linkage of fluorescent labels to protein (Thompson *et al.*, 1974). PAS staining of glycoproteins is notoriously insensitive, capricious and fades rapidly (Fairbanks *et al.*, 1971).

A final note of caution regarding breakdown of proteins during analytical manipulations. It is well-known that sodium dodecyl sulphate can enhance proteolysis (Pringle, 1970) and hence special measures should be taken to avoid this during electrophoresis (Weber *et al.*, 1972), during peptide mapping (Bray & Brownlee, 1973) and from the initial preparation of homogenates (Willard *et al.*, 1974). These measures mainly consist of blockade of serine proteates by an initial covalent linkage with PMFS (see Weber *et al.*, 1972) and inhibition of divalent cation dependent enzymes using chelators such as orthophenanthroline or EDTA. Enzymes with sulfhydryl groups at the catalytic site can be blocked by carboxymethylation. Prolonged extraction with EDTA and mercaptoethanol showed breakdown of membrane proteins (Trayer *et al.*, 1971).

2. OTHER POTENTIALLY SPECIFIC ANTIGENS

Specific antisera to proteins have generally been raised in one of two ways. Either a known protein with an identified function has been purified and separated from all other proteins, or extracts of brain have been compared with other tissues, and those proteins unique to brain have been used as immunogen.

In the search for a biochemical basis of nervous system specificity, several investigators have studied the monomeric protein profiles of different cellular and subcellular membrane constituents of the brain. Some of these preparations will be mentioned here as a perspective for the various proteins discussed above and because they may contain much useful information regarding other possible candidates for specific proteins of the nervous system. The use of sodium dodecyl sulphate electrophoresis allows not only excellent separation of individual proteins, but also yields them in a form most suited for immunogenesis (Lazarides & Weber, 1974). Subcellular fractions of brain have been prepared and interesting unique protein profiles were noted (Waehneldt *et al.*, 1973).

The preparation of neuronal versus glial fractions has been a thorny issue, but some differences of individual proteins have been noted in these fractions in spite of many similarities (Karlsson *et al.*, 1973). The proteins of synaptic membranes have attracted special interest (Banker *et al.*, 1972). This has been extended to glycoproteins and synaptic vesicles (Morgan *et al.*, 1973 and 1974) and by differential solubilization of the post-synaptic density (Banker *et al.*, 1974).

Axons can be obtained for polypeptide analysis without myelin or other contaminants which usually occur in subcellular fractionation, by using garfish olfactory nerve (Grefarth & Reynolds, 1973). Developmental studies of myelin and other membrane fractions have given helpful data on the ontogenetic aspect of the individual proteins and their combinations (Waehneldt & Neuhoff, 1974). Some most intriguing results have come from the study of proteins which are transported down the axon (Willard et al., 1974). It is seen that discrete proteins move at 4 rates (240, 50, 6 and 3 mm per day).

General glycoproteins have been studied using different extraction methods (Margolis & Margolis, 1973) or by using labelled sugars as precursors (Dutton & Barondes, 1970).

It is worth sounding a note of caution, namely, that cerebroside may also be identified as a soluble brain specific protein. This shows how general attention to the biochemistry of analysis of the antigen can be so important: because of the micellar behaviour of lipids, it is important to characterize all findings by reliable methods of chemical analysis (Tremblay et al., 1974).

3. PEPTIDE MAPPING

Sometimes knowledge of the total amino acid composition can be helpful, but often proteins can be closely related simply due to common evolutionary origin, e.g. tubulin, ATPase, actin, myosin and dynein (Weltman & Dowben, 1973). In these cases peptide mapping can be more useful to distinguish small differences in the face of many similarities. The traditional two-dimensional method can be most informative; but often the same data can be obtained, indeed on lesser amounts of protein, from one-dimensional runs at two different pH values (Bray & Brownlee, 1973). The method lends itself admirably in pilot run capacity and may well provide the definitive answer required (see Chapter 1).

B. IMMUNIZATION

Various methods have been employed to render some of these antigens immunogenic, as they often represent highly conserved proteins in the evolutionary sense.

One method has been to mix the antigen with methylated bovine serum albumin (Dannies & Levine, 1971). Although this was used as part of the adjuvant for the 14-3-2 protein, antigen alpha, which is immunologically indistinguishable, did not include this particular adjuvant.

The amount of dried tubercle bacillus which is used to 'complete' the Freunds adjuvant is crucial for production of antibody as well as the full syndrome or experimental allergic encephalitis (Kies et al., 1973). Absence of tubercle bacillus is, of course, part of the regimen to induce tolerance in animals (Teitelbaum et al., 1972).

It has recently been noted that denaturation in sodium dodecyl sulphate renders antigens quite immunogenic (Lazarides & Weber, 1974). Indeed, proteins can be removed from the polyacrylamide and used directly for immunization. It should be noted that acrylamide on its own can produce a peripheral neuropathy.

One interesting point regarding possible tolerance is to give the animals a fairly long interval between boosts and/or bleeds. A striking example is choline acetylase (Rossier *et al.*, 1973). Also, tubulin, which has been a notoriously poor immunogen was made an effective immunogen by using a 10-week pause in the scheme of immunization (Kowit & Fulton, 1974).

Glycolipids usually exist as micelles and some have been used with and without auxiliary protein (Czlonkowska & Leibowitz, 1974) or have been chemical coupled to synthetic co-polymers (Sela, 1969).

A final point on animals. As rabbits are not inbred, their response is not as predictable as that from lines of rats, mice, guinea pigs or other genetically defined strains.

As there has been much progress in the characterization of the acetylcholine receptor protein (nicotinic), so one hopes for similar advances relating to receptors for other neurotransmitters. Encouraging starts have been made for the following:

The muscarinic acetylcholine receptor has been found enriched in cholinergic areas of the brain (Yamamura & Snyder, 1974; Hiley & Burgen, 1974), the glycine receptor has been shown to bind strychnine at inhibitory synapses (Young & Snyder, 1974), the morphine receptor may parallel the distribution of brain cholinergic areas (Pert & Snyder, 1973; Lowney *et al.*, 1974), the dopamine receptor has been followed due to the increase in cyclic GMP which the neurohormone elicits (Miller & Iversen, 1974; Clement-Cormier *et al.*, 1974), as the histamine receptor has, which acts via increased cyclic AMP levels in brain (Baudry *et al.*, 1975).

VIII Conclusion

Unlike other tissues, that of the nervous system is comprised of an exceedingly complex assembly of different cell types with varying functions. It is thus not surprising to find that certain specific proteins and lipids have characteristic cellular localizations.

Since in many diseases of the nervous system there is selective damage (pathoclisis) to groups of neurones or attack on a single cell type, recognition of such specific macromolecules should in the future be of immense value. Just as the increased transaminase activity has been of use in the diagnosis of myocardial infarction, so it is conceivable that one or more of the brain specific antigens could aid in the differential diagnosis of stroke. Immunological methods provide one of the most sensitive techniques for the identification and quantitation of traces of specific antigens. There are, as indicated in this chapter, an astonishingly large number and variety of immunogenically active macromolecules. Specific antisera have been successfully used in neurobiological research, but so far relatively little use has been made of antisera in anaylsis of body fluids in neurological disease or in research on the pathogenesis of these conditions. For example, in the case of different tumours of the central nervous system specific products may be detectable either in the circulation generally, in urine, or, more probably, in the cerebrospinal fluid. The various tumour lines have been found to synthesize many early recognizable antigens. Specific antibody coupled to a gamma emitter could be used in diagnosis or an aid in post-operative 'follow-up', especially where recurrence and/or metastases of the tumour were suspected. Indeed a start has been made in the case of antibody to the S-100 protein (Haglid et al., 1973; Stavrou et al., 1971). Other tumours should now be assayed as well. The question of therapy poses an intriguing consideration. It is possible that antibody directed against specific tumour products could be quite useful in enhancing the effects of cytotoxic drugs upon the tumour in question. Some indication of the potential value of this has already been published (Rubens, 1974; Rubens & Dulbecco, 1974).

Although cytotoxic drugs can be coupled by covalent linkage to specific antibody, the effect can be achieved in some cases without such linkage. It would appear that the antibody thus sensitizes the tumour cells in some fashion such that they show an increased affinity for the drug. This is obviously a fascinating new concept in chemotherapy of tumours but, again, what is urgently required is further work on specific antibody.

Another intriguing therapy based on immunological methods has been the attempts to desensitize patients with multiple sclerosis, using myelin basic protein, (Campbell et al., 1973). Because of the well-known clinical variability in remission and relapse, results must be viewed critically. Likewise, immuno-suppression has been tried by various investigators (Waksman, 1974), as has immuno-enhancement, giving transfer factor (Jersild et al., 1973). A most interesting recent finding has been the association of this disorder with specific haplotypes and hence by implication immune response genes (Lehrich et al., 1974). It is also of note that the state of activity in multiple sclerosis appears related to ability of patients' white cells to produce inhibition of macrophage migration (Sheremata et al., 1974).

Further understanding of the pathogenesis of neurological diseases such as multiple sclerosis, in which there is an important immunological component, will be aided by progress in basic research on the subject. Apart from the many immunochemical questions already discussed in this chapter, there is the problem of the dynamics of membrane antigens. Movement of a specific antigen was noted first in lymphocytes (Taylor et al., 1971), but has since been shown to be a general phenomenon in tissues throughout the body. The acetylcholine receptor has been shown to normally have restricted mobility (Anderson & Cohen, 1974). This is, of course, altered in the case of denervation. The question of mobility of concanavalin A binding sites has been investigated for subcellular fractions with membranes of synaptosomes and myelin (Matus et al., 1973).

It is of interest to follow the parallel with the glycoprotein of CNS myelin. Altered membrane dynamics is of importance in relation to transformed versus non-transformed cells (via agglutination), as is the finding of differences in the glycolipid pattern (Fishman *et al.*, 1974).

We can safely conclude that the understanding of how the mechanism of antibody reacts with specific membrane molecules is of considerable importance to questions of pathogenesis of many diseases. Moreover, the appropriate use of such markers will, in the near future, undoubtedly bring useful new information about the normal nervous system as well as helping in the diagnosis and treatment of neurological diseases as a whole.

ACKNOWLEDGEMENTS

It is a pleasure to acknowledge many helpful discussions with Dr. M.C. Raff and my debt to Mrs. M. Wilson for her careful attention to the manuscript.

References

ALMON R.R., ANDREW G.G. & APPEL S.H. (1974) Serum globulin in myasthenia gravis: Inhibition of alpha-bungarotoxin binding to acetylcholine receptors. *Sci.* **186**, 55–57.

ANDERSON M.J. & COHEN M.W. (1974) Fluorescent staining of acetylcholine receptors in vertebrate skeletal muscle. *J. Physiol.* **237**, 385–400.

ASKARI A. (1974) The effects of antibodies to (Na^+, K^+) on the reactions catalysed by the enzyme, *Ann. N.Y. Acad. Sci.* **242**, 372–388.

AUNIS D., ALLARD, D., MIRAS-PORTUGAL, M-T. & MANDEL, P. (1975) Bovine adrenal medullary chromogranin A: Studies on the structure and further evidence for identity with Dopamine-β-Hydroxylase subunit. *Biochem. Biophys. Acta*, **393**, 284–295.

BANERJEE S.P., SNYDER S.H., CUATRECASAS P. & GREENE L.A. (1973) Identification of specific nerve growth factor (NGF) binding in superior cervical ganglia. *Fed. Proc.* **32**, 1721.

BANIK N.L. & DAVISON A.N. (1974) Lipid and Basic protein interaction in myelin. *Biochem. J.* **143**, 39–45.

BANKER G.B., CHURCHILL L. & COTMAN C.W. (1974) Proteins of the post synaptic density *J. Cell. Biol.* **63**, 456–465.

BANKER G., CRAIN B. & COTMAN C.W. (1972) Molecular weights of the polypeptide chains of synaptic plasma membranes *Brain. Res.* **42**, 508–513.

BAUDRY M., MARTRES M.P. & SCHWARTZ J.C. (1975) H_1 and H_2 receptors on the histamine-induced accumulation of cyclic AMP in guinea pig brain slices. *Nature London* **253**, 362–364.

BENDA P., TSUJI S., DAUSSANT J. & CHANGEUX J.P. (1970) Localization of acetylcholine esterase by immunofluorescence in eel electroplax. *Nature, London,* **225**, 1149–1150.

BENNETT G.S. (1974) Immunological and electrophoretic identity between nervous system specific proteins antigen alpha and 14-3-2 *Brain Res.* **68**, 365–369.

BENNETT G.S. & EDELMAN G.M. (1968) Isolation of an acidic protein from rat brain *J. Biol. Chem.* **243**, 6234–6241.

BERL S., PUSZKIN, S. & NICKLAS W.J. (1973) Actomyosin-like proteins in brain. *Sci.* **179**, 441–446.

BIGNAMI A. & DAHL D. (1973) Differentiation of astrocytes in the cerebellar cortex and the pyramidal tracts of the newborn rat. An immunofluorescent study with antibodies to a protein specific to astrocytes. *Brain Res.* **49**, 393–402.

BIGNAMI A. & DAHL D. (1974) Astrocyte-specific protein and radial glia in the cerebral cortex of newborn rat. *Nature, London,* **252**, 55–56.

BISSELL M.G., RUBENSTEIN L.J., BIGNAMI A. & HERMAN M.M. (1974) Characteristics of the rat C-6 glioma maintained in organ culture systems. Production of glial fibrillary acidic protein in absence of bliofibrillogenesis. *Brain. Res.* **82**, 77–89.

BJERRE B., BJÖRKLUND A. & STENEVI U. (1974) Inhibition of the regenerative growth of central noradrenergic neurones by intracerebrally administered anti-NGF serum. *Brain Res.* **74**, 1–18.

BLITZ A.L. & FINE R.E. (1974) Muscle-like contractile proteins and tubulin in synaptosomes, *Proc. Nat. Acad. Sci.* **71**, 4472–4476.

BOGOCH S. (1970) Glycoproteins of the brain of the training pigeon. In Lajtha A. (Ed.) *Protein Metabolism of the Nervous System*, pp. 555–569.

BRAY D. (1972) Cytoplasmic actin: A comparative study. *Cold Spring Harbor Symp. Quant. Biol.* **37**, 567–571.

BRAY D. (1973) Model for membrane movements in the neural growth cone. *Nature*, **244**, 93–96.

BRAY D. & BROWNLEE S.M. (1973) Peptide mapping of proteins from acrylamide gels *Anal. Biochem.* **55**, 213–221.

BRECKENRIDGE W.C., GOMBOS G. & MORGAN I.G. (1972) The lipid composition of adult rat brain synaptosomal plasma membranes *Biochim. Biophys. Acta* **266**, 695–707.

BROSTOFF S., BURNETT P., LAMPERT P. & EYLAR E.H. (1972) Isolation and characterization of a protein from sciatic nerve myelin responsible for experimental allergic neuritis. *Nature New Biol.* **235**, 210–212.

BRUNNGRABER E.G., SUSZ K.P. & WARECKA K. (1974) Electrophoretic analysis of human brain—specific proteins obtained by affinity chromatography. *J. Neurochem.* **22**, 181–182.

CALISSANO P. & COZZARI C. (1974) Interaction of nerve growth factor with the mouse brain neurotubule protein(s) *Proc. Nat. Acad. Sci.* **71**, 2131–2135.

CALISSANO P., MOORE B.W. & FRIESEN A. (1969) Effect of calcium ion on S-100, a protein of the nervous system *Biochem.* **8**, 4318–4326.

CAMPAGNONI A.T. & MAGNO C.S. (1974) Molecular weight estimation of mouse and guinea-pig myelin basic proteins by polyacrylamide gel electrophoresis in the presence of sodium dodecyl sulphate: Influence of ionic strength. *J. Neurochem.* **23**, 887–890.

CAMPBELL B., VOGEL P.J., FISHER E. & LORENZ R. (1973) Myelin basic protein administration in multiple sclerosis. *Arch. Neurol.* **29**, 10–15.

CHANG C.M. & GOLDMAN R.D. (1973) The localization of actin-like fibres in cultured neuroblastoma cells as revealed by heavy meromyosin binding. *J. Cell. Biol.* **57**, 867–874.

CLEMENT-CORMIER Y.C., KEBABIAN J.W., PETZOLD G.L. & GREENGARD P. (1974) Dopamine-sensitive adenylate cyclase in mammalian brain: A possible site of action of antipsychotic drugs. *Proc. Nat. Acad. Sci.* **71**, 1113–1117.

CUATRECASAS P. (1973) Gangliosides and membrane receptors for cholera toxin. *Biochem.* **12**, 3558–3566.

CZLONKOWSKA A. & LIEBOWITZ S. (1974) The effect of homologous and heterologous carriers on the immunogenicity of the galactocerebroside hapten. *Immunol.* **27**, 1117–1126.

DAHL D. & BIGNAMI A. (1973) Glial fibrillary acidic protein from normal human brain. Purification and properties. *Brain Res.* **57**, 343–360.

DAHL D. & BIGNAMI A. (1974) Heterogeneity of the glial fibrillary acidic protein in gliosed human brains. *J. Neurol. Sci.* **23**, 551–563.

DANNIES P.S. & LEVINE L. (1971) Structural properties of bovine brain S-100 protein *J. Biol. Chem.* **246**, 6276–6283.

DAVISON P.F. & WINSLOW B. (1974) The protein subunit of calf brain neurofilament. *J. Neurobiol.* **5**, 119–133.

DRISCOLL B.F., KRAMER A.J. & KIES M.W. (1974) Myelin basic protein: location of multiple antigenic regions. *Sci.* **184**, 73–75.

DUDAI Y. & SILMAN I. (1974) The molecular weight and subunit structure of acetylcholinesterase preparations from the electric organ of the electric eel, *Biochem. & Biophys. Res. Comm.* **59**, 117–124.

DUTTON G.R. & BARONDES S.H. (1970) Glycoprotein metabolism in developing mouse brain. *J. Neurochem.* **17**, 913–920.

DUTTON G.R. & BARONDES S.H. (1972) Macromolecular behaviour of gangliosides on electrophoresis in sodium dodecyl sulphate. *J. Neurochem.* **19**, 559–562.

ENG L.F., VANDERHAEGHEN J.J., BIGNAMI A. & GERSTL B. (1971) An acidic protein isolated from fibrous astrocytes. *Brain Res.* **28**, 351–354.

ENG L.F., UYEDA C.T., CHAO L.P. & WOLFGRAM F. (1974) Antibody to bovine choline acetyltransferase and immunofluorescent localization of the enzyme in neurones. *Nature, London* **250**, 243–245.

ESSELMAN W.J. & MILLER H.C. (1974) Brain and thymus lipid inhibition of antibrain-associated theta cytoxicity. *J. Exper. Med.* **139**, 445–450.

EVERLY J.L., BRADY R.O. & QUARLES R.H. (1973) Evidence that the major protein of rat sciatic nerve myelin is a glycoprotein. *J. Neurochem.* **21**, 329–334.

EYLAR E.H., BROSTOFF S., JACKSON J. & CARTER H. (1972) Allergic encephalomyelitis in monkeys induced by a peptide from the A1 protein. *Proc. Nat. Acad. Sci.*, **69**, 617–619.

FAIRBANKS G., STECK T.L. & WALLACH D.F.H. (1971) Electrophoretic analysis of major polypeptides of the human erythrocyte membrane. *Biochem.* **10**, 2606–2617.

FAMBROUGH D.M., DRACHMAN D.B. & SATYAMURTI S. (1973) Neuromuscular junction in myasthenia gravis: decreased acetylcholine receptors. *Sci.* **182**, 293–295.

FELTKAMP T.E.W., VAN DEN BERG-LOONEN P.M., NIJENHUIS L.E., ENGELFRIET C.P., VAN ROSSUM A.L., VAN LOGHEM J.J. & OOSTERHUIS H.J.G.H. (1974) Myasthenia Gravis, Autoantibodies and HL-A Antigens. *Brit. Med. J.* **1**, 131–133.

FILIPOWICZ W., VINCENDON G., MANDEL P. & GOMBOS G. (1968) Topographical distribution of fast and slow migrating fractions of beef brain S-100 protein fraction. *Life Sci.* **7**, 1243–1250.

FINE R.E. (1971) Heterogeneity of tubulin. *Nature New Biol.* **233**, 283–284.

FINE R.E. & BRAY D. (1971) Actin in growing nerve cells. *Nature New Biol.* **234**, 115–8.

FINE R.E., BLITZ A.L., HITCHCOCK S.E. & KAMINER B. (1973) Tropomyosin in brain and growing neurones. *Nature New Biol.* **245**, 182–186.

FISHMAN P.H., BRADY R.O., BRADLEY R.M., AARONSON S.A. & TODARO G.J. (1974) Absence of a specific ganglioside galactosyltransferase in mouse cells transformed by murine sarcoma virus. *Proc. Nat. Acad. Sci.* **71**, 298–301.

FRAZIER W.A., ANGELETTI R.H. & BRADSHAW R.A. (1972) Nerve growth factor and insulin. *Sci.* **176**, 482–488.

FRAZIER W.A., BOYD L.F. & BRADSHAW R.A. (1973) Interaction of nerve growth factor with surface membranes: biological competence of insolubilized nerve growth factor, *Proc. Nat. Acid. Sci.* **70**, 2931–2935.

FRY J.M., WEISSBARTH S., LEHRER G.M. & BORNSTEIN M.B. (1974) Cerebroside antibody inhibits sulfatide synthesis and myelination and demyelinates in cord tissue cultures. *Sci.* **183**, 540–542.

GASKIN F., KRAMER S.B., CANTOR C.R., ADELSTEIN R. & SHELANSKI M.L. (1974) A dynein-like protein associated with neurotubules. *FEBS Lett.* **40**, 281–286.

GEFFEN L.B., LIVETT B.G. & RUSH R.A. (1969) Immunohistochemical localization of protein components of catecholamine storage vesicles. *J. Physiol.* **204**, 593–605.

GOLDS E.E. & BRAUN P.E. (1974) Asymmetric location of the basic protein in the intact myelin membrane *Fed. Proc.* **33**, 1038.

GOROVSKY M.H., CARLSON K. & ROSENBAUM J.L. (1970) Simple method for quantitative densitometry of polyacrylamide gels using fast green. *Analyt. Biochem.* **35**, 359–370.

GREAVES M.F., OWEN J.J.T. & RAFF M.C. (1973) T & B Lymphocytes: origins, properties and roles in immune responses *Excerpta Medica*, Amsterdam, pp. 113–48.

GREENE L.A., SYTKOWSKI A.J., VOGEL Z. & NIRENBERG M.W. (1973) Alpha-Bungarotoxin used as a probe for acetylcholine receptors of cultured neurones. *Nature, London*, **243**, 163–166.

GREENFIELD S., BROSTOFF S., EYLAR E.H. & MORELL P. (1973) Protein composition of myelin of the peripheral nervous system. *J. Neurochem.* **20**, 1207–1216.

GREFARTH S.P. & REYNOLDS J.A. (1973) Polypeptide composition of an excitable plasma membrane. *J. Biol. Chem.* **248**, 6091–6094.

GREFART S.P. & REYNOLDS J.A. (1974) The molecular weight of the major glycoprotein of human erythrocyte membranes. *Proc. Nat. Acad. Sci.* **71**, 3913–3917.

HAGLID F.G., STAVROU D., RÖNNBÄCK L., CARLSSON C.A. & WEIDENBACH W. (1973) The S-100 protein in water-soluble and pentanol-extractable form in normal human brain and tumours of the human nervous system: A quantitative study *J. Neurol. Sci.* **20**, 103–111.

HAMBERGER A. & SVENNERHOLM L. (1971) Composition of gangliosides and phospholipids of neuronal and glial cell enriched fractions. *J. Neurochem.* **18**, 1821–9.

HARTMAN B.K. & LIM R. (1974) Differentiation of axon systems using a specific protein marker, *Abstracts: Soc. for Neurosci.* **4**, 283,

HENDERSON R.M., RAUCH H.C. & EINSTEIN E.R. (1973) Immuno-electron microscopic localization of the encephalitogenic basic protein in myelin. *Immunol. Comm.* **2**, 163–172.

HENDRY I.A., STACH R. & HERRUP K. (1974) Characteristics of the retrograde axonal transport system for nerve growth factor in the sympathetic nervous system. *Brain Res.* **82**, 117–128.

HERSCHMAN H.R. & LERNER M.D. (1973) Production of a nervous system specific protein (14-3-2) by human neuroblastoma cells in culture. *Nature New Biol.* **241**, 242–244.

VAN HEYNINGEN W.E. (1974) Gangliosides as membrane receptors for tetanus toxin, cholera toxin and serotonin. *Nature, London*, **249**, 415–417.

HILEY C.R. & BIRD E.P. (1974) Decreased muscarinic receptor concentration in postmortem brain in Huntington's chorea. *Brain Res.* **80**, 355–358.

HILEY C.R. & BURGEN A.S.V. (1974) The distribution of muscarinic receptor sites in the nervous system of the dog. *J. Neurochem.* **22**, 159–162.

HOELDTKE R., LLOYD T. & KAUFMAN S. (1974) An immunochemical study of the induction of tyrosine hydroxylase in rat adrenal glands. *Biochem. Biophys. Res. Comm.* **57**, 1045–1053.

HÖRTNAGL H., WINKLER H. & LOCHS H. (1972) Membrane proteins of chromaffin granules: dopamine beta hydroxylase, a major constituent. *Biochem. J.* **129**, 187–195.

HUNEEUS F.C. & DAVISON P.F. (1970) Fibrillar proteins from squid axons. I. Neurofilament protein. *J. Molec. Biol.* **52**, 415–428.

HYDEN H. & MCEWAN B. (1966) A glial protein specific for the nervous system. *Proc. Nat. Acad. Sci.* **55**, 354–358.

IQBAL K., WISNIEWSKI H.M., SHELANSKI M.L., BROSTOFF S., LIWNICZ B.H. & TERRY R.D. (1974) Protein changes in senile dementia. *Brain Res.* **77**, 337–343.

JEAN R.H., ALBERS R.W. & KOVAL G.J. (1974) Immunochemical properties of the subunits of lubrol solubilized (Na^+, K^+)-ATPase from electric eel. *Abstracts: Soc. for Neurosci.* **4**, 323.

JERSILD C., HANSEN G.S., SVEJGAARD A., FOG T., THOMSEN M. & DUPONT B. (1973) Histocompatibility determinants in multiple sclerosis, with special reference to clinical course. *Lancet* **ii**, 1221–1225.

JOHNSON L.S. & SINEX F.M. (1974) On the relationship of brain filaments to microtubules. *J. Neurochem.* **22**, 321–326.

KARLSSON J.O., HAMBERGER A. & HENN F.A. (1973) Polypeptide composition of membranes derived from neuronal and glial cells. *Biochim. Biophys. Acta*, **298**, 219–229.

KIES M.W., DRISCOLL B.F., SEIL F.J. & ALVORD, JR., E.C. (1973) Myelin inhibition factor: dissociation from induction of experimental allergic encephalomyelitis. *Sci.* **179**, 689–690.

KIRSHNER A.G. & KIRSHNER N. (1969) A specific soluble protein from the catecholamine storage vesicles of bovine adrenal medulla. II. Physical characterization. *Biochim. Biphys. Acta.* **181**, 219–225.

KLETT R.P., FULPIUS B.W., COOPER D., SMITH M., REICH E. & POSSANI L.D. (1973) The acetylcholine receptor: I. Purification and characterization of a macromolecule isolated from *Electrophorous electricus*. *J. Biol. Chem.* **248**, 6841–6853.

KOLBER A.R., GOLDSTEIN M.N. & MOORE B.W. (1974) Effect of nerve growth factor on expression of colchicine-binding activity and 14-3-2 protein in an established line of human neuroblastoma *Proc. Nat. Acad. Sci.* **71**, 4203–4207.

KOWIT J.D. & FULTON C. (1974) Purification and properties of flagellar outer doublet tubulin from *Naegleria gruberi* and a radioimmune assay for tubulin. *J. Biol. Chem.* **249**, 3638–3646.

KYTE J. (1974) The reaction of sodium and potassium ion-activated adenosine triphosphatase with specific antibodies. Implications for mechanism of active transport. *J. Biol. Chem.* **249**, 3652–3660.

LAZARIDES E. & WEBER K. (1974) Actin antibody: The specific visualization of actin filaments in non-muscle cells. *Proc. Nat. Acad. Sci.* **71**, 2268–2272.

LEHRICH J.R., ARNASON G.W., FULLER T.C. & WRAY S.H. (1974) Parainfluenza, histocompatibility and multiple sclerosis. *Acta. Neurol. Scand.* **50**, 183–193.

LENNON V.A. & DUNKLEY P.R. (1974) Humoral and cell-mediated immune responses of Lewis rats to syngeneic basic protein of myelin. *Int. Arch. Allergy*, **47**, 598–608.

LEVI-MONTALCINI R. & ANGELETTI P.U. (1968) Nerve growth factor. *Physiol. Rev.* **48**, 534–569.

LIM R. & GOEDKEN M.P. (1973) A partially purified membrane protein specific to the brain. *Biochim. Biophys. Acta.* **322**, 359–371.

LIVETT B.G., UTTENTHAL L.O. & HOPE D.B. (1971) Release of neurophysin together with vasopressin by a Ca^{++} dependent mechanism *Phil. Trans. Roy. Soc. B.* **261**, 379–380.

LONGO A.M. & PENHOET E.E. (1974) Nerve growth factor in rat glioma cells. *Proc. Nat. Acad. Sci.* **71**, 2347–2349.

LOWEY S. & STEINER L.A. (1972) An immuno-chemical approach to the structure of myosin and the thick filament. *J. Mol. Biol.* **65**, 111–126.

LOWNEY L.I., SCHULZ K., LOWERY P.J. & GOLDSTEIN A. (1974) Partial purification of an opiate receptor from mouse brain. *Sci.* **183**, 749–753.

MARGOLIS R.K. & MARGOLIS R.U. (1973) Extractability of glycoproteins and mucopolysaccharides of brain. *J. Neurochem.* **20**, 1285–1288.

MATTHIEU J.M., DANIEL A., QUARLES R.H. & BRADY R.O. (1974) Interactions of Concanavalin A and other lectins with CNS myelin. *Brain Res.* **81**, 348–353.

MATUS A., DE PETRIS S. & RAFF M.C. (1973) Mobility of Concanavalin A receptors in myelin and synaptic membranes. *Nature New Biol.* **244**, 278–280.

MCLAUGHLIN B.J., BARBER R., SAITO K. & ROBERTS E. (1974) Fine structural localization of glutamate decarboxylase in synaptic terminals of rat spinal cord. *Abstracts: Soc. for Neurosci.* **4**, 452.

MILEDI R. & POTTER L.T. (1971) Acetylcholine receptors in muscle fibres. *Nature, London* **233**, 599–603.

MILLER R.J. & IVERSEN L.L. (1974) Effect of psychoactive drugs on dopamine (3,4-Dihydroxy-phenethylamine) sensitive adenylate cyclase activity in corpus striatus of rat brain. *Trans. Biochem. Soc.* **2**, 256–259.

MOORE B.W. (1965) A soluble protein characteristic of the nervous system. *Biochem. Biophys. Res. Comm.* **19**, 739–744.

MOORE B.W. & MCGREGOR D. (1965) Chromatographic and electrophoretic fractionation of soluble proteins of brain and liver. *J. Biol. Chem.* **240**, 1647–1653.

MORGAN I.G., ZANETTA J.P., BRECKENRIDGE W.C., VINCENDON G. & GOMBOS G. (1973) The chemical structure of synaptic membranes. *Brain Res.* **62**, 405–411.

MORGAN I.G., ZANETTA J.P., BRECKENRIDGE W.C., VINCENDON G. & GOMBOS G. (1974) Adult rat brain synaptic vesicles: protein and glycoprotein composition. *Trans. Biochem. Soc.* **2**, 249–252.

MORELL P., LIPKIND R. & GREENFIELD S. (1973) Protein composition of myelin from brain and spinal cord of several species. *Brain Res.* **58**, 510–514.

NAGATSU I., HARTMAN B.K. & UDENFRIEND S. (1974) The anatomical characteristics of dopamine-beta-hydroxylase accumulation in ligated sciatic nerve. *J. Histochem. Cytochem.* **22**, 1010–1018.

VAN NIEUW AMERONGEN A. & ROUKEMA P.A. (1974) GP-350, A sialoglycoprotein from calf brain: Its subcellular localization and occurrence in various brain areas. *J. Neurochem.* **23**, 85–89.

VAN NIEUW AMERONGEN A., ROUKEMA P.A. & VAN ROSSUM A.L. (1974) Immunofluorescence study of the cellular localization of GP-350, a sialoglycoprotein from brain. *Brain Res.* **81**, 1–19.

OGER J., ARNASON B.G.W., PANTAZIS N., LEHRICH J. & YOUNG M. (1974) Synthesis of nerve growth factor by L and 3T3 cells in culture. *Proc. Nat. Acad. Sci.* **71**, 1554–1558.

OLMSTEAD J.B., CARLSON K., KLEBE R., RUDDLE F. & ROSENBAUM J. (1970) Isolation of microtubule protein from cultured mouse neuroblastoma cells. *Proc. Nat. Acad. Sci.* **65**, 129–136.

OLSSON J.E., BLOMSTRAND C. & HAGLID K.G. (1974) Cellular distribution of beta-trace protein in CNS and brain tumours. *J. Neurol. Neurosurg. & Psychiat,* **37**, 302–311.

PALMER & DAWSON (1969) Complex-formation between triphosphoinositide and experimental allergic encephalitogenic protein. *Biochem. J.* **111**, 637–646.

PATERSON P.Y. (1966) Experimental allergic encephalomyelitis and autoimmune disease. *Adv. Immunol.* **5**, 131–208.

PEREZ V.J. & MOORE B.W. (1968) Wallerian degeneration in rabbit tibial nerve: changes in amount of the S-100 protein. *J. Neurochem.* **15**, 971–977.

PEREZ V.J. & MOORE B.W. (1970) Biochemistry of the nervous system in ageing. *Interdiscipl. Topics Geront.* **7**, 22–45.

PEREZ V.J., OLNEY J.W., CICERO T.J., MOORE B.W. & BAIN B.A. (1970) Wallerian degeneration in rabbit optic nerve: cellular localization in the central nervous system of the S-100 and 14-3-2 proteins. *J. Neurochem.* **17**, 511–519.

PERT C.B. & SNYDER S.H. (1973) Opiate receptor: demonstration in nervous tissue. *Sci.* **179**, 1011–1014.

PFEIFFER S.E., HERSCHMAN H.R., LIGHTBODY J.E., SATO G. & LEVINE L. (1971) Modification of cell surface antigenicity as a function of culture conditions. *J. Cell. Physiol.* **78**, 145–152.

PICKEL V.M., JOH T.H. & REIS D.J. (1974) Immunohistochemical localization of tyrosine hydroxylase by light and electron microscopy. *Abstracts: Soc. for Neurosci.* **4**, 532.

PITT-RIVERS R. & IMPIOMBATO F.S.A. (1968) The binding of sodium dodecyl sulphate to various proteins. *Biochem. J.* **109**, 825–830.

PRINGLE J.R. (1970) The molecular weight of the undegraded polypeptide chains of yeast hexokinase. *Biochim. Biophys. Res. Comm.* **39**, 46–52.

PROHASKA J.R., CLARK D.A. & WELLS W.W. (1973) Improved rapidity and precision in the determination of brain 2', 3'-cyclic nucleotide 3'-phosphohydrolase. *Anal. Biochem.* **56**, 275–282.

PUSZKIN S. & KOCHWA S. (1974) Regulation of neurotransmitter release by a complex of actin with relaxing protein isolated from rat brain synaptosomes. *J. Biol. Chem.* **249**, 7711–7714.

PUSZKIN S., NICKLAS W.J. & BERL S. (1972) Actomyosin-like protein in brain: subcellular distribution. *J. Neurochem.* **19**, 1319–1333.

QUARLES R.H., EVERLY J.L. & BRADY R.O. (1973) Evidence for the close association of a glycoprotein with myelin in rat brain. *J. Neurochem.* **21**, 1177–1191.

RAKIC P. (1971) Neurone-glia relationships during granule cell migration in developing cerebellar cortex. A Golgi and electronmicroscopic study in *Maccacus rhesus. J. Comp. Neurol.* **141**, 283–312.

REYNOLDS J.A. & TANFORD C. (1970) Binding of dodecyl sulphate to proteins at high binding ratios. Possible implications for the state of proteins in biological membranes. *Proc. Nat. Acad. Sci.* **66**, 1002–1007.

REYNOLDS J.A. & GREEN H.O. (1973) Polypeptide chains from porcine cerebral myelin. *J. Biol. Chem.* **248**, 1207–1210.

RIGHETTI P.G., PERRELLA M., ZANELLA A. & SIRCHIA G. (1973) The membrane abnormality of the red cell in paroxysmal nocturnal haemoglobinuria. *Nature New Biol.* **245**, 273–276.

ROGERS A.W., DARZYNKIEWICZ Z., SALPETER M.M., OSTROWSKI K. & BARNARD E.A. (1969) Quantitative studies on enzymes in structures in striated muscles by labelled inhibitor methods. *J. Cell. Biol.* **41**, 665–685.

ROSSIER J., BAUMAN A. & BENDA P. (1973) Antibodies to brain choline acetyltransferase: species and organ specificity. *FEBS Lett.* **36**, 43–48.

RUBENS R.D. (1974) Antibodies as carriers of anticancer agents. *Lancet* **i**, 498–499.

RUBENS R.D. & DULBECCO R. (1974) Augmentation of cytotoxic drug action by antibodies directed at cell surface *Nature, London* **248**, 81–82.

Rush R.A., Thomas P.E., Nagatsu T. & Udenfriend S. (1974) Comparison of human serum dopamine-beta-hydroxylase levels by radioimmunoassay and enzymatic assay. *Proc. Nat. Acad. Sci.* **71**, 872–874.

Saito K., Barber R., Wu J.Y., Matsuda T. & Roberts E. (1974) Immunohistochemical localization of glutamate decarboxylase in rat cerebellum. *Proc. Nat. Acad. Sci.* **71**, 269–273.

Sarkar S., Sketer F.A. & Gergely J. (1971) Light chains of myosins from white, red and cardiac muscles. *Proc. Nat, Acad. Sci.* **68**, 946–950.

Schmitt F.O. (1968) Fibrous proteins—neuronal organelles. *Proc. Nat. Acad. Sci.* **60**, 1092–1101.

Schubert D., Heinemann S., Carlisle H., Tarikas H., Kimes B., Patrick J., Steinbach H.H., Culp W. & Brandt B.L. (1974) Clonal cell lines from the rat central nervous system. *Nature, London*, **249**, 224–227.

Sela M. (1969) Antigenicity: Some molecular aspects. *Sci.* **166**, 1365–1374.

Shelanski M.L., Albert S., Devries G.H. & Norton W.T. (1971) Isolation of filaments from brain. *Sci.* **174**, 1242–1245.

Sheremata W., Cosgrove J.B.R. & Eylar E.H. (1974) Cellular hypersensitivity to basic myelin (A1) protein and clinical multiple sclerosis. *New Eng. J. Med.* **291**, 14–17.

Shooter E.M. & Roboz-Einstein E.R. (1971) Proteins of the nervous system. *Am. Rev. Biochem.* **40**, 635–652.

Sierra M.F. & Tzagoloff A. (1973) Assembly of the mitochondrial membrane system. Purification of a mitochondrial product of the ATPase. *Proc. Nat. Acad. Sci.* **70**, 3155–3159.

Smith A.D. (1971) Secretion of proteins (Chromogranin A and dopamine-beta-hydroxylase) from a sympathetic nerve. *Phil. Trans. Roy. Soc. Lond. B.* **261**, 363–370.

Stavrou D., Haglid K.G. & Weidenbach W. (1971) The brain specific proteins S-100 and 14-3-2 in experimental brain tumours of the rat. *Z. ges. exp. med.* **156**, 237–242.

Sugiyama H., Benda P., Meunier J.C. & Changeux J.P. (1973) Immunological characterization of the cholinergic receptor protein from *Electrophorous electricus*. *FEBS Lett.* **35**, 124–128.

Swanson L.W. & Hartman B.K. (1974) A systematic immunofluorescence study of the central dopamine-beta-hydroxylase (DBH) neurones and the distribution of their processes in the rat. *Abstracts: Soc. for Neurosci.* **4**, 673,

Taylor R.B., Duffus P.H., Raff M.C. & de Petris S. (1971) Redistribution and pinocytosis of lymphocyte surface immunoglobulin molecules induced by anti-immunoglobulin antibody. *Nature New Biol.* **233**, 225–229.

Teitelbaum D., Webb C., Meshorer A., Arnon R. & Sela M. (1972) Protection against experimental allergic encephalomyelitis. *Nature, London* **240**, 564–566.

Thompson E.J., Daroga B.M., Quick J.M.S., Shortman R.C., Hughes B.P. & Davison A.N. (1974). Quantitative determination of fluorescent labelled proteins using dansyl chloride or fluorescamine with application to polyacrylamide gels. *Trans. Biochem. Soc.* **2**, 989–990.

Trams E.G. (1973) A rapid fluorometric assay for 2′, 3′-cyclic adenosine monophosphate 3′-phosphoester hydrolase. *J. Neurochem.* **21**, 995–997.

Trayer H.R., Nozaki Y., Reynolds J.A. & Tanford C. (1971) Polypeptide chains from human red blood cell membranes. *J. Biol. Chem.* **246**, 4485–4488.

TREMBLAY J., SIMON M. & BARONDES S.H. (1974) Cerebroside may be falsely identified as a soluble 'brain specific protein'. *J. Neurochem.* **23**, 315–318.
ULMAR G. & WHITTAKER V.P. (1974) Immunological approach to the characterization of cholinergic vesicular protein. *J. Neurochem.* **22**, 451–454.
UTTENTHAL L.O. & HOPE D.B. (1970) The isolation of three neurophysins from porcine posterior pituitary lobes. *Biochem. J.* **116**, 899–909.
VOGEL Z., DANIELS M.P. & NIRENBERG M. (1974) Localization of acetylcholine receptors during synaptogenesis in retina. *Fed. Proc.* **33**, 1431.
WAEHNELDT T.V. & BEUHOFF V. (1974) Membrane proteins of rat brain: compositional changes during postnatal development. *J. Neurochem.* **23**, 71–77.
WAEHNELDT T.V., RALSTON H.J. & SHOOTER E.M. (1973) The protein composition of subcellular fractions from mouse brain. *Neurobiol.* **3**, 215–224.
WAKSMAN B.H. (1974) Immunologic mysteries of multiple sclerosis. *New England J. Med.* **291**, 45–46.
WARECKA K., MÖLLER H.J., VOGEL H.M. & TRIPATZIS I. (1972) Human brain-specific alpha 2-glycoprotein: Purification by affinity chromatography and detection of a new component; localization in nervous cells. *J. Neurochem.* **19**, 719–725.
WEBER K. & GROESCHEL-STEWART U. (1974) Antibody to myosin: The specific visualization of myosin-containing filaments in non-muscle cells. *Proc. Nat. Acad. Sci.* **71**, 4561–4564.
WEBER K., PRINGLE J.R. & OSBORN M. (1972) Measurement of molecular weights by electrophoresis on SDS-acrylamide gel *Meth. in Enzymol.* **26**(C), 3–27.
WEEDS A.G., TRENTHAM D.R., KEAN C.J.C. & BULLER A.J. (1974) Myosin from cross-reinnervated cat muscles. *Nature, London* **247**, 135–139.
WELTMAN J.K. & DOWBEN R.M. (1973) Relatedness among contractile and membrane proteins: evidence for evolution from common ancestral genes. *Proc. Nat. Acad. Sci.* **70**, 3230–3234.
WESSELLS N.K., SPOONER B.S., ASH S.F., BRADLEY M.O., LUDUENA M.A., TAYLOR E.L., WRENN J.T. & YAMADA K.M. (1971) Microfilaments in cellular and developmental processes. *Sci.* **171**, 135–143.
WHITTAKER V.P., DOWDALL M.J., DOWE G.H.C., FACINO R.M. & SCOTTO J. (1974) Proteins of cholinergic synaptic vesicles from the electric organ of *Torpedo*: Characterization of a low molecular weight acetic protein. *Brain Res.* **75**, 115–131.
WILLARD M., COWAN W.M. & VAGELOS P.R. (1974) The polypeptide composition of intra-axonally transported proteins: evidence for four transport velocities. *Proc. Nat. Acad. Sci.* **71**, 2183–2187.
WILLINGHAM M.C., OSTLUND R.E. & PASTAN I. (1974) Myosin is a component of the cell surface of cultured cells. *Proc. Nat. Acad. Sci.* **71**, 4144–4148.
WOOD J.G. & DAWSON R.M.C. (1973) A major myelin glycoprotein of sciatic nerve. *J. Neurochem.* **21**, 717–719.
WOOD J.G. & DAWSON R.M.C. (1974a) Some properties of a major structural glycoprotein of sciatic nerve. *J. Neurochem.* **22**, 627–630.
WOOD J.G. & DAWSON R.M.C. (1974b) Lipid and protein changes in sciatic nerve during Wallerian Degeneration. *J. Neurochem.* **22**, 631–635.
YAMADA K.M., SPOONER B.S. & WESSELLS N.K. (1971) Ultrastructure and function of growth cones and axons of cultured nerve cells. *J. Cell Biol.* **49**, 614–635.
YAMAMURA H.I. & SNYDER S.H. (1974) Muscarinic cholinergic binding in rat brain. *Proc. Nat. Acad. Sci.* **71**, 1725–1729.
YOUNG A.B. & SNYDER S.H. (1974) Strychnine binding in rat spinal cord membranes associated with synaptic glycine receptor: co-operativity of glycine interactions. *Molec. Pharmacol.* **10**, 790–809.

Index

Abetalipoproteinaemia 57, 69
Abiotrophies 4, 12, 39, 56
Acetylcholine 132
   receptor 307
   receptor protein (Nicotinic) 285–288, 306
Acetylcholinesterase 21–23, 288
Acetylsalicylic acid
   retina in 80
Acid protease
   retina in 72
Acid proteinase (Cathepsin D) 20, 27
   plaques in 29
Actin 33, 290
   peptide mapping and 305
   -relaxing protein complex and 292
Actinomyosin D 7
Adrenaline 177, 288
Ageing 10–14
   theory 32
Alanine 148
Alcohol dehydrogenase
   retina in 64–65
Allyglycine 137
Alumina foci 126
Aluminium
   ion 13
   oxide 120
Alzheimer's disease 12
   cells in 262
   neurofibrillary changes in 39
   neuronin S-6 in 34
Amines
   brain in 182
   hypoxia in 247–248
Amino acids 123, 127 et seq., 147–149
   brain in 195
   CSF content of 143
   changes, content of freeze-focus in 141
   changes in brain 137
   changes, metabolic implications 137
   epileptogenic foci in 129, 136
   experimental seizures in 139
   metabolism and mental dysfunction 207
   neurotransmission and 135
   plasma concentration in 145
   retina, content of 59
   taurine influence on 152
γ-Aminobutyric acid (GABA) 127, 135–136, 140–143
   enzymes and 143
   Huntington's chorea in 196
   hypoxia in 246
   -mediated neurones 5
   metabolism of 40
   retina in 57
   shunt 137, 144, 249
   -transaminase (GABA-T) 143
α-Amino oxyacetic acid 144
di-(p-Aminophenoxy)-pentane 93–94
Ammonia 136, 246, 264, 267
Ammonium
   ions 22
   salts, injection of 265
Ammon's horn 4
Amyloid 16

Amyotrophic lateral sclerosis 5
Anaesthetics 230
Anoxia 19–22, 35–37, 235, 239–244
Anticonvulsants 116, 121, 124–126, 132
  folate and 153
Antigen
  alpha (14:3:2) 33, 38, 283
  neuronal 285
  tissue 28
Antipyridoxal agents 123, 137
Apomorphine 193
Arterio-venous differences
  hypoxia in 245
Aspartate 127, 136
  retina in 57
Astrocytes 14, 290, 293, 295
Astrofilaments 293–294
Atherosclerosis 10
ATPase
  retina, changes in 65
Audiogenic
  models 126
  seizures 134
Autoimmune
  diseases 285, 299
  reactions 32
Autism
  infantile 206
Autoxidation 17
Axonal flow 288, 293, 305
Axons 304
  lesions of 22
  transport 305

Baboon
  photosensitive 123, 134
  seizures, light-induced of 133
Basal ganglia
  biochemistry of 185
  brain decarboxylases and 39, 40
Batten's disease 12, 17
Bicuculline 151
Blindness
  abnormalities in 54
Blood-brain barrier 25, 117, 126, 146, 153
Blood glucose
  hypoglycaemia in 257
Brain
  amines in 182
  amino acid changes in 137
  amino acids 195
Bufotenine 172

Canine distemper 29
Carbamylcholine 285
Carbon monoxide (CO)
  poisoning 21
Catalase 29
Catecholamines 170 *et seq.*
  hypoxia in 247–249
  metabolism of 173
Cathepsin D (*see* acid proteinase)
Cations 127
  epilepsy and 131
Cell
  abnormalities of 22
  damage in 253
  death of 4, 17
  Kupffer 8
  loss 10
  -mediated reactions 24
  nervous system in 4
  neural 14
  photoreceptor 73 *et seq.*
  Purkinje 4
  senile changes in 15
  T, lymphocytes and 299
  visual, organization of 75
Central nervous system (CNS) 24
  cells in 4
  myelin from rat 300
Cerebellar hypoplasia 26
Cerebrosides 28, 301
Cerebrospinal fluid (CSF) 143, 148, 179
  extracellular 135
  measles antigen and 31
  metabolites in 180, 192, 203
  multiple sclerosis and 31
Cholera toxin 284
Cholesterol ester 8, 29, 31
Choline acetylase 306
Choline acetyltransferase 288
Chromatolysis 6
Chromogranin A 290
Chromomembranin A 288
Clinical pathology
  epilepsy and 116
Cobalt 25, 119, 128–129, 148
Colchine 293

Coma
    biochemistry of   230 *et seq.*
    development and   267
    enzymes, effect on   21
    GAD activity and   42
    hepatic   261
    metabolic causes of   231
Convulsants   126, 134, 137
    ammonium ion   267
    chemical   126, 138
    α-oxoglutaramate in   266
Copper   5
Creutzfeldt-Jacob disease   23
    animal virus infection and   26
    'slow-virus' and   12
Cuprizone (biscyclohexanone oxaldihydrazone)   24
2',3'-Cyclic phosphodiesterase   301
3',5'-Cyclic adenosine monophosphate cyclic (AMP)
    retina in   83, 84

Dark adaptation   95
Deafferentiation   126
Degeneration   5, 9
    light deprivation and   77, 78
    neurofibrillar   14–16
    optic nerve in   54
    pigmentary   54
    retina in   54, 88
    spongiform   23
Demyelination
    antibody to basic protein and   301–302
    axon and nerve cell of   9
    axonal damage   10
    experimental   23
    tissue culture in   299
Deoxyglucose
    hypoglycaemia and   254–256
    metabolism of   255
Deoxyribonucleic acid (DNA)
    retina in   58
Depression
    biochemistry of   199
    catechol amines and   202
    glutamate and   150
    indoleamines and   200
    spreading   150
Diabetes
    alloxan by   97

biochemistry of   98
fatty acid metabolism in retina in   99
insulin in   97
membrane of kidney, lens and retina changes in   101
mucopolysaccharides in retina in   100
platelet abnormalities in   103
retina in   99
Diabetic retinopathy   54
    biochemical aspects of   97
    drug therapy of   103
Diaminodiphenoxypentane
    retina in   94
Diaschisis   4
Dihydroxyphenylalanine (L-DOPA) 170 *et seq.*
    therapy   189
Dihydroxyphenylalanine (L-DOPA) decarboxylase (DOPAD)   23, 251
    carbon monoxide poisoning and   21
    changes in activities   41
    senile dementia i   33–35, 38–39
Diphtheria toxin   23
n-Dipropyl acetate (Epilim)   144
Dopamine   20, 173–175, 185
    metabolism of   40
    receptor   306
Dopamine β-hydroxylase   173–174, 288
Down's syndrome   9, 206
    development of   12
Drugs
    anti-inflammatory in retinopathy   78, 80
    therapy of diabetic retinopathy   103
'Dying back' process   9, 23
Dynein   295
    peptide mapping   305
Dysfunction
    visual cycle and   70
Dyskinesias
    drug provoked   197
    Huntington's chorea   12, 35, 194
    management of   199
Dystrophy
    animals, behavioural studies in   91
    inherited retinal   54
    lipid metabolism and   58

Electrical
    features   125

Electrical—(contd)
  stimulation 124
Electroencephalogram (EEG) 117 et seq.
Electroretinogram (ERG) 56
  coma in 230
  epilepsy and 119
  glutamine formation and 265
  hypoglycaemia in 253
  sleep in 230
Enzymes
  activity 21
  GABA and 143
  glutamate and 146
  retina in 82
  retinal dystrophy and 83
  retinal lysosomes, release of 74, 80
Epilepsy
  acetylcholine and 132
  agents causing 119
  aluminium oxide and 120
  amino acids and 127, 143–145, 148
  amino acid changes and 129
  anticonvulsants and 124
  audiogenic models of 126
  biochemical changes in 126
  blood-brain barrier and 117, 153
  cations and 131
  cobalt and 119, 128, 147–148
  cobalt-induced 120
  convulsants and 126
  experimental 118, 127
  experimental models 118, 126
  focus 25, 124, 130, 136, 153
  folate and 153
  freeze-lesions in 121
  GABA and 136
  gliosis and 124
  glycolysis and 130
  histopathology and 116
  hypocalcaemia and 132
  inhibitors and 133
  intermediary metabolism and 137
  mechanism of 127
  metals and 119
  monoamines and 133
  neuropathology of 117, 119, 126
  neurotransmission and 135
  ouabain and 132
  pathology of 116
  penicillin and 122
  photogenic models of 126
  reflex 120–123
  sclerosis in 116
  seizures in 116, 133
  spread of 127
  taurine and 127, 151–152
  temporal lobe and 127
  tungstic acid and 122
Epilim (see n-Dipropyl acetate) 144
Eserine 285
Etaretin 57
Excitation 122, 142
Experimental allergic encephalitis
  immunization and 305
Experimental allergic encephalomyelitis (EAE) 24
  antibodies in 301–302
  myelin basic protein in 299
Experimental allergic neuritis (EAN)
  glycoprotein in 300
Experimental models 19
  epilepsy of 118

Fatty acid
  metabolism in diabetic retina 99
Filarin 15, 293
'Floating fraction' 8, 23
Focal epilepsy
  agents producing 119
Foci
  epileptic, amino acid content of 136
  experimental 125
Focus
  concept of 124
Folate
  epilepsy and 153
  molecule and 154
Freeze-focus
  amino acid content of 141
Freeze-lesions
  epilepsy in 121

β-Galactosidase 23, 27
  retina in 79
Gangliosides 284
Gitter cells 8
Glioma 295
  beta trace protein and 299
Gliosis 8, 27, 117, 124
Glucoreceptors 259

Glucose
  retina in 99
Glucose transport 257–258
β-Glucuronidase 20, 23, 27, 29
  retina in 80
Glutamate 23, 127, 142, 148
  cobalt-focus release from 147–148
  decarboxylase (GAD) 23, 288
  depression and 150
  enzymes and 146
  Huntington's chorea and 196
  hypoxia and 247
  metabolism of 135–136, 140–141, 146, 267
  retina in 57
Glutamic acid decarboxylase (GAD) 23, 288
  epilepsy and 137
  senile dementia and 35, 38
Glutamine 23, 136, 266
Glycine 135
Glycolipids
  auxiliary protein and 306
Glycolysis 130, 233, 238
  anaerobic in retina 56
  ischaemia and 237
Glycoprotein 288
  alpha 2 297
  10 B 299
  CNS and 300
  CNS myelin and 308
  GP350 290
  PAS staining and 304
Golgi complex 75
Grand-mal 121
Granulovacuoles 12
Granulovacuolar degeneration 16
Guam-Parkinsonism dementia 12, 33

Hamartoma 124
Hepatic coma 261
  animal studies in 261
Hepatolenticular degeneration (Wilson's disease) 5
Hexachlorophene 23
Hexokinase 19, 245
Hexose monophosphate shunt (HMP) pathway
  retina in 63–64
Hippocampus 16
Hirano bodies 14

Histochemistry
  fluorescence methods in 177
Histamine
  receptor 306
Histological changes 16
Histopathology
  degenerative brain diseases and 6
  epilepsy and 116
Homovanillic acid (HVA) 173, 184, 198
Huntington's chorea 5, 12, 35, 39, 194, 288
Hydrazides 137
Hydranencephaly 26
5-Hydroxyindolylacetic acid (5HIAA) 172–173
L-5-Hydroxytryptamine (5HT) 171–173, 176, 207
L-5-Hydroxytryptophan (5-HTP) 170–173
Hyperammonaemia 22, 262, 263
Hyperexcitability 117
Hypertensive
  patients 288
Hypocalcaemia 132
Hypoglycaemia 116, 252–259, 269
  blood glucose in 257
  deoxyglucose induced 254–256
  glucose efflux and 260
  insulin induced 252–254
Hypoxia 19, 116, 244–252, 271
  glycolysis in 244
  glycolytic enzymes in 245
  respiration in 244

IgG 31
Immunogenesis 304–305
Indoleamine
  metabolism of 173, 200, 208, 251
Inflammatory response 8
Inhibition 142
Inhibitors 133–135
  monoamine oxidase 134
  peripheral decarboxylase 193
Insulin
  diabetes in 97
  hypoglycaemia and 252–254
  retinopathy in 99
Interperiod line 299–300
Intra-neuronal changes 14

Ischaemia 4, 19, 21, 232–239
  cell damage in 253
  glycolysis in 237
  lactate production in 242

'Kindling effect' 124
Kupffer cells 8
Kuru 23, 26

Lactate dehydrogenase
  retina in 82
Laurence-Moon-Biedl syndrome 54
Learning
  pigeons in 299
Lesions
  experimental in brain 193
Levine preparation 242
Light deprivation
  retinal degeneration and 77
  retinitis pigmentosa and 77
Lipidoses 7
Lipofuscin 12, 15–17, 32
Lymphocytes
  'killer' 9
  T 285
Lysosomes 15–18, 30
  'hypothesis' 73
  retinal 78, 80

Macrophages 29
Manganese 199
  dyskinesia and 199
Markers
  cellular 27–29
Mechanical damage 121
Membranes
  synaptic plasma 289, 292
Meningioma
  actin and 295
Mental dysfunction
  amino acid metabolism in 207
  catechol amines in 206
Metabolic
  aminoa acid changes 137
  coma, causes of 231
  implications 137
Metabolism
  carbohydrate 130
  energy 130
  intermediary in epilepsy 137
  retinal
    aerobic glucose 57
    amino acid 59
    arachidonate 57
    fatty acid 99
    glucose 56
    lipid 57
    nucleic acid 58
    visual pigment 66
Metals 25
Methionine sulphoximine 137
3-Methoxy 4-Hydroxy-mandelic acid (VMA) 173–174
Metrazole 134
Micrencephaly 26
Microcephaly 26
Microfilaments 292–294
Microglia 14, 29
Microtubules 15, 294
Microvacuoles 242
Migraine 208
  amine changes 209
  foods and amines in 210
  precipitants of 210
'Mirror-foci' 121
Mitotic spindle 293
Molecular
  glycoprotein weights 303
Mongol 12
Monoamine oxidase 172
  inhibitors 134
  platelet 211
Monoamines 132–133
Morphine
  receptor 306
Motor neurone disease 12
Mucopolysaccharides
  diabetic retina of 100
Multiple Sclerosis
  animal virus infection and 26
  axon and 10
  beta trace protein and 299
  epidemiology 28
  immunological methods in 307
  macroscopic changes in 11
  oligodendrocytes in 295
  plaque tissue in 26
Myasthenia gravis 285
Myasthenic syndrome 285
Myelin
  axons 304
  degeneration 9
  glycoprotein in CNS 300

Myelin—(contd)
  lamellae fragmentation of 23
  loss 7, 24
  PNS in 300
  sheath 15, 27
Myosin 292
  peptide mapping and 305

Nerve
  growth factor 281
  ligation 288–289
Neurin 292
Neuritic plaques 12, 17
Neuroblastoma 283
Neurofilaments 293–294
Neurones
  ageing of 14
  loss of 10
Neuronin 284
  S-6 33 *et seq.*
Neurophagia 5
Neuronophagia 10
Neuropathy 293
Neurophysins 290
Neuropil
  changes in 17
Neurotransmission 23, 135
  amino acids and 135
Nissl substance 6, 15
Nodes of Ranvier 7
Noradrenaline 20, 173–175, 288–289

Oligodendrocytes 14, 292
  agent acting on 23
  degeneration of 24
  multiple sclerosis and 295
  proteins in myelin and 300
Opsin 66
Optic
  atrophy 54
  nerve 54
  pathway 55
Ouabain 122, 132
Oxygen
  brain and 19
  deprivation 19, 243
  hyperbaric 134
  K_m value 250
  supply 19
α-oxyglutaramate 266

Parkinson's disease 5, 32–35, 183

  clinical and morphological changes in 12
  neuropathology of 185
  neurotransmitter metabolism in 39, 185
  'on-off' effects in 192
  treatment of 183, 186
Paroxysmal depolarization shift (PDS) 125, 151–155
Pathoclisis 5, 22
  cellular 20
Pellagra 207
Penicillin
  epilepsy and 122
Peripheral neuropathy 306
Perivascular cuffing 9
Peroxisomes 17–18
Petit-mal 118, 146
Phagocytes
  CNS in 8
Phagocytosis 8, 77
Phenothiazines 198
Phenylalanine 207
Phenylethylamine
  migraine in 210
Phenytoin 132
Phosphocreatine 21
Phosphodiesterase
  retina in 83
Phosphofructokinase 19, 245–247
Photochemical
  rhodopsin and 62
Photogenic
  models 126
Photoreceptor
  cells 73 *et seq.*
  renewal–removal mechanism 74
Phytanic-acid storage disease (Refsum's disease) 57
Pick's disease 12
Plaques 11, 17, 28
Plasma
  amino acid 145
  epilepsy, changes in amino acids 145
Platelets
  adhesives 104
  diabetes, abnormalities in 103
  Down's syndrome in 206
  5HT in 206
Pollution
  heavy metals by 25

Porphyropsin
  retina in 61
Portocaval anastomosis 23, 262
Post-mortem
  changes 28
Post-synaptic
  densities 293, 304
Potassium (K$^+$) 127–128
  ATPase 284, 305
Protein
  14:3:2 283, 297
  acetylcholine receptor (Nicotinic) 285–288
  acidic 34
  actin-relaxing complex 292
  AP22 295
  band (M.W.50,000) 32
  beta trace 297
  glial fibrillary acidic 8, 295
  myelin basic (encephalitogenic) 24, 31, 299, 307
  neuronin S-6 33, 37
  proteolipid 300–301
  retina, content in 59, 73
  S-100 283–284, 295, 297, 307
  synthesis 74, 137
  Wolfgram 300–301
Proteolipid
  proteins 300
Psychological disorders 199

Q analysis 38

Receptors
  post-synaptic 142
Reflex epilepsy
  models of 123
Refsum's disease (phytanic-acid storage disease) 57
Retina
  acetylsalicylic acid in 80
  acid protease in 72
  alcohol dehydrogenase in 64
  amino acid metabolism in 59
  di-(p-aminophenoxy)-pentane in 93–94
  ATPase in 65
  biochemistry of 71, 98
  carbohydrate metabolism in 64
  chloroquine in 92, 96
  chlorpromazine in 97
  cyclic AMP 84
  cyclic GMP 84
  dithizone in 92
  dystrophy of 54 *et seq.*
  enzymes in 82
  glucose in 99
  glycine in 60
  hexose monophosphate (HMP) pathway in 63
  insulin in 99
  inherited dystrophy of 54
  lactate dehydrogenase in 82
  lipid metabolism in 57
  nucleic acid in 58
  phosphodiesterase in 83
  pigment
    degeneration of 54
    metabolism of 66
  primary biochemical lesion of 71
  protein metabolism of 59
  rhodopsin in 62
  RNA content in 80
  toxic agents sensitivity and 96
  vitamin A metabolism in 68
Retinal degeneration 54 *et seq.*
  studies of 88
Retinal dystrophy
  animal behavioural studies 91
  cortical effects of 88
  inherited 54
Retinitis pigmentosa
  in man 54, 57, 77
  light deprivation effect and 77
  retinol, action of 78
Retinotoxic substances 92
Rhodopsin 61–62
Ribonucleic acid (RNA) 7
  retina in 58, 79, 89

Schilder's disease 26
Schizophrenia 204
  catechol amines and 204–205
  dopaminergic overactivity and 205
Sclerosis
  Ammons horn 116
Schwann cell 7
  proteins in myelin in 300
Seizures 118, 146
  audiogenic 20, 134
  cerebral metabolism during 131
  experimental 139

Seizures—(contd)
  initiation 132–133
  light-induced 133
  photogenic 123
  propagation 132
Senile dementia 10, 31 et seq.
  abnormal structural elements of 32
  biochemical changes in 33, 37
  clinical and morphological changes in 12
  decarboxylase activities in 35, 36, 37
  histological changes in 16
  macroscopic changes in 11
  metabolism of dopamine and GABA in 40
  pathogenesis of 38
Senile plaques 17
Serotonin (5-HT) 285
Shock
  electroconvulsive 134
Sleep
  electroencephalogram (EEG) and 230
Sodium (Na$^+$) 128
  ATPase 284, 305
  -pump 122, 127, 132
Sodium pump
  rods in 63
Status epilepticus 117
Stroke
  beta trace protein and 299
Strychnine 151, 306
Subacute sclerosing panencephalitis (SSPE) 12
  animal virus infection and 26
Succinic semialdehyde dehydrogenase (SSDH) 143
Superfusion 133
Synaptic
  plasma membranes 289–292, 304
  population 14
  receptor 126
  vesicles 289–292, 294, 304
Synaptosomes 290

Tangles 12
Taurine 127, 135
  amino acid content 152
  epilepsy and 120, 136, 145, 151
Temporal lobes
  epilepsy in 116, 127

senile dementia in 37
Tetanus toxin 285
Thiol groups 92
Topistic unit 5
Toxic agents
  retinal sensitivity to 96
Transmitters 132
  anatomical pathways 176–177
  energy consumed in synthesis 250
  rare 174
Transport
  amino acid metabolites in 181, 207
  ion 122, 132
Tricarboxylic acid
  cycle 249
Triethyl tin 23
Tri-ortho cresyl phosphate 9
Tropomyosin
  brain in 289
Tryptamine 285
Tryptophan 170, 202
  depression in 200–201
Tubulin 282, 293, 306
  -dynein 293
  peptide mapping 305
Tungstic acid
  epilepsy and 122
Tyramine
  migraine in 210
Tyrosine hydroxylase 289

Uptake system
  high-affinity 135
Urease
  injection of 262

Vesiculin 290
Vinblastin 293–295
Viruses 26
Visual
  cycle 61
  pigment 56, 66
Vitamin
  A 61
  A metabolism of retina 68
  B6 deficiency 123

'Waelsch effect' 135
Wallerian degeneration 7, 21
Wilson's disease 5

Zygology
  method 38